Solid State Gas Sensing

Elisabetta Comini · Guido Faglia ·
Giorgio Sberveglieri
Editors

Solid State Gas Sensing

Editors
Elisabetta Comini
Sensor Laboratory
CNR-INFM and Brescia University
Via Valotti, 9
25133 Brescia
Italy

Guido Faglia
Sensor Laboratory
CNR-INFM and Brescia University
Via Valotti, 9
25133 Brescia
Italy

Giorgio Sberveglieri
Sensor Laboratory
CNR-INFM and Brescia University
Via Valotti, 9
25133 Brescia
Italy

ISBN: 978-1-4419-3507-6 e-ISBN: 978-0-387-09665-0
DOI: 10.1007/978-0-387-09665-0

Printed on acid-free paper

springer.com

Preface

Gas sensor technology has advanced remarkably during the past few decades and is becoming an essential technology. Many gas sensors are now commercially available and researchers are making efforts, thanks to pioneering novel ideas, to develop next-generation gas sensors, having all the necessary requirements, such as small size, low production costs and power consumption.

This book covers all gas sensor topics, a research field with an increasing interest in the last few years due to the demands of reliable, inexpensive and portable systems for environmental monitoring, indoor air quality, food quality control and many other applications. The goal of this book is to provide a critical assessment of the new trends in the gas sensor field, by describing the working principle and the applications related to the different types of sensors and paying attention to the recent impact of nanotechnology on the field.

Nanotechnology is a new field that will dramatically change solid-state gas sensing. In the last few decades, the study of one-dimensional (1D) materials has become a primary focus in nanoscience and nanotechnology. With reduction in size, novel electrical, mechanical, chemical and optical properties have been introduced, which are largely believed to be the result of surface and quantum confinement effects. For example, nanowire-like structures are the ideal system for studying the transport process in one-dimensionally (1D) confined objects, which are of benefit not only for understanding the fundamental phenomena in low dimensional systems but also for developing new generation nanodevices and gas sensors with high performances.

The first chapter is devoted to micro-fabrication for gas sensors. Following chapters deal with the subject on the base of the transduction principle as electrical, permittivity, field effect, electrochemical, optical, thermometric and mass (both quartz and cantilever types) based.

The book is characterized by a methodical and thorough treatment of the subject matters. The chapters are logically related, and each has its own introduction and bibliography, in order to make it accessible to any reader notwithstanding his background on related subjects.

Since the last years have seen an enormous amount of activity in the field of gas sensor systems, this book represents a valuable and accessible guide and reference for researchers with up-to-date examples and state-of-the-art results.

Italy
May 2008

Elisabetta Comini
Guido Faglia
Giorgio Sberveglieri

Contents

Contributors

István Bársony
Research Institute for Technical Physics and Materials Science – MFA
Hungarian Academy of Sciences, Budapest, Hungary

Ilaria Cacciari
CNR, "Nello Carrara" Institute of Applied Physics (IFAC), 50019 Sesto Fiorentino, Italy

Elisabetta Comini
Sensor Laboratory, CNR-INFM and Brescia University, Via Valotti, 9, 25133 Brescia, Italy, comini@ing.unibs.it

Csaba Dücső
Research Institute for Technical Physics and Materials Science - MFA
Hungarian Academy of Sciences, Budapest, Hungary

Perumal Elumalai
Art, Science and Technology Center for Cooperative Research, Kyushu University, Kasuga-shi, Fukuoka, 816-8580, Japan

Guido Faglia
Sensor Laboratory, CNR-INFM and Brescia University, Via Valotti, 9, 25133 Brescia, Italy, guido.faglia@ing.unibs.it

Péter Fürjes
Research Institute for Technical Physics and Materials Science - MFA
Hungarian Academy of Sciences, Budapest, Hungary

Andreas Helwig
EADS Innovation Works, D81663 München, Germany

Samuel J. Ippolito
RMIT University, Department of Applied Chemistry, Melbourne, Victoria Australia, sipp@ieee.org

Hans Peter Lang
National Center of Competence for Research in Nanoscale Science, Institute of Physics of the University of Basel, Klingelbergstrasse 82, 4056 Basel Switzerland, Hans-Peter.Lang@unibas.ch

Norio Miura
Art, Science and Technology Center for Cooperative Research, Kyushu University, Kasuga-shi, Fukuoka, 816-8580, Japan
miurano@astec.kyushu-u.ac.jp

Gerhard Müller
EADS Innovation Works, D81663 München, Germany
Gerhard.mueller@eads.net

Lars Ojamäe
Department of Physics, Chemistry and Biology, Linköping University
SE-581 83 Linköping, Sweden, lars@ifm.liu.se

Vladimir V. Plashnitsa
Art, Science and Technology Center for Cooperative Research, Kyushu University, Kasuga-shi, Fukuoka, 816-8580, Japan

David A. Powell
Australian National University, Research School of Physical Sciences and Engineering, Canberra, ACT, Australia, david.powell@ieee.org

Giancarlo C. Righini
CNR, "Nello Carrara" Institute of Applied Physics (IFAC), 50019 Sesto Fiorentino, Italy; CNR, Department of Materials and Devices, 00185 Roma, Italy
g.c.righini@ifac.cnr.it

Yoshihiko Sadaoka
Department of Materials Science and Biotechnology, Graduate School of Science and Engineering, Ehime University, Matsuyama 790-8577, Japan
sadaoka@eng.ehime-u.ac.jp

Giorgio Sberveglieri
Sensor Laboratory, CNR-INFM and Brescia University, Via Valotti, 9, 25133 Brescia, Italy, sbervegl@ing.unibs.it

Olaf Schulz
EADS Innovation Works, D81663 München, Germany

Magnus Skoglundh
Competence Centre for Catalysis, Chalmers University of Technology
SE-412 96 Göteborg, Sweden, skoglund@chalmers.se

Jan Spannhake

EADS Innovation Works, D81663 München, Germany

Anita Lloyd Spetz
Department of Physics, Chemistry and Biology, Linköping University
SE-581 83 Linköping, Sweden, spetz@ifm.liu.se

Adrian Trinchi
CSIRO Division of Materials Science and Engineering, Melbourne, Victoria
Australia, avt@ieee.org

Taro Ueda
Interdisciplinary Graduate School of Engineering Sciences, Kyushu University
Kasuga-shi, Fukuoka, 816-8580, Japan

Masahiro Utiyama
National Institute for Environmental Studies, Tsukuba-shi, Ibaraki, 305-8506
Japan

Ryotaro Wama
Interdisciplinary Graduate School of Engineering Sciences, Kyushu University
Kasuga-shi, Fukuoka, 816-8580, Japan

Wojtek Wlodarski
RMIT University, School of Electrical and Computer Engineering, Melbourne
Victoria, Australia, ww@rmit.edu.au

Chapter 1
Micro-Fabrication of Gas Sensors

Jan Spannhake, Andreas Helwig, Olaf Schulz, and Gerhard Müller

1.1 Introduction

Gas sensors are increasingly used in the growing markets of automotive [1, 2, 3], aerospace [2, 3, 4, 5, 6, 7], and logistic [8, 9, 10] applications. Within these domains, gas sensors play important roles in providing comfort and safety or in enabling process control or smart maintenance functionalities. Future important markets are likely to emerge in the fields of safety and security [11]. With regard to the sensitivity and selectivity of gas detection these various applications require very different levels of sensor performance. Very often the different applications also impose highly varying price, size, weight, and power consumption constraints on an acceptable sensor solution. In practice, therefore, a whole range of different gas sensors and gas-sensing principles need to be employed [8, 9, 10, 11]. Although the sensitivity of gas detection is not normally a major concern, selectivity is much harder to attain. Usually selectivity is obtained at the expense of an increased system complexity [11]. Depending on the degree of selectivity required, gas sensors can take the form of simple low-cost solid-state devices with broadband sensitivity or the form of desktop analytical instruments in case high selectivity is required.

In the field of low-cost solid-state gas sensors, metal oxide sensors have found widespread commercial application [12, 13, 14, 15]. Such sensors consist of gas-sensitive metal oxide (MOX) materials screen-printed onto ceramic heater substrates. To date such sensors are being produced in quantities of several millions per year, mainly for applications in the automotive and domestic markets. In the automotive industry, MOX gas sensors are used to control the air ventilation flaps in an attempt at keeping toxic air contaminants or highly odorous gases to a minimum in the car interior [3]. In the domestic market, MOX gas sensors are being used to detect natural gas leaks and to set off alarms before any explosive air–gas mixture might have developed.

G. Müller
EADS Innovation Works, D81663 München, Germany
e-mail: Gerhard.mueller@eads.net

E. Comini et al. (eds.), *Solid State Gas Sensing*,
DOI: 10.1007/978-0-387-09665-0_1, © Springer Science+Business Media, LLC 2009

Such inexpensive gas sensors provide high sensitivity to a wide variety of gases. However, they have also become ill-famed for their cross-sensitivity and the resulting high level of false alarms. Considering their high sensitivity and low complexity, considerable attempts have been undertaken to improve the selectivity of MOX gas sensors. In order to achieve this, many different kinds of MOX materials have been studied [15]. Rather than arriving at materials with narrow cross-sensitivity profiles, a range of materials with broad, but distinctly different cross-sensitivity profiles has evolved. Attempts at arriving at higher selectivity with MOX gas sensors therefore have concentrated on forming arrays of several MOX gas sensors with different cross-sensitivity profiles [12, 16, 17, 18, 19, 20]. In this case pattern recognition techniques are applied to the multi-sensor output to arrive at an acceptable level of gas discrimination. It has become popular to refer to such arrays as "electronic nose" devices [18]. Application areas of electronic nose devices range from the food industry up to supply chain monitoring in the automotive industry.

High-performance gas sensors, on the other hand, usually employ spectroscopic principles such as optical absorption spectrometry [21], gas chromatography [22] or ion mobility spectrometry [23]. Examples of optical spectrometers are dispersive and non-dispersive infrared (IR) absorption gas detectors [16, 24]. Such devices are selective in that they identify the molecules of interest by their specific molecular vibration modes. IR-based devices therefore produce fewer false alarms than MOX-based gas detectors in similar applications. However, they cannot usually compete with the superior gas sensitivity of MOX gas sensors. The range of applications of IR-based gas sensors is therefore limited to a certain range of air quality monitoring and leak detection applications. In case higher sensitivity is required – as in security applications – it has become popular to employ chromatographic and/or ionization based detection principles [23]. This latter range of devices allows tiny traces of explosives, illicit drugs and other contraband materials to be detected in a huge background of less interesting air constituents or commonly occurring air contaminants.

In the past two decades micro-electro-mechanical systems (MEMS) miniaturization techniques have started to penetrate all these various fields of gas-sensing technologies. Examples of MEMS-based sensing devices now range from low-power MOX gas sensors [19, 20, 25, 26] up to miniature mass spectrometers including on-chip vacuum generation [27]. The driving force in all these miniaturization attempts is arriving at small-size low-cost devices, either for the use in handheld instruments or as miniaturized low-power consumption devices in distributed and bus-connected sensor networks. In all these applications MEMS batch fabrication technologies are not only interesting because of their miniaturization potential but also because of their potential of providing high performance at low cost in mass production scenarios.

In a limited size review, like the present, it is impossible to do justice to the whole range of MEMS developments that have been undertaken so far. In the following we will therefore concentrate on development work carried out in the author's laboratory. In this work the focus has been building up a MEMS toolkit [28] around thermal microstructures to allow miniaturized gas-sensing systems to be set-up in a flexible manner. Using this kit, both MOX and IR-based gas-sensing microsystems have been realized and tested in various application scenarios. In the following the individual components in this toolkit will be described and several examples of micro-sensor systems will be pointed out. Last-not-least, industrialization issues are discussed that need to be dealt with to introduce MEMS gas-sensing systems into high-tech end-user markets.

1.2 Gas Sensors and MEMS Miniaturization Techniques

During the past five decades, silicon microelectronics technology has been the driver for the so-called microelectronics revolution. During that time electronic micro-miniaturization techniques have smoothly developed into a highly developed top-down nanotechnology with minimum feature sizes now approaching the 45 nm region [29, 30]. Later on the silicon base technology was extended to allow three-dimensional mechanical structures to be formed out of mono-crystalline silicon and/or silicon-compatible thin-film materials. This latter field is now generally referred to as MEMS [31, 32, 33]. In this latter form the silicon technology has also started to contribute to the evolution of chemical sensors and chemical sensor systems [25, 26]. Before we enter into a detailed discussion of MEMS gas sensors and MEMS gas-sensing systems, we take a brief look at the silicon technology and the silicon semiconductor material itself to see how both can contribute to the development of micro-gas-sensors and micro-gas-sensor systems.

1.2.1 Silicon as a Sensor Material

Whereas mono-crystalline silicon is widely recognized as the base material for the microelectronics industry, it is also well-known that silicon supports a variety of solid-state effects that make it interesting as a sensor material [17]. Transduction effects that are useful for the fabrication of micro-sensors serving the five main signal domains are indicated in Fig. 1.1.

Making use of these effects, a wide variety of sensor devices can be produced, which, in addition to the very sensing function, also contain important down-stream electronic functionalities such as preamplification and temperature compensation. The second important feature of silicon sensors is related to the fact that both the sensing and the electronic functions can be realized using a common set of processing steps, which can all be carried out on wafer-scale, i.e., in a batch process, which allows large economic benefits in mass production.

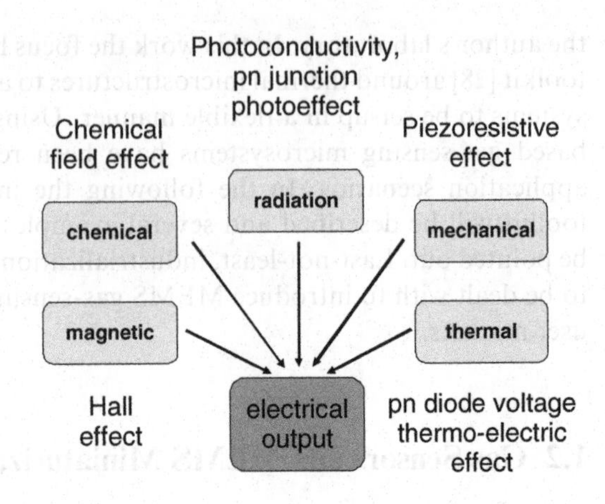

Fig. 1.1 Sensor signal domains and silicon solid-state effects that allow non-electrical input signals to be transformed into an electrical output signal using silicon sensor devices

Considering Fig. 1.1, however, it is also relevant to note that silicon as a semiconductor material is only of limited use for the fabrication of chemical sensors in general and for gas sensors in particular. Examples of silicon-based chemical sensors are pH- and ion-sensitive field-effect transistors and catalytic-gate MOS (metal oxide semiconductor) gas sensors. Whereas the first kind of device was introduced by Bergveld et al. [34], the latter was first described by the Lundström group [35, 36, 37]. In the meantime, all these devices have been developed to a considerable degree of sophistication. The functioning of both kinds of devices is based on the interaction of the analyte ion or analyte gas species with SiO_2 surfaces, i.e., with the surface of those oxide layers that naturally form on silicon surfaces. A severe constraint of catalytic-gate MOS gas sensors is that, in the normal operating temperature range of silicon devices, such sensors are almost completely selective to H_2. They therefore do not provide a sufficiently generic approach towards micro-gas-sensors. The field of semiconductor gas sensors, therefore, has very much remained a field of specialized transducer materials such as MOXs [15, 38], polymers [39], and other kinds of gas-sensitive materials [40].

1.2.2 Thermal Sensors and Actuators

A breakthrough in the field of silicon-based gas sensors occurred once silicon-based MEMS technologies became available [31, 32, 33]. Evolving out of the traditional microelectronic silicon processing technology, MEMS technology includes additional processing steps such as bulk- and surface-micromachining processes as well as silicon and glass wafer bonding techniques. In this way it has become possible to produce miniaturized devices, which combine mechanical, electrical, and thermal functionalities within a single piece of silicon. In addition, the portfolio of MEMS processes also includes an increasing variety of silicon-compatible functional materials to enable sensor and actuator functionalities that are impossible to realize using the silicon base material itself. Important representatives of such functional materials are thin- and thick-film gas-sensitive layers.

Whereas MEMS fabrication techniques have first made a major impact in the field of mechanical sensors such as pressure transducers, accelerometers, angular rate sensors, and so forth [17, 41], MEMS micro-miniaturization techniques soon thereafter have started to contribute to the development of miniaturized gas sensors and gas-sensing devices as well [25, 26]. In this latter context, the capability of MEMS technologies to arrive at thermal microstructures has proven to be most useful. Such microstructures are the basic building blocks in thermal sensor and actuator devices [36].

Thermal sensors (Fig. 1.2a) represent a special class of sensors that convert the input signals of interest into an intermediate thermal signal – usually a temperature change – that is then electrically detected by performing a temperature measurement. Figure 1.2b further shows that by inverting the direction of the energy flow in a thermal transducer, thermal sensors can be converted into thermal actuators. In this latter case the non-electrical output energy form is generated by dissipating electrical energy. The thermal energy in turn is converted into the output energy form.

The usefulness of thermal microstructures is revealed if we consider some specific cases. Considering the case of IR gas detection first, we note that here a thermal actuator may be used to produce the thermal IR radiation that may be absorbed by the analyte gas molecules and that the transmitted IR radiation in turn may then be detected by a bolometer or a thermopile, i.e., a thermal IR detector. Thermal sensors and actuators therefore, form integral parts of various kinds of infrared gas-sensing microsystems [16, 24, 42]. Considering other kinds of gas-sensing technologies, thermal microstructures often play the dual

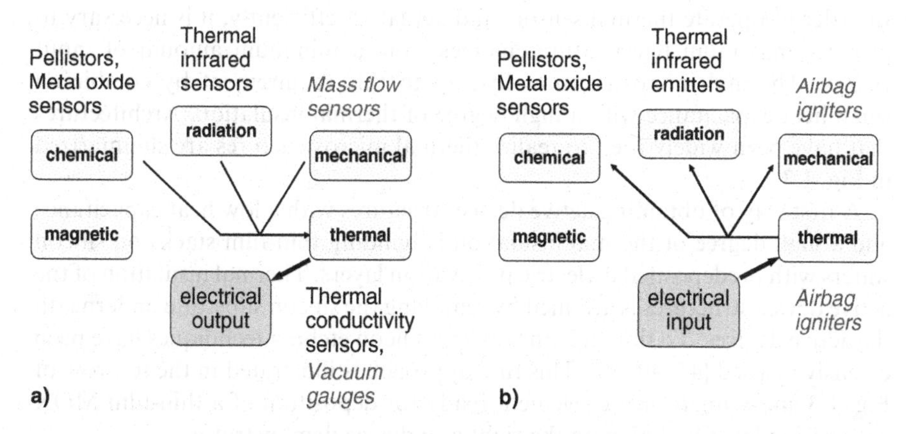

Fig. 1.2 Functional principle of thermal sensors (**a**) and of thermal actuators (**b**). For the sake of illustration some typical application examples are mentioned. Non-gas-sensing applications with mass production potential are indicated by italic letters

role of a sensor and an actuator within one and the same device. Such a situation, for instance, arises in the case of thermal conductivity gas sensors [43]. Such sensors consist of a heated membrane, which is cooled by the ambient atmosphere. As an actuator this microstructure produces the heat that is carried away by the molecules in the ambient atmosphere. As a sensor this microstructure senses the heat that is being carried away and thus provides information concerning the molecular composition of the ambient atmosphere. This dual role can also be observed in catalytic gas sensors, i.e., pellistors [44]. Such devices first produce the heat that is required to initiate chemical reactions of the analyte molecules at a catalytically active surface and secondly they detect the amounts of heat that are generated as the analyte molecules are being burnt at the sensor surface. Once heated to the reaction threshold, micro-heaters may also enable conductance changes in MOX materials [45, 46, 47, 48] that had been deposited onto such micro-heaters. Thermal sensors and actuators also do play a role in non-gas-sensing applications as for instance in the automotive industry. Important mass-driver applications are thermal mass-flow sensors and air-bag igniters.

These examples show that thermal microstructures can be used in a variety of ways to arrive at micro-gas-sensors. Furthermore there is a potential that their production can be aligned with the production of non-gas-sensing devices to arrive at production lots that are sufficiently large to benefit from the MEMS mass production capabilities. Before discussing specific application examples, we still need to consider some general architectural and technological details relevant to thermal microstructures.

1.2.3 Thermal Microstructures

In order to operate thermal sensors and actuators efficiently, it is necessary to produce maximum temperature changes from a minimum amount of input energy. Thermal microstructures live up to this requirement by combining small heat capacitance with a high degree of thermal insulation. Architectures that have been widely used to realize thermal microstructures are summarized in Fig. 1.3.

A first way of obtaining active device structures with a low heat capacitance and a high degree of thermal insulation is building thin-film stacks on silicon wafers with predeposited dielectric passivation layers. Thermal insulation of the active device structures is attained by removing the silicon substrate underneath the active devices. To this end, anisotropic silicon etching techniques have been extensively used [45, 49, 50]. This first approach is illustrated in the top row of Fig. 1.3 and a micro-heater element ready for deposition of a thin-film MOX gas-sensing layer is shown on the right as a device demonstrator.

Standard materials for forming such suspended membrane structures are silicon dioxide (SiO_2) and silicon nitride (Si_3N_4) – either in the form of layer

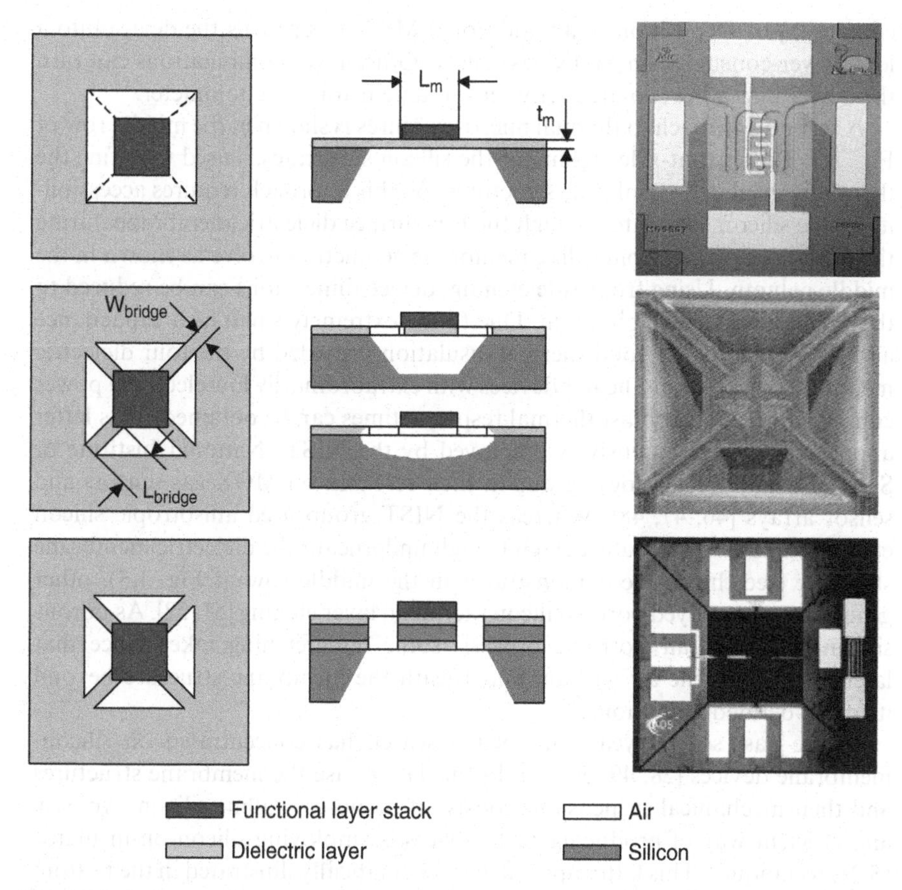

Fig. 1.3 Principle architectures of thermal microstructures. *Left row*: schematics top view; *middle row*: schematics cross-section; *right row*: realized device demonstrators. The active area in the membrane centre usually contains a thin-film stack consisting of a Pt meander that can serve both as a heater and temperature sensing element, a dielectric passivation layer and additional functional layers that depend on the very device application. Geometrical parameters that critically determine the device performance are indicated in the schematics

stacks or in the form of thin-film alloys – to arrive at a small level of tensile mechanical stress inside the suspended membrane. Forming a Pt meander on top of such a membrane, the structure can be used both as a thermal sensor and an actuator device. This double function is enabled by the fact that Pt exhibits a positive and almost linear temperature coefficient of resistivity (TCR). Pt meanders, therefore can serve both as active heating as well as a passive temperature sensing elements. In order to avoid catalytic interactions with the heated Pt meander, the meander is coated with a thin, chemically inert layer of SiO_2. Very often, such passivation layers form the substrate for further functional layers that may be required to arrive at a particular sensor or actuator function. In the demonstrator case shown, one can recognize an interdigital contact layer. Deposition of an additional MOX layer turns the device into a low-power-consumption MOX gas sensor. Other layer combinations can turn the device into a heat conductivity sensor, a pellistor, or a bolometer.

A second approach to thermal microstructures is shown in the middle row of Fig. 1.3. There front-side etching of the silicon substrates is used to define the thermally insulated membrane structures. As this approach requires accessibility of the silicon substrate through the top-surface dielectric membrane during the device processing, spider-like membrane geometries evolve as shown in the middle column. Using front-side etching, device dimensions can be reduced to the smallest necessary minimum. Due to the extremely small heat capacitance and the exceptionally good thermal insulation provided by the four dielectric membrane suspensions, heater devices with extraordinarily low electrical power consumption and very fast thermal response times can be obtained. This latter approach has been intensively employed by the NIST (National Institute of Standards and Technology) group in their research on MOX gas sensors and sensor arrays [46, 47, 48]. Whereas the NIST group used anisotropic silicon etching [49, 50] to generate an etch trough underneath the dielectric membrane structure (see the device demonstrator in the middle row of Fig. 1.3), other groups have employed porous silicon sacrificial layer etching [51, 52]. As porous silicon etching is an isotropic process, some underetching takes place that laterally extends the air cushion underneath the membrane structure beyond its mask-defined limitations.

In the past several years our own research has concentrated on silicon-membrane devices [28, 49, 50, 51]. In this latter case the membrane structures and their mechanical suspensions consist of mono-crystalline silicon. An easy and efficient way of producing such devices is employing silicon-on-insulator (SOI) techniques. This latter approach is schematically illustrated in the bottom row of Fig. 1.3. A more detailed cross-section through such a device is shown in Fig. 1.4 together with some clues concerning their technological realization.

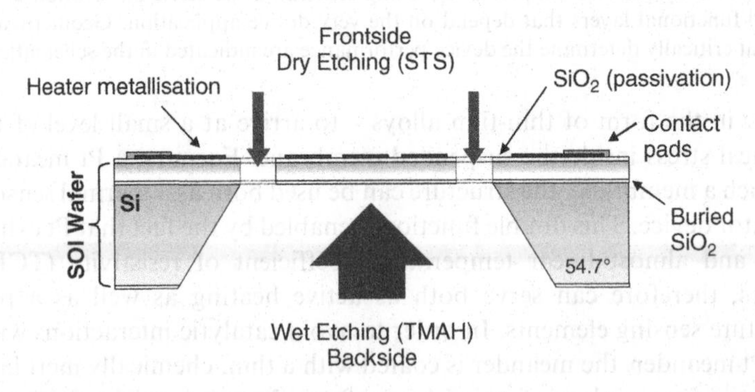

Fig. 1.4 Cross-section through a SOI-based micro-heater device. The membrane structure consists of a thin mono-crystalline silicon layer that is thermally, electrically and mechanically separated from a thick supporting handle wafer

Because the thermal conductivity of bulk silicon is roughly one hundred times higher than that of SiO_2 or Si_3N_4, ultra-low-power-consumption devices cannot be made with this kind of SOI technology. The disadvantage of high thermal conductivity of the silicon membrane material, however, is offset by a number of advantages:

(i) Superior mechanical stability with regard to dielectric membranes [28],
(ii) Integration of active semiconductor devices into the membrane material [53, 54, 55, 56],
(iii) Direct resistance heating of the membrane material [50, 53],
(iv) Enhanced high-temperature stability (T~1000°C) due to the reduced electro-migration in semiconductor as compared to noble metal heaters [50, 57],
(v) Usefulness of the SOI base technology also for other kinds of membrane-type devices such as silicon microphones [58, 59, 60] or flexural plate wave devices [61], which are all of considerable interest in the field of chemical sensors [62].

Quantitatively the performance of the above microstructures depends on material properties, device dimensions and the conditions of device operation. An efficient tool for analyzing the gross performance of such devices is the method of thermal equivalent circuits. Mathematical details concerning this approach have been presented elsewhere [63, 64]. Qualitatively the idea behind this approach is illustrated by the thermal equivalent circuits presented in Fig. 1.5.

In this equivalent circuit it is assumed that heated membrane devices, like the ones shown in Figs. 1.3, can be modeled assuming a homogeneously heated membrane center, heated to a temperature $T_m > T_a$. For simplicity it is further assumed that heat is not generated in the periphery of the membrane that

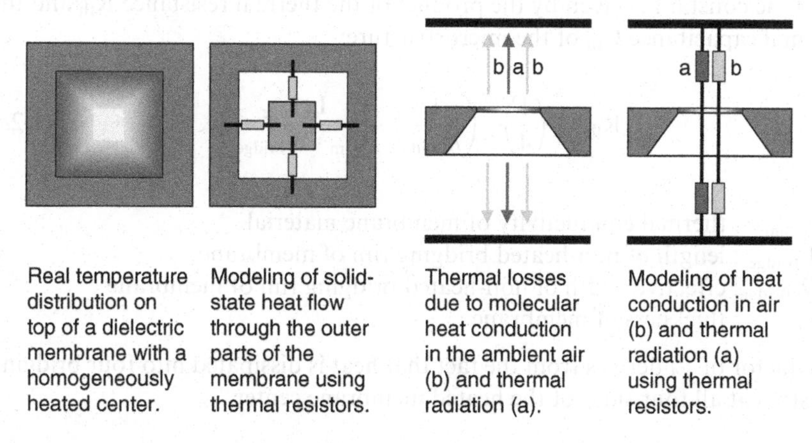

b a b
a b
Real temperature distribution on top of a dielectric membrane with a homogeneously heated center.
Modeling of solid-state heat flow through the outer parts of the membrane using thermal resistors.
Thermal losses due to molecular heat conduction in the ambient air (b) and thermal radiation (a).
Modeling of heat conduction in air (b) and thermal radiation (a) using thermal resistors.

Fig. 1.5 Thermal equivalent circuit describing the flow of heat from a homogeneously heated membrane towards its neighboring heat sinks

bridges the heated membrane center with the cold silicon rim, maintained at the ambient temperature T_a. Heat conduction from this membrane center can then occur via three independent processes: (i) solid-state heat conduction through the membrane, (ii) heat conduction (convection) through the ambient air, and (iii) heat radiation. Each of these processes can be associated with equivalent thermal resistors whose magnitude depends on the temperature difference ($T_m - T_a$), as well as on the thermal and geometrical properties of the heat transporting medium that bridges the gap between the heated membrane center and its cold surroundings [65, 66]. As in electrical circuits, the overall thermal resistance R_{th} of the membrane structure arises from the parallel combination of the individual thermal resistances. Further, using the equivalent of Ohm's law for thermal equivalent circuits the temperature of the membrane center can be obtained as a function of the thermal power that is input into the membrane center:

$$T_m - T_a = R_{th}(T_m - T_a) \cdot P_{in}. \tag{1.1}$$

This latter result shows that a high thermal resistance minimizes the input power P_{in} for attaining a certain membrane temperature T_m. Furthermore this result shows that the actual value of T_m depends on T_a, i.e., on the ambient temperature in which the micro-heater is being operated in. The temperature baseline value of T_a therefore corresponds to the zero-potential baseline in the equivalent electrical circuit.

Neglecting heat conduction, convection and thermal radiation, which can be the dominant thermal losses from a heated membrane, the thermal response time τ_{th} of the microstructure can easily be estimated. This latter time constant measures the rate with which the temperature of the membrane centre can follow changes in the thermal input power P_{in}. Again, as in the electrical case, this time constant is given by the product of the thermal resistance R_{th} and the thermal capacitance C_{th} of the microstructure:

$$R_{th} = \left(\frac{1}{4}\right) \cdot \left(\frac{1}{\lambda_{solid}}\right) \cdot \left(\frac{L_{bridge}}{t_m \cdot W_{bridge}}\right) \tag{1.2a}$$

λ_{solid} : thermal conductivity of membrane material
L_{bridge} : length of non-heated bridging rim of membrane
W_{bridge}: effective width of non-heated bridging rim of membrane
t_m : thickness of membrane

The factor of ¼ derives from the fact that heat is dissipated into four bridging resistors at all four sides of the heated membrane center.

$$C_{th} = \rho_{solid} \cdot C_{P\_solid} \cdot L_m^2 \cdot t_m \tag{1.2b}$$

ρ_{solid} : mass density of membrane material
C_{p_solid}: heat capacity at constant pressure of membrane material.

With the latter two formulas, the thermal time constant can be obtained as a function of the thermal properties of the membrane material and the geometrical dimensions of the membrane structure:

$$\tau_{th} = \left(\frac{\rho_{solid} \cdot C_{p\_solid}}{\lambda_{solid}}\right) \cdot \left(\frac{L_{bridge}}{W_{bridge}}\right) \cdot \left(L_m^2\right) \tag{1.3}$$

Inserting typical values, time constants in the range of milliseconds are revealed. On the other hand, this latter formula also reveals a trade-off. Whereas small heated areas (L_m^2) clearly minimize τ_{th}, there is a conflict when we simultaneously want to obtain good insulation and a small time constant: whereas forming long and narrow bridges yields large thermal resistances and good thermal insulation, the reduced heat flow through such suspensions also leads to long time constants. Similar considerations also apply to the heat conductivity of the membrane materials. This conflict is the thermal analog of the well-known gain–bandwidth limitation in conventional electrical engineering.

1.3 Specific Sensor Examples

In this chapter we should like to discuss how the base structures described above can be used to realize actual device functionalities. The examples discussed include heat conductivity sensors, MOX and field-effect gas sensors as well as thermal infrared emitters. This range of devices is far from being exhaustive. However, it is large enough to demonstrate some of the main features and applications of thermal microstructures.

1.3.1 Heat Conductivity Sensors

Among the thermal microstructures shown in Fig. 1.3 those with dielectric membranes feature the highest degree of thermal insulation and the smallest thermal mass. The high thermal insulation results from the very low heat conductivity of the membrane materials and the low thermal mass from the small thickness of the dielectric membranes ($t_m \sim 0.2\ \mu m$). Operating such a device in ambient air, the membrane center can be heated to about 400°C at the expense of an electrical power input of about 20 mW into the Pt heater meander. Operating this same device in vacuum, the power consumption is reduced to about 2–3 mW. The individual contributions to the heat loss as estimated from the above thermal resistor model are shown as a function of the membrane temperature in Fig. 1.6a. Figure 1.6b shows how the thermal time constant of this same device varies as the temperature of the membrane center is raised.

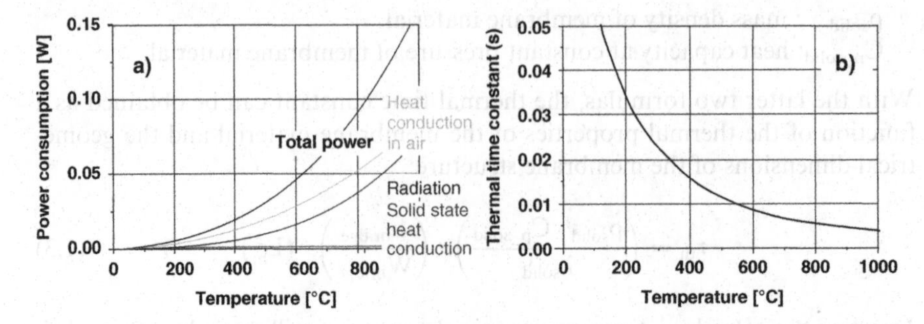

Fig. 1.6 **a**) Calculated heating power consumption of the dielectric-membrane-type device shown in the top row of Fig. 1.3. The individual contributions due to dissipation into solid-state heat conduction, heat transport in air and thermal radiation are indicated; (**b**) thermal time constant as a function of the membrane temperature. The time constant drops with increasing membrane temperature due to the increasing thermal losses due to heat conduction in air and due to thermal radiation

These latter data vividly demonstrate that dielectric-membrane-type heaters make very low-power-consumption and fast-responding devices whose power dissipation is largely determined by heat conduction processes in the ambient air. As a result, the main application of such microstructures to date is in thermal mass flow sensors, which regulate the air intake into car engines [43, 67, 68]. There the convective cooling of flowing air determines the electrical power consumption of the device and, due to their small thermal mass; such devices can follow changes in the air mass flow with single-cylinder resolution. This kind of a non-gas-sensing application, in fact, is a very useful MEMS mass driver application, on which a variety of architecturally related MEMS gas-sensing devices can be based.

In order to operate dielectric-membrane-type micro-heaters as gas sensors, the heaters need to be operated in stationary air. This can be achieved by a packaging as indicated in Fig. 1.7. Very often the access holes in such packages are additionally closed with porous seals that prevent air flows across the heated membrane but allow analyte molecules in the ambient air to diffuse into the sensor package.

In order to see how such as sensor can be used to obtain information about the ambient gas atmosphere, we consider the formula for the heat conductivity λ_{air} of an ideal gas [64, 69]:

$$\lambda_{air}(P, L_c, M_{rel}, T) = \frac{5}{3} \cdot \left(\frac{2 \cdot k_B^3 \cdot T}{\pi \cdot M_{rel} \cdot M_0 \cdot \sigma_s} \right) \cdot \left(\frac{P}{P + P_c} \right) \tag{1.4a}$$

$$P_c = \frac{k_B \cdot T}{\sigma_s \cdot L_c} \tag{1.4b}$$

k_B Boltzmann constant
T air temperature

Fig. 1.7 Principle architecture of a heat conductivity sensor. A heated membrane is enclosed in a package with the membrane at a distance L_{wall}. At normal ambient pressure it can be used to detect the presence of combustible gases (H_2, CH_4) in air. At low pressures, in which the mean free path is constrained by L_{wall}, it can be used to measure the vacuum pressure. *Upright positioning* of the sensor accelerates the exchange of air inside the package by thermal convection

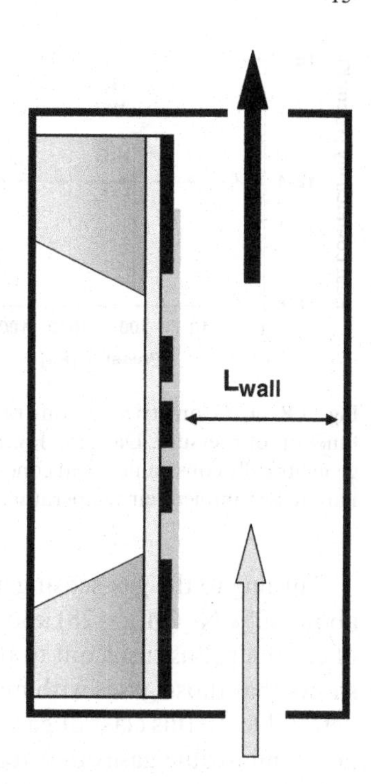

P air pressure
M_{rel} average molecular mass in atomic mass units
M_0 atomic mass unit (amu)
σ_s molecular collision cross-section

Here, P_c is a pressure parameter that characterizes geometrically constrained situations, characterized by a critical length L_c. Such situations arise, whenever the molecular mean free path becomes comparable to the geometrical dimensions of the device or its packaging. In Fig. 1.8 the above formula has been evaluated for the packaging situation indicated in Fig. 1.7. There it has been assumed that the motion of the air molecules is confined to the depth of a typical etch trough inside a thermal microstructure, i.e. $L_c = L_{wall} \sim 300\ \mu m$. As revealed from Fig. 1.8 this constraint becomes relevant when the air pressure P is reduced to about 100 Pa, i.e., to about one thousandth of the normal air pressure. Above this pressure the heat conductivity is independent of P but dependent on the average molecular mass (Fig. 1.8a) and on the gas temperature (Fig. 1.8b). Below about 10 Pa, the heat conductivity becomes linearly dependent on P.

The above results can be used in two ways: in the range of normal ambient pressures the dependence of the average molecular mass can be used to detect gases in air whose molecular mass differs from the average molecular mass of air [43]. In the low-pressure range, heated membrane devices can be used to measure the pressure inside vacuum vessels [70].

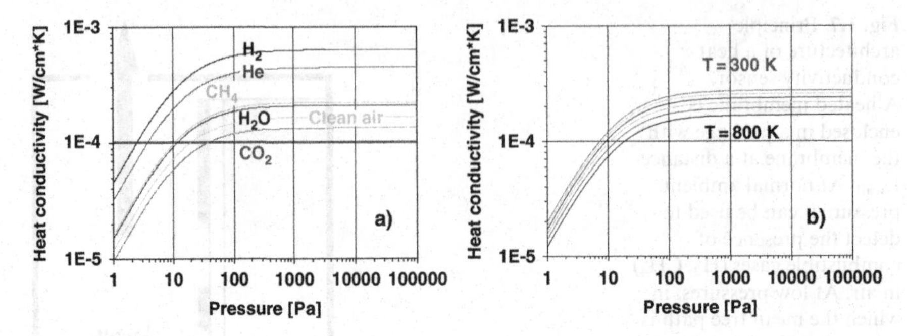

Fig. 1.8 (**a**) Geometrically constrained heat conductivity of clean air at T = 300 K as a function of pressure. Data for H_2, He, CH_4, H_2O, and CO_2 are shown for comparison; (**b**) geometrically constrained heat conductivity of clean air ($M_{rel} = 28.8$) as a function of pressure. Parameter: ambient air temperature

Turning to the gas-sensing application first we note that clean air consists of about 80% N_2 ($M_{rel} = 28$) and 20% O_2 ($M_{rel} = 32$); the average molecular mass of clean air thus turns out to be $M_{rel} \sim 28.8$. Reference to the data of Fig. 1.8a shows that those gases with molecular masses lighter than this should easily be detectable. In this class of gases we find combustible gases such as H_2 and CH_4, i.e., combustible gases that represent safety hazards when their concentration exceeds their lower explosive limits (LEL) of about 4% in air. For detection a heated membrane device is used as in Fig. 1.7 but with the gas inlet and outlet ports being closed by porous membranes. This porous sealing allows for a diffusive gas exchange but it does not allow an ignition process to initiate should the LEL concentration be exceeded. For detection the membrane device is heated above the ambient temperature and kept constant by an electronic control circuit. With this kind of control being established and gases of low molecular weight entering the sensor package, more electrical power needs to be fed into the membrane heater to maintain a constant surface temperature. This excess power consequently can be used as a sensor output signal. The data in Fig. 1.9a show that in this way H_2 can be detected in concentrations far below the LEL of H_2 of 4% or 40,000 ppm. A likely application scenario in the near future will be H_2 driven cars, where safety requires all components of the power train to be monitored for H_2 leakages.

The data of Fig. 1.9b point out a problem that is commonly encountered in the field of chemical sensors, i.e., cross-sensitivity. It shows that heat conductivity sensors not only detect H_2 but also a range of other gases provided two conditions are met: (i) the gas concentration needs to exceed the minimum detectable concentration of about 1000 ppm and the gases need to have a molecular mass that is different from the average mass of clean air, i.e., $M_{rel} \sim 28.8$. Such conditions are particularly true in the case of H_2O vapor. As the heat conductivity is further dependent on temperature (see Fig. 1.8b), it is evident that heat conductivity sensors require both temperature and humidity

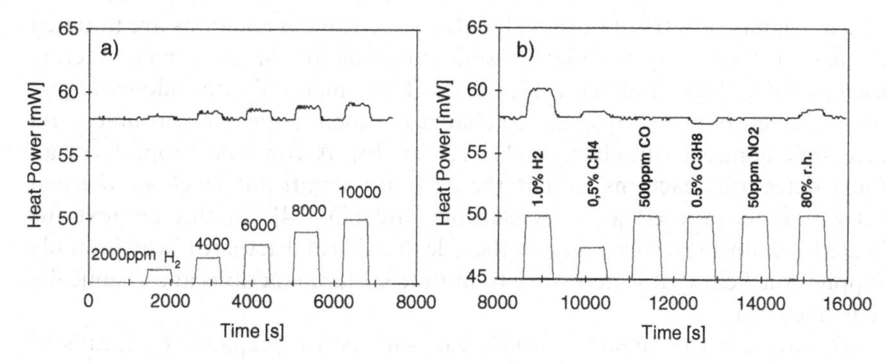

Fig. 1.9 (**a**) Output signal of a heated-membrane-type sensor in response to increasing concentrations of H_2; (**b**) Response of a heat conductivity sensor towards a range of gases that might interfere with H_2 detection

compensation in order to be useful. Discrimination of several gases becomes possible employing sensor array concepts [59].

Turning to the second application, namely vacuum pressure sensors, Fig. 1.8 shows that this is possible at pressures at which the mean free path becomes constrained by the dimensions of the micro-heater device. In the example considered in Fig. 1.7 we assumed that the distance between the heated membrane and the package is on the order of a typical silicon wafer thickness ($L_{wall} \sim 300$ μm). Such macroscopic air gaps can also be realized without invoking MEMS technologies. Macroscopic versions of heat conductivity sensors are currently in wide-spread use in the field of vacuum technology and they are well known under the name of Pirani gauges [70]. Making use of MEMS technologies and realizing very small gaps in the order of 1 μm or less between the heated membrane and a neighboring cold wall, the condition of geometrically constrained mean free paths can already be met in the range of normal ambient pressures [71, 72]. MEMS thermal conductivity sensors therefore can also be used for measurements in the meteorologically relevant range of air pressures. For such sensors all the above considerations concerning cross-sensitivity are relevant. As in a meteorological measurement the ambient temperature and the humidity are being measured alongside with the air pressure, temperature, and humidity compensation are of no concern in this particular application.

1.3.2 Metal-Oxide-Based Gas Sensors

The thermal conductivity sensors discussed above are physical sensors in the sense in that there is no chemical reaction at the heated membrane surface that leads to the detection reaction. A true chemical detection is enabled in case the heated membranes are fitted with gas-sensitive semiconductor materials, which generate electrical output signals once chemical reactions are initiated at their surface. Very popular sensing materials are MOX semiconductors such as SnO_2, TiO_2, and WO_3 [13, 14, 15]. These materials are wide-band-gap semiconductors, which respond to changes in their gaseous environment via reversible conductivity changes [11, 12, 15, 16]. A common property of all these detection reactions is that they require significant levels of thermal activation to proceed at a measurable rate [73, 74]. In this context the heated-membrane device takes on the role of a thermal actuator, which simply supplies the heat that is necessary to initiate surface reactions at a chemically sensitive layer.

To date most commercial MOX gas sensors are prepared by means of screen printing onto bulk ceramic substrates [13, 14]. Due to their bulky nature, ceramic heater substrates consume heating powers on the order of 0.5–1 W per sensor element for attaining the required operation temperatures of about 400°C. This high level of power consumption represents a severe drawback when MOX gas sensors are to be used in bus-connected sensor networks or when sensor arrays with higher levels of gas discrimination are to be realized.

In order to improve this situation, a lot of research has been performed to arrive at much lower-power-consumption devices. This research has involved both the development of low-power-consumption heater devices as well as MOX deposition technologies that are compatible with these much more fragile micro-heaters [75, 76, 77, 78, 79, 80, 81, 82]. With regard to micro-heaters, most of the previous research had been dedicated to dielectric membrane-type devices [45, 46, 47, 48]. In contrast, most of our recent work was devoted to SOI-based heater technologies [28, 49, 50, 54, 55]. This move was largely motivated by the desire to gain processing stability on the level of the silicon MEMS technology as well as by providing substrates with higher mechanical stability and post-processing capabilities that allow innovative sensing layer technologies to be combined with micro-heater technologies [75, 76, 77, 78, 79, 80, 81, 82, 83, 84, 85, 86]. We will come back to this subject in Chapter 5 where we discuss such technologies with a special focus on industrialization aspects.

The SOI-based micro-heater devices developed in our lab are shown in Fig. 1.10 [28, 49, 50]. The top SOI layer in these array heater devices is shaped into thin silicon bridges that are suspended within a rigid chip frame. The heater current is passed directly through the heavily doped top silicon (Si:B) layer, which is about 6 μm thick (Fig. 1.10a). Due to the good thermal insulation, the center regions of the bridges can be heated to typical sensor operation temperatures of about 400°C by electrical power inputs of about 50 mW. On these center regions high-temperature stable and catalytically inactive SnO_2:Sb contacts have been formed to provide electrical access to the sensitive layers [57]. In order to enable different kinds of gases to be distinguished, the left-hand-side bridges carry pure tin dioxide (SnO_2) and the right-hand-side ones Pt-catalyzed SnO_2 gas-sensing layer (see Fig. 1.10b). The data in Fig. 1.11 show that the first

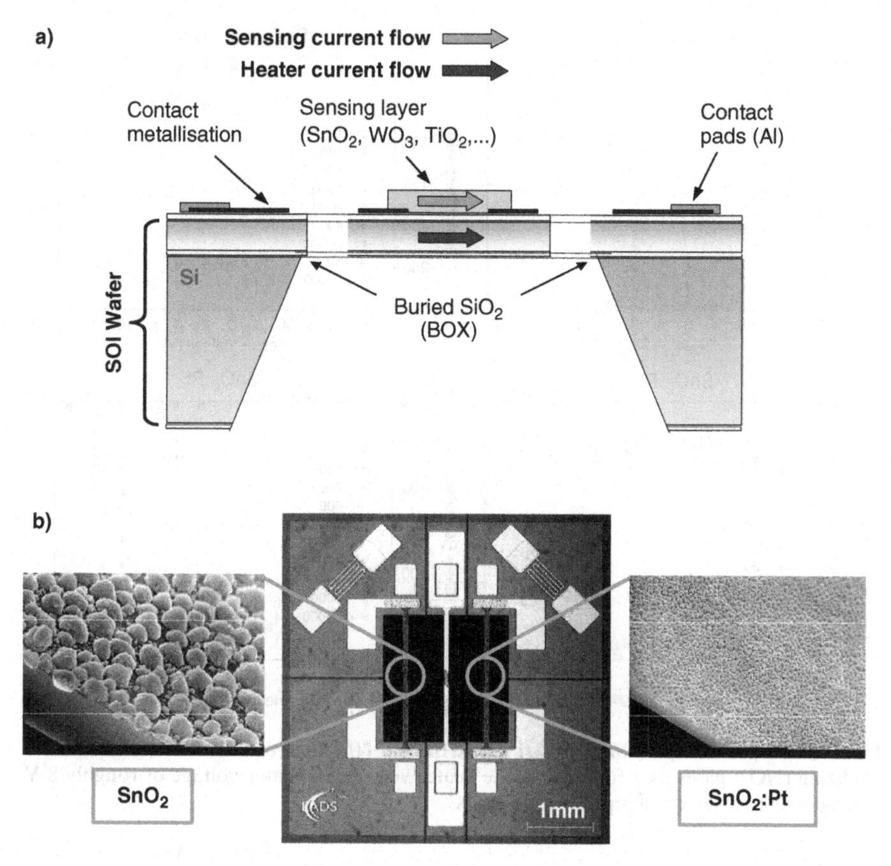

Fig. 1.10 (**a**) Device cross-section of the SOI-based micro-heater arrays; (**b**) SOI-based gas sensor array featuring two MOX gas sensors (outer bridges) and a fast, thermally insulated Pt thermometer on the centre bridge

kind of sensing layers predominantly responds to oxidizing gases (NO_2, O_3), whereas the latter one preferably responds to reducing gases such as H_2 and hydrocarbons [38, 73, 74]. The Pt thermometer in the middle can be used for reference purposes.

A biasing option afforded by the SOI hot-plate technique is that an electric field can be applied in a direction vertical to the current flow in the gas sensitive layer. Several authors have shown that in this way the gas sensitivity of the MOX layers can be changed by means of the chemical field effect [87, 88, 89, 90, 91]. The corresponding biasing arrangement and some representative data are shown in Fig. 1.12. Experiments quite generally reveal that it is the sensitivity to oxidizing gases, such as NO_2, that is most sensitively influenced by the chemical field effect. This latter effect therefore represents another device operation option that can be applied to enhance the discrimination power of thin-film MOX gas sensors.

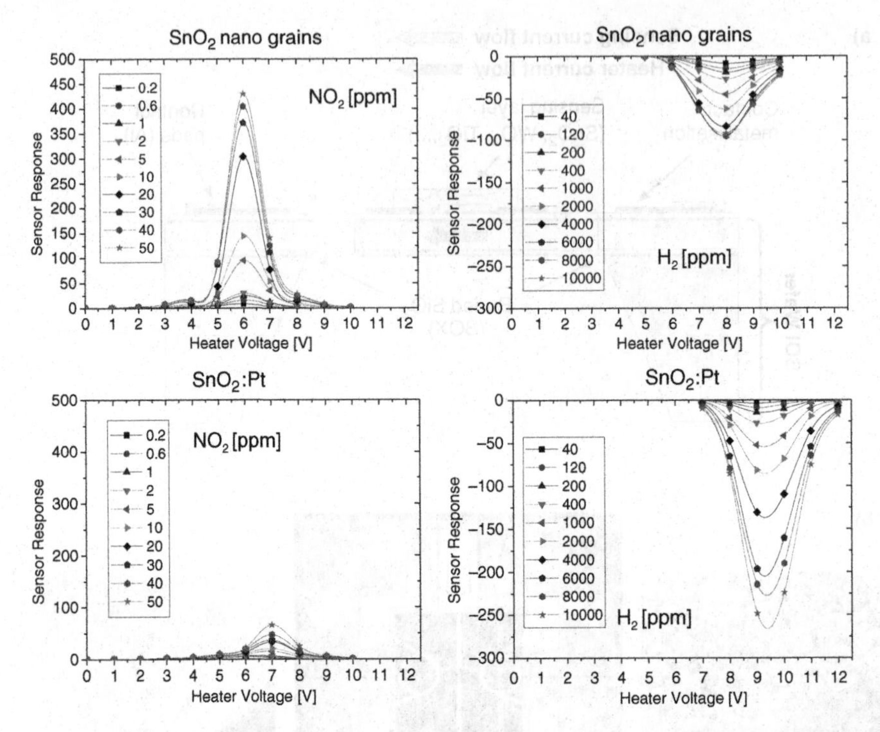

Fig. 1.11 Response of pure SnO_2 (*top row*) and SnO_2: Pt (*bottom row*) to reducing (H_2) and to oxidizing (NO_2) gases as a function of the heater voltage. A heater voltage of roughly 8 V corresponds to a surface temperature of 400°C

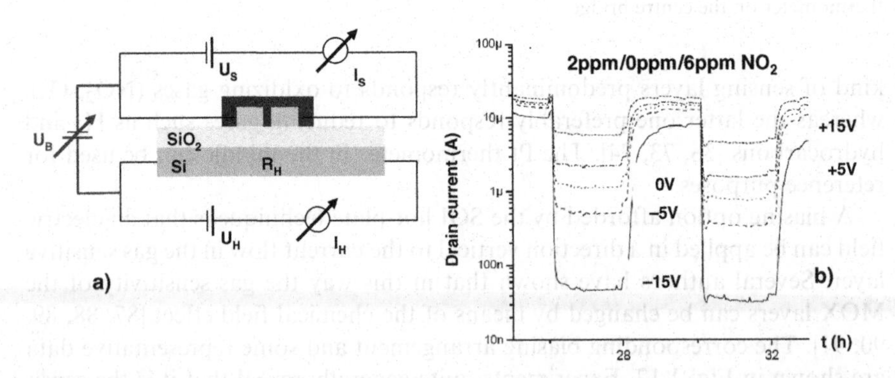

Fig. 1.12 (**a**) Cross-section through a heater bridge demonstrating the biasing options afforded by SOI hot plates. By applying a voltage U_B, the gas response of the MOX layer can be affected by the chemical field effect; (**b**) Sensor response towards two different NO_2 concentrations as observed at a surface temperature of 200°C and after applying different gate bias potentials [89]

1.3.3 Field-Effect Gas Sensors

Building thermal microstructures on SOI material offers the very important advantage that the supporting mechanical structures are made out of mono-crystalline silicon. This allows additional electronic functionalities to be integrated into membrane- and micro-strip heaters. An application that we have been investigating is integrating catalytic-gate field-effect gas sensors [54, 55, 56]. Such devices have first been demonstrated by Lundström and coworkers [35, 36, 37]. With the devices being kept at room temperature or slightly above, such sensors are almost completely selective H_2 sensors. Later work on SiC devices has shown that a much wider variety of hydrocarbon gases can be detected in case sensor operation temperatures in the order of 400–600°C can be applied [92, 93, 94]. As demonstrated in Fig. 1.13, silicon-MOS (SiMOS) capacitors rapidly loose their electronic performance at temperatures approaching 200°C. This so far has limited the widespread adoption of SiMOS sensors to gas-sensing applications.

Building on the thermal properties of SOI-based micro-heaters we have developed a novel operational principle for silicon MOS gas sensors. The principle idea behind this operational mode is applying rapid temperature changes to the SiMOS structures: in a first step a high temperature (HT) is applied. During this high-temperature period catalytic surface reactions are enabled at the noble metal gate and H atoms split-off from hydrocarbon species are able to diffuse as protons to the noble metal – SiO_2 interface (Fig. 1.14). Applying an additional positive high-temperature bias (HTB) during this heating period, the protons can diffuse deeper into the MOS structures up to the SiO_2:Si interface, where they can become trapped and frozen in when the device is rapidly cooled towards room temperature (Fig. 1.15). The data in Fig. 1.16

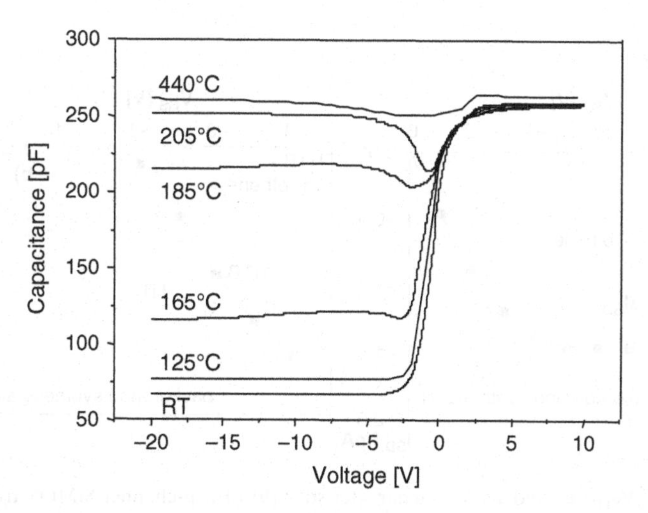

Fig. 1.13 Impact of increasing operation temperature on the C-V characteristics of a silicon-MOS (SiMOS) capacitor

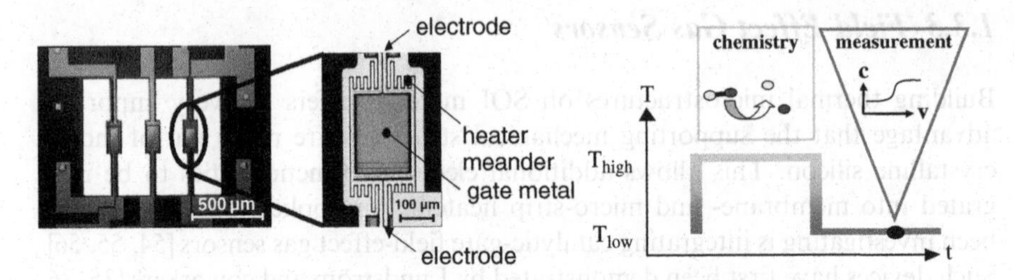

Fig. 1.14 (*left*) Sensor chip with three hot plates; (*right*) Principle idea of a time-sequential operation of chemically sensitive SiMOS devices

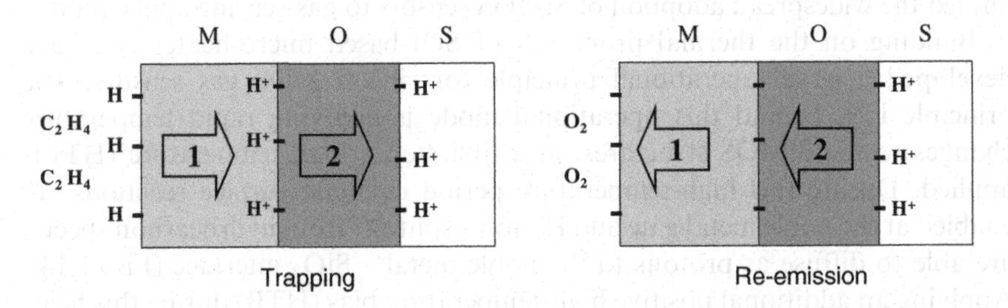

Fig. 1.15 Trapping and re-emission of atomic hydrogen inside SiMOS devices with a catalytically active noble metal electrode

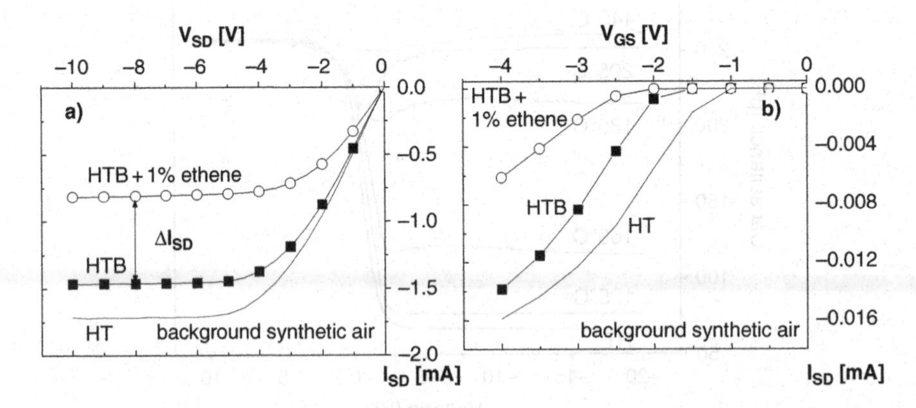

Fig. 1.16 $I_{SD}V_{SD}$- (**a**) and $I_{SD}V_{SG}$ -characteristics (**b**) of a p-channel SiMOS transistor in response to high-temperature cycling (HT), high-temperature-/biascycling (HTB; (X)) and to ethene exposure during high-temperature-/bias-cycling (HTB + gas; (O))

demonstrate that under such conditions changes in the I-V characteristics can be observed that have originated from the previous high-temperature gas exposure. Tests have shown that in this way a whole range of saturated and unsaturated hydrocarbon species can be detected that previously have only been accessible to silicon carbide devices [54, 55, 56].

1.3.4 Thermal Infrared Emitters

Building on the superior high-temperature performance of SOI-based micro-heaters, thermal IR emitter devices have been developed [50, 57]. The emitters shown in Fig. 1.17 use the same kind of SOI base technology as in the MOX and catalytic-gate field-effect gas sensors above. Due to the different actuator requirements, however, these latter devices exhibit a much different geometry. Whereas tiny hot spots are sufficient to heat MOX gas-sensing layers to their appropriate operation temperature, the heated area in IR-emitting devices needs to be made much larger to generate a sufficiently large radiation output. The heated area, on the other hand, cannot be increased beyond limits as larger heated areas increase the thermal response time of the emitters and thus reduce their maximum modulation speed. Requirements for a sufficiently high modulation speed arise from the request of obviating bulky and expensive mechanical choppers in NDIR gas-sensing systems. Both conflicting requirements lead to relatively large hot plates with heated areas of $1.5 \times 1.5\ mm^2$. Although being very similar to MOX gas sensor arrays from the point of view of the silicon technology, very different requirements arise on the level packaging and assembly techniques [95, 96].

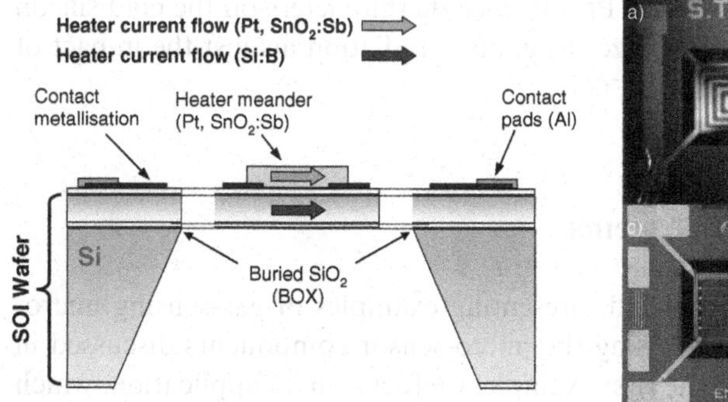

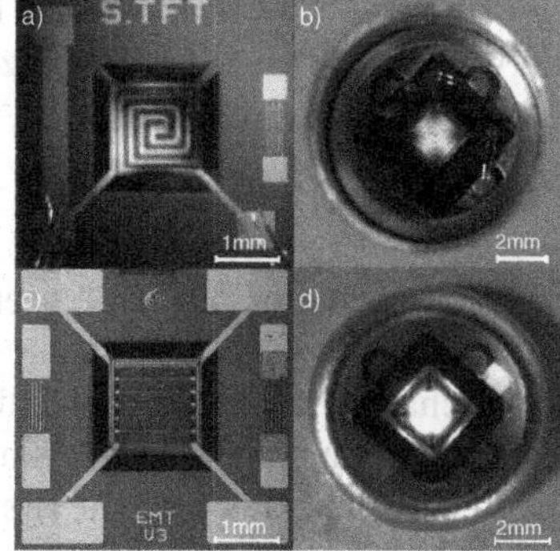

Fig. 1.17 (*left*) Device cross-section of SOI-based thermal IR emitters. The heater current can either be passed through the heavily doped silicon hot plate or through a specific heater metallization deposited on top of the hot plate; (*right*) different versions of SOI-based thermal emitter devices: (**a**) chip with Pt heater meander; (**b**) chip with SnO_2:Sb strip heaters; (**c**) mounted Pt emitter device operated at around 800°C; (**d**) mounted Pt emitter device emitting from the higher-emissivity rear surface of the membrane

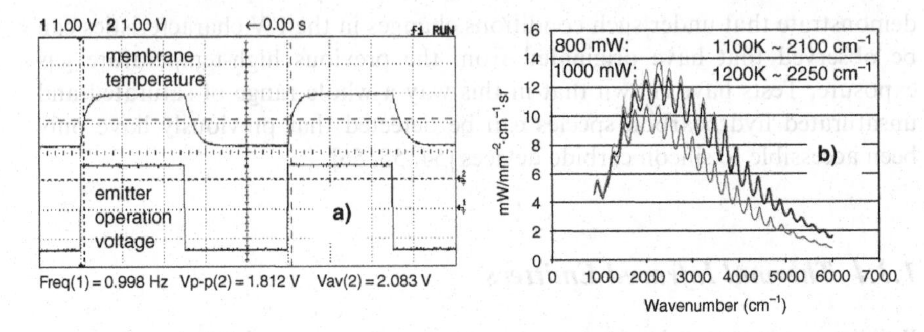

Fig. 1.18 (**a**) Surface temperature change of the hot-plate IR emitters in response to rectangular drive voltage pulses. Maximum modulation frequencies are on the order of 10 Hz; (**b**) Emitted IR power as a function of the wave number. Interference fringes are due to reflections at the upper and lower silicon–air boundaries of the Si:B emitter hot plates

Figure 1.18 presents some device characteristics of the emitters shown in Fig. 1.17. Figure 1.18a shows how the surface temperature of the hot plates responds to rectangular changes in the heater driver voltage. These data reveal that thermal response times are on the order of 100 ms, i.e. possible modulation speeds go up to 10 Hz, which is enough for operating conventional and photo-acoustic NDIR gas-sensing systems in an efficient manner [42, 97]. Figure 1.18b, on the other hand shows that the emitted radiation peaks at around 4–5 μm, thus allowing for efficient NDIR detection in the 3–5 μm atmospheric window. An important point to note is that SOI-based IR emitters exhibit a vastly improved long-term stability with regard to earlier dielectric membrane-type devices: accelerated degradation tests have indicated that SOI emitters can exhibit lifetimes up to 10 years when operated at temperatures approaching 1000°C [50, 57]. The chip layouts in Fig. 1.17 further show Pt reference thermometers on the cold silicon rim, which can be used to stabilize the emitted radiation against the impact of ambient temperature changes [63, 66].

1.4 Gas-Sensing Microsystems

In this chapter we move forward, presenting examples of gas-sensing microsystems that can be built up using the micro-sensor components discussed in Chapters 1.3.2 to 1.3.4. In the first example, we focus on an application, which makes use of a miniaturized electronic nose device to augment the functionality of a conventional smoke detector. It is shown, how the so-formed multi-criteria fire detectors can live up to the low false-alarm-rate requirements of the aeronautic industry. In the second example, we present a novel high-selectivity spectroscopic gas-sensing technique that may find multiple applications in the fields of air quality monitoring, leak-, and fire detection.

1.4.1 Low False-Alarm-Rate Fire Detection

A fire event that may be harmless on ground might become catastrophic when it occurs during flight inside a passenger airplane. For this reason tight authority regulations [98] exist, which require that any area in a passenger airplane, that is inaccessible during flight, needs to be equipped with photoelectric fire detectors. Authority regulations further require that any fire alarm – whether true or not – needs to be answered by appropriate safety measures. These safety measures firstly consist in fire suppression by halon injection into the affected area and secondly in an emergency landing. It is evident that such measures are associated with considerable cost and that – in the case of a false alarm – such actions also cause a loss in the confidence in the aircraft – both on the side of the passengers as well as on the side of the airline operators. Statistics show that the number of such incidents tends to increase along with the general increase in the air traffic. In terms of absolute numbers, such events amount to about 200–300 per year and thus are not negligibly small [99]. Currently, this false-alarm rate is on the order of 1 in 10^5 flight hours and airlines and authorities request that this should be decreased to 1 in 10^7 flight hours, at least for long overseas flights and flights across arctic areas [100].

Currently used fire detectors in commercial and passenger aircraft are based on light-scattering arrangements as shown in Fig. 1.19. These detect the smoke particles that are emitted in the case of a fire event and thereby trigger a fire alarm. As smoke particles represent only one out of a large number of fire criteria, scattered light detectors suffer from a number of severe drawbacks:

- smoke only occurs after a fire has already started. Scattered light detectors therefore do not possess an early fire warning capability;
- condensing humidity and dust particles cause light scattering and therefore are likely to be misinterpreted as fire indicators;
- some fires burn with little or no smoke release and thus go undetected.

Dealing with these drawbacks and trying to decrease false-alarm rates, fire detectors are being investigated that allow measuring more than one single fire characteristic. An approach towards arriving at a multi-criteria fire detector is

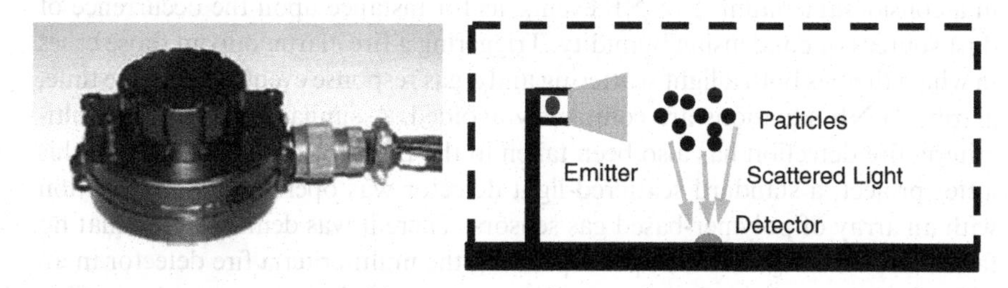

Fig. 1.19 (*left*) State of the art smoke detector; (*right*) light-scattering principle of smoke detection

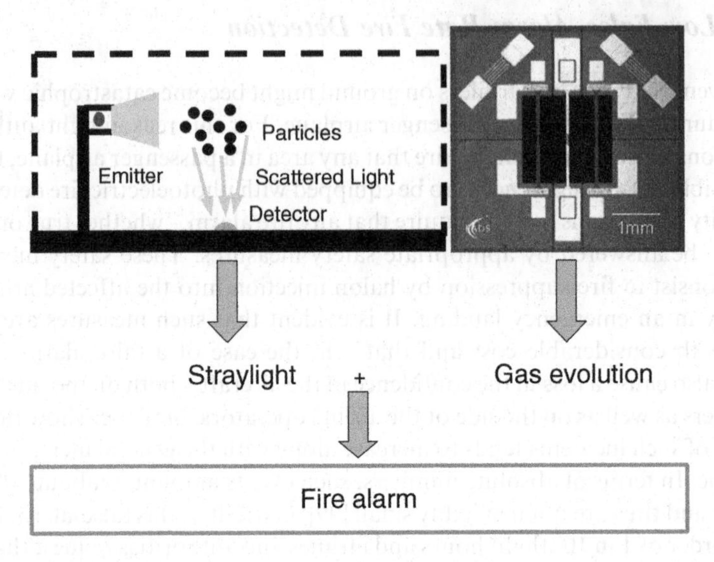

Fig. 1.20 Coincidence detection of smoke particles and fire gases to reduce false-alarm rates in fire detection [93, 94]

illustrated in Fig. 1.20. In such a detector, a conventional smoke detector is combined with a MOX gas sensor array such as the ones shown in Figs. 1.10 and 1.11 above. The innovative principle of this detector is that smoke and gas detection are combined in the form of a logical AND, which means that a smoke signal is only taken seriously once it is accompanied by the presence of fire gases.

We have fabricated several such coincidence detectors and tested them in the Trauen fire test facility of DLR (Deutches Zentrum für Luft- und Raumfahrt). In these tests, the multi-criteria detectors were subjected to a range of test-fire (TF) and non-fire (NF) events as illustrated in Fig. 1.21 below.

In a first series of tests, we have been investigating the possibility of false-alarm suppression. For this reason, the multi-criteria detector was subjected to a series of fire and non-fire events and both sensor outputs, i.e., the smoke sensor and the individual gas sensor outputs were simultaneously monitored. Figure 1.22 shows that using the scattered light detector alone would have produced fire alarms also in a considerable number of NF events, as for instance upon the occurrence of dust sources or condensing humidity. Triggering a fire alarm, only in those cases in which there is both a light scattering and a gas response event at the same time, alarms to NF situations are completely avoided. A similar approach to multi-criteria fire detection has also been taken in the EU project IMOS [102]. In this latter project, a standard scattered-light detector was operated in combination with an array of polymer-based gas sensors. There it was demonstrated that no false alarms were generated upon operating the multi-criteria fire detector in an office building, where cigarette smoke was generated, or in a wood recycling centre, where a lot of dust was produced.

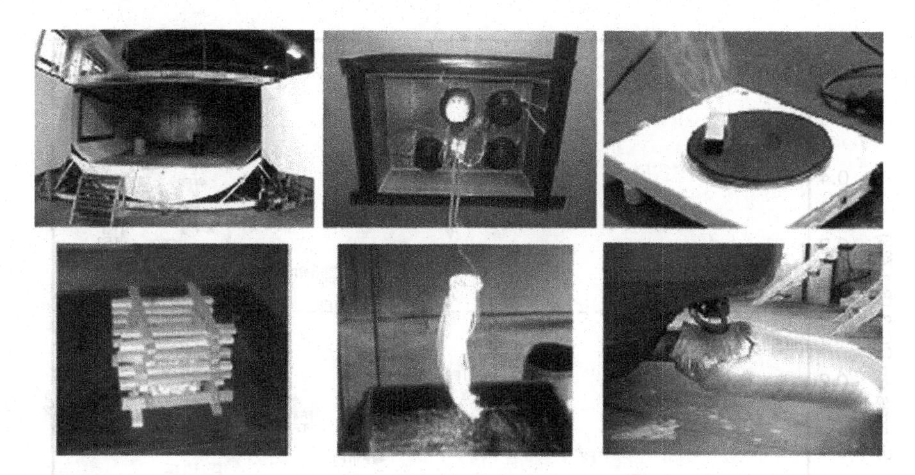

Fig. 1.21 Generating open and smoldering fires as well as non-fire events inside an aircraft environment

False alarms may still be generated by such multi-criteria detectors, in case there are independent sources of dust or condensing humidity and sources of gas, such as air contaminants drawn in with the outside air or odorous events inside the aircraft. In order to distinguish such situations from true fire events, the output of the gas sensor array needs to be considered in greater detail. In order to allow gases and gas mixtures to be distinguished, the two sensors in the array were fitted with two kinds of sensitive layers with distinctly different cross-sensitivity profiles as shown in Fig. 1.11. In this array, the first sensor predominantly responded to oxidizing gases such as NO_2 and O_3 and the second to combustible gases, such as H_2, hydrocarbons, and CO. With such an array,

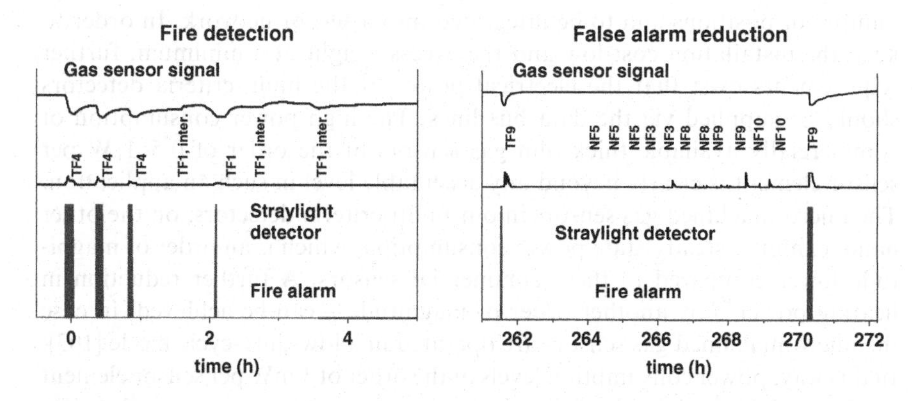

Fig. 1.22 Fire detection and false-alarm reduction using coincidence detection of smoke particles and fire gases [101]

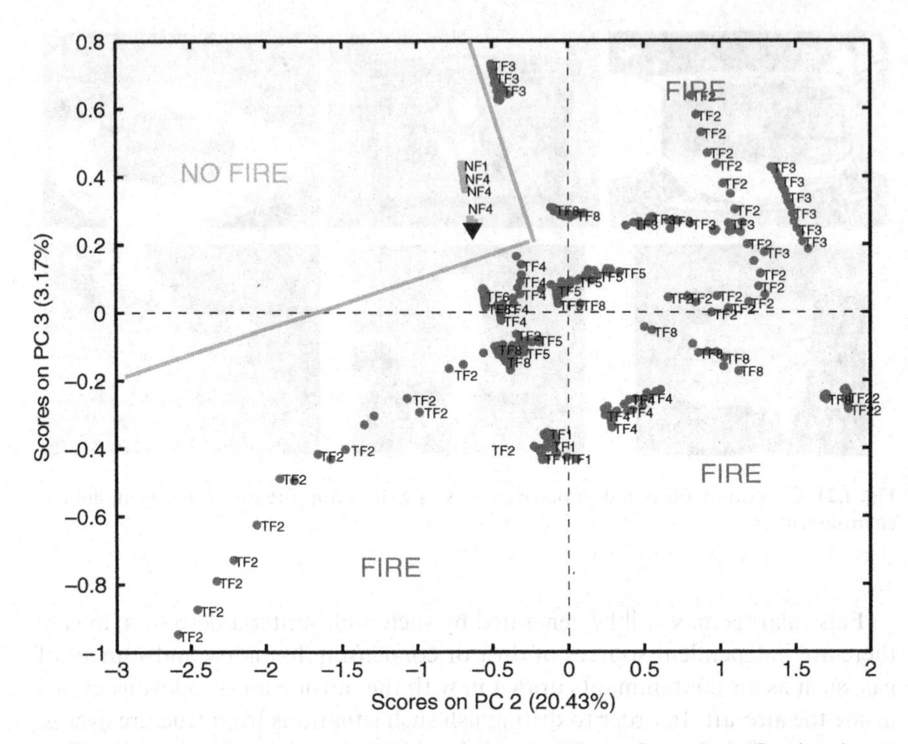

Fig. 1.23 Response of a two-sensor MOX array towards different fire and non-fire situations. TF and NF stand for well-defined TF and NF situations as generated in the standardized test procedures indicated in Fig. 1.21

different fire events can be distinguished on account of their kinds and concentrations of fire gases emitted. This latter possibility of fire distinction is illustrated in Fig. 1.23.

In an airplane, such multi-criteria detectors need to be positioned in a number of positions and to be integrated into a sensor network. In order to keep the installation cost low and the excess weight at a minimum, further requirements exist that the electrical power of the multi-criteria detectors should be supplied via the data bus lines. The high power consumption of commercially available thick-film gas sensors of the order of 0.5–1 W per sensor element is clearly beyond any acceptable level in such an application. The micro-machined gas sensors in our multi-criteria detectors, on the other hand, exhibit a steady-state power consumption, which is an order of magnitude lower compared to these commercial sensors. A further reduction in heating power – by another order of magnitude – can be achieved, in case the micro-machined gas sensors are operated in a low-duty-cycle mode [103]. In this way, power consumption levels of the order of 1 mW per sensor element can be reached. Such levels of electrical power are easy to transmit via data bus lines.

1.4.2 Air Quality Monitoring and Leak Detection

The MOX based gas-sensing microsystems discussed above are unspecific in the sense that they do not allow identifying the analyte gas molecules as such. MOX gas sensors and gas sensor arrays rather classify different kinds of analyte gas molecules with regard to their oxidizing or reducing power relative to the oxygen ions that abound on heated MOX surfaces.

A spectroscopic technique that can be built on principally the same kinds of thermal microstructures as MOX gas sensor arrays, and that is capable of identifying analyte gas molecules as such, is non-dispersive infrared (NDIR) absorption. The principle architecture of an NDIR system is shown in Fig. 1.24. Such systems consist of an electrically modulated thermal IR emitter, an optical absorption path, and a thermal IR detector, which is made wavelength selective by an appropriate narrow-band optical filter. The optical absorption path is an air gap, typically a few centimeters long, which can communicate with the ambient air and which therefore can contain a certain amount of the analyte gas to be detected. The narrow-band IR filter, on the other hand, is tuned to an IR wavelength that coincides with a strong vibrational absorption mode of the target analyte molecule. The number of analyte gas molecules in the optical absorption path can then be determined from the attenuation of the IR radiation within the surveillance path.

NDIR gas detection technology, in general, is best suited to the detection of gases, which have characteristic absorption lines in the 3–5 μm or 8–12 μm ranges, i.e., outside those spectral regions where broad and strong water-related absorption features prevail. Potentially interesting molecules that satisfy this condition are: CO_2, CO, N_2O, and a wide range of hydrocarbon species (represented in Fig. 1.25 by the example of CH_4). Also included in this group are hydrogenated fluorocarbon gases such as tetrafluoroethane (CF_3CH_2F), which still are important cooling agents with a potentially high impact on the UV

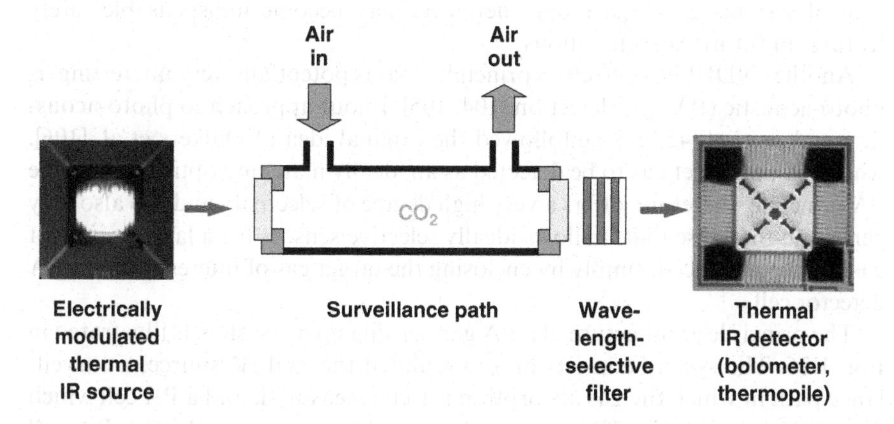

Fig. 1.24 Principle architecture of a NDIR gas-sensing microsystem

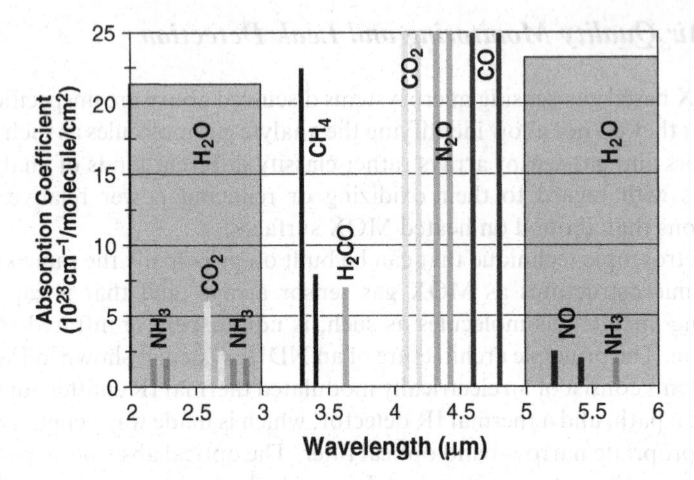

Fig. 1.25 Infrared absorption spectra of a number of relevant gas species in the area between 2 and 6 μm. This is a simplified graphic presentation to show the importance of the individual absorption lines, as well as the shadowing effect of water vapor

shielding properties of the stratospheric ozone layer. Examples that do not satisfy this condition are NO_x species and NH_3. Within the aeronautic field, the largest interest in these species is on CO_2, CO and hydrocarbons. CO_2 monitoring is interesting for the reason of air-quality monitoring and realizing ventilation-on-demand strategies. The interest in CO is due to CO being a potential fire indicator and hydrocarbon species are interesting for detecting fuel and hydraulic leaks in those areas of a commercial aircraft, which are not readily accessible by maintenance personnel. In the field of automotive applications, CO_2 detection is very interesting because low concentrations of CO_2 lead to reduced driver attention, drowsiness, and eventually "micro-sleep," High concentrations of CO_2 are also potential safety hazards in cars with CO_2 climate control systems. CO_2 detectors, therefore, may become indispensable safety features in future car generations.

Another NDIR gas detection principle, that is potentially very interesting, is photo-acoustic (PA) gas detection [104, 105]. In our approach to photo-acoustic gas detection [42, 97] we followed the original idea of Ohlkers et al. [106], who used the target gas to be detected as an ideally matching optical filter. The PA principle, therefore, offers a very high degree of selectivity and it is also very generic in the sense that it allows ideally selective sensors for a large variety of gases to be produced, simply by enclosing the target gas of interest inside a PA detector cell.

The principle architecture of a PA gas-sensing micro-system is illustrated in Fig. 1.26. The system consists of a modulated thermal IR source, a surveillance path in which the gas absorption is to be measured, and a PA cell which forms the gas-specific IR sensing element. In our approach, the PA cell

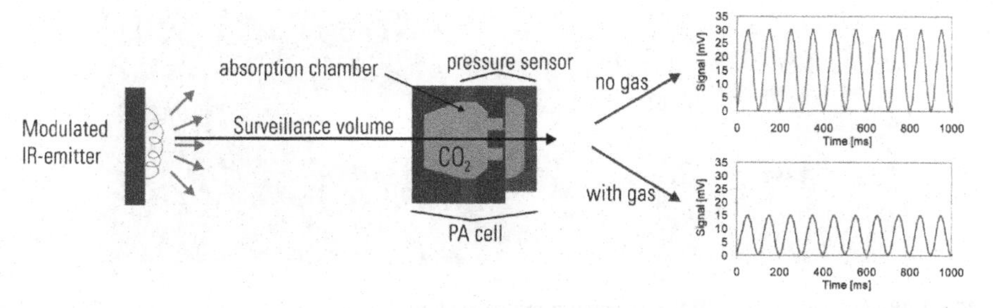

Fig. 1.26 Principle architecture of a PA gas-sensing micro-system. The output signals with and without target gas in the surveillance volume are shown in the *right*

contains a lifetime filling of the gas to be detected. When such a PA cell is irradiated with pulsed thermal IR light, the enclosed gas absorbs incident radiation within its own characteristic absorption spectrum as indicated in Fig. 1.27. The vibrational-rotational energy stored in the target gas molecules is rapidly converted into kinetic energy, i.e., into heat. Further, as the target gas is enclosed in a fixed volume, these temperature changes generate acoustic pressure pulses that can be detected with a sensitive microphone [58,59,60]. The strength of this acoustic output signal depends on the incoming light intensity and the number of target gas molecules enclosed within the absorption chamber. Target gas molecules crossing the surveillance volume, i.e., the volume between the IR emitter and the PA absorption cell, absorb light within the characteristic absorption feature of the target gas and thus reduce the strength of the acoustic output signal. The measured attenuation of the

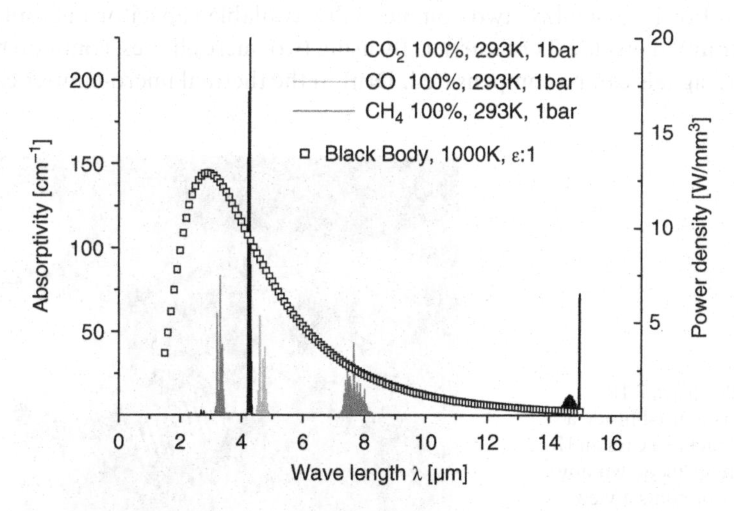

Fig. 1.27 Molecular absorption spectra of CO_2, CO, CH_4 and the emission spectrum of a 1000 K black body source

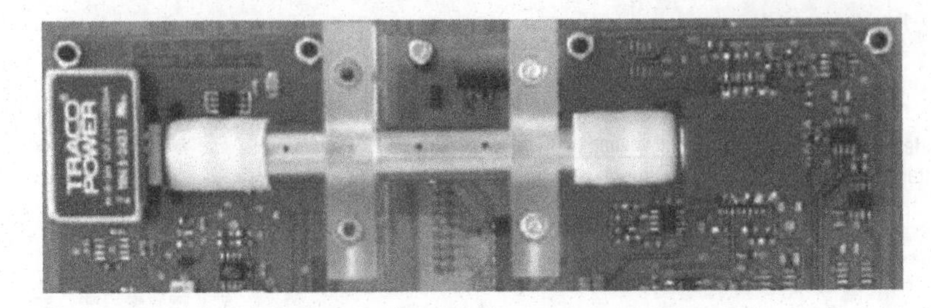

Fig. 1.28 *Top view* onto a PA gas-sensing micro-system

acoustic output signal therefore provides a direct measure of the target gas concentration within the surveillance volume. Moreover, as the PA cell also produces an output signal when the target gas is not present, the PA detection principle also contains a built-in self-test feature.

A hardware realization of a PA gas-sensing system is shown in Fig. 1.28. The surveillance path consists of a 10 cm perforated stainless steel tube. The printed circuit board (PCB) on the left contains the thermal IR emitter and the associated driver electronics. The PCB on the right contains the receiver electronics including the PA cell itself. This latter cell is mounted inside a plastic cube which – besides the PA cell – also contains the preamplifier circuits for the microphone and an optional thermopile reference detector inside the PA cell. Close-up views onto the thermal micro-emitters and onto the PA detector cells are presented in Figs. 1.29–1.31.

Figure 1.29 shows the thermal IR emitters already discussed in Chapter 1.3.4. Further innovative micro-components are the hybrid and monolithically assembled PA cells shown in Figs. 1.30 and 1.31. Figure 1.30 shows a hybrid PA cell. For its assembly, two commercially available capacitor microphones were mounted inside a TO-8 carrier. By using two microphones, common mode acoustic signals can be compensated. Drift in the thermal micro-emitter can be

Fig. 1.29 Thermal IR emitter as housed inside a ceramic package with an IR-transparent silicon window. The inset presents a view onto the glowing surface of an un-housed emitter chip

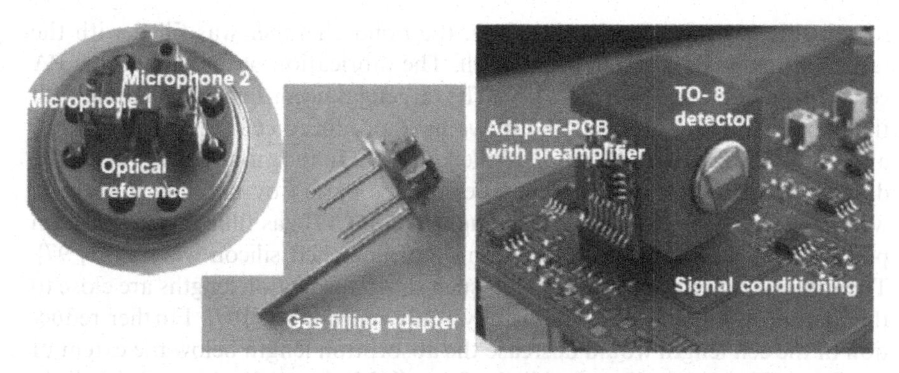

Fig. 1.30 Hybrid-assembled PA cell. *Left:* two microphones and a reference thermopile as integrated into a TO-8 carrier; *Center:* TO-8 carrier showing the metal tube for gas filling and PA cell sealing. *Right:* fully assembled and gas-filled hybrid PA cell as mounted inside a plastic cube containing preamplifier circuits for the microphones and the thermopile detector

monitored by the reference thermopile mounted in the space in between the two microphones. The center panel in Fig. 1.30 shows a metal pipe at the bottom of the TO-8 carrier. After assembly of the microphones and the thermopile reference detector on the TO-8 socket and after sealing the lid with an integrated IR-transparent silicon window on top, the hybrid PA cell can be evacuated and refilled with the target gas to be detected. Instead of the target gas itself, this type of cell can also be filled with gas mixtures consisting of the target gas and suitably chosen buffer gases that enhance the vibrational translational energy relaxation [107]. The hybrid PA cell affords a high degree of flexibility, both with regard to the type of gas filling as well as with regard to possible electronic compensation features.

A higher degree of miniaturization was achieved, using the monolithic assembly approach shown in Fig. 1.31. The left hand side of Fig. 1.31 shows a PA cell that was formed from an anodically bonded glass–silicon–glass stack. This stack contains as a central layer a piezoresistive microphone, fabricated using the SensoNor MultiMEMS foundry service [108]. The PA cell was formed by sealing the microphone glass–silicon–glass stack via anodic bonding to a

Fig. 1.31 Monolithic PA cell [58]. *Left:* Single PA cell and cross-section. *Center:* PA cell as mounted in a ceramic carrier. *Right:* Ceramic carrier mounted inside a plastic cube

fourth silicon wafer. During bonding, the bond chamber was filled with the target gas to be enclosed in the PA cell. The fabrication of the monolithic PA cell is further detailed elsewhere [58]. The target gas needs to be compatible with the anodic bonding process. So far, we have produced cells that contain CO_2 and N_2O, i.e., two gases of major applied interest. In the monolithic PA cell, the dimensions of the absorption chamber can be varied by bonding bottom Si wafers with differently-sized etch troughs. With CO_2 gas fillings, an optimum performance was obtained by bonding flat, unetched silicon wafers [42, 97]. Theoretical considerations show that such short absorption lengths are close to the limit of miniaturization that is physically reasonable [107]. Further reduction of the cell length would decrease the absorption length below the extent of the thermal boundary layer inside the PA cell. Miniaturizing beyond that limit would counteract the building up of a significant PA cell pressure.

The gas-sensing performance of a system with a CO_2-filled, monolithic PA cell is shown in Fig. 1.32. The diagram in the upper part of Fig. 1.32 shows, how the sensor signal varies as the PA system is exposed to a train of CO_2 pulses of decreasing amplitude. This figure clearly shows, that in the absence of any CO_2, a large output signal is generated due to the un-attenuated absorption of the thermal IR radiation inside the PA cell. Upon application of an intense CO_2 gas pulse, the output signal is reduced as part of the radiation that can be absorbed by the PA cell is now absorbed in the surveillance path. Decreasing the amplitude of the CO_2 gas pulse increases the output of the PA cell until eventually the clean-air output signal is approached again. A more in-depth analysis of these data shows that the minimum detectable CO_2 concentration is on the order of 20 ppm. The lower diagram of Fig. 1.32, on the other hand, addresses the

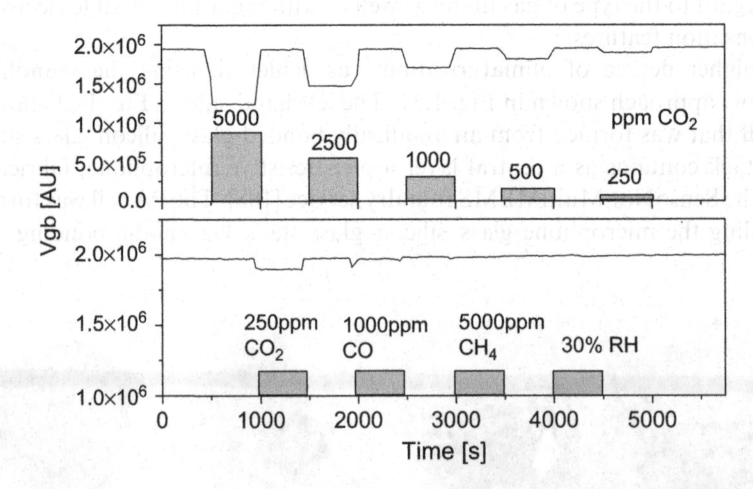

Fig. 1.32 (*top*) Digitally processed sensor output signal V_{gb} in response to different concentrations of CO_2; (*bottom*) cross-sensitivity of a PA cell towards 1000 ppm CO, 5000 ppm CH_4, and 30% humidity. The individual sensor responses to 250 ppm CO_2 are shown for comparison

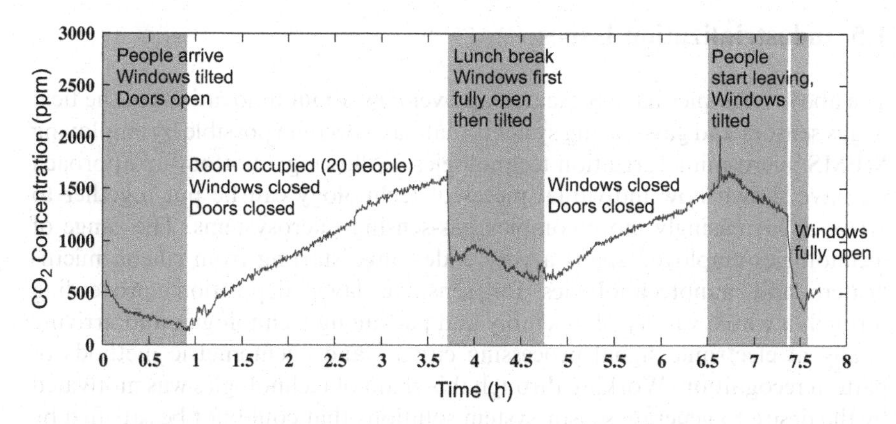

Fig. 1.33 Variation of the CO_2 concentration inside a big meeting room during a 1-day meeting with a maximum occupation of 20 persons. During lunchtime and after the end of the meeting windows were opened to admit fresh outside air

important issue of cross-sensitivity. In this latter experiment, sensor systems with CO_2-filled monolithic PA cells were exposed to pulses of CO_2, CO, CH_4, and water vapor. It is clearly seen that a response is only produced in those cases in which the analyte gas in the surveillance path is identical to the gas inside the PA cell.

Figure 1.33, finally, shows the result of an application test. In this latter case the sensor system was positioned inside a large meeting room, as about 20 people were gathering for a project review. As indicated by the remarks in this figure, the sensor signal clearly traces the production of CO_2 by the respiration of the participants and the effect of ventilation actions.

Whereas, due to safety reasons, monolithic cells could only be filled with CO_2 and N_2O so far, hybrid cells allowed for a much wider range of gas fillings. Due to their larger geometric size, these latter cells exhibited somewhat smaller gas sensitivity than the monolithic ones. With this type of cells, however, the flexibility and the potentially very wide range of gas-sensing applications, accessible to the PA gas-sensing technique, could be demonstrated (Table 1.1).

Table 1.1 Minimum detectable gas concentrations using hybrid PA cells

Cell type	Gas filling	Gas sensitivity [ppm]
Hybrid	40% CH_4, 60% Ar	100 ±10
Hybrid	40% CO, 60% Ar	50 ±10
Hybrid	40% CO_2, 60% Ar	50 ±5

1.5 Industrialization Issues

The above examples have presented an overview about innovations in the field of gas sensors and gas-sensing systems that have become possible by employing MEMS micro-miniaturization technologies. Following a bottom-up approach we have shown how individual pieces of technology can be put together to arrive at increasingly more complex gas-sensing microsystems. The range of technologies employed spans a very wide range starting from silicon micro-heaters and nanotechnologies for sensitive layer deposition, proceeding through a whole variety of assembly and packaging technologies and arriving finally at electronic signal processing circuits and chemometric methods of pattern recognition. Working through this chain of technologies was motivated by the desire to generate sensor system solutions that could not be satisfied by sensor systems readily available on the supplier markets. In this final chapter we should like to take a more top-down view on the above MEMS-based sensor technologies. Considering the example of low false-alarm-rate fire detection we should like to consider in some detail the kind of work share that needs to be established, both on the level of research and development as well as in the commercialization, to successfully introduce such products into the market. The challenges along this way are discussed under the subheadings below.

1.5.1 Initiating a System-Level Innovation

As already mentioned, the base motivation to employ novel pieces of technology is solving problems that require a solution and that cannot be satisfied with commercial sensor products. In the case of low false-alarm-rate fire detection the requirement is developing fire detectors which feature small-size, light-weight, low-power consumption and which are capable of network integration. In order to arrive at the required low level of false alarms, a multi-sensor solution needs to be sought that allows more fire criteria to be measured than just smoke. All these requests clearly call for a micro-system solution. Considering fire gases as the most important fire criterion next to smoke, there is a clear request to develop miniaturized gas sensor arrays that can be combined with conventional smoke detectors and that can fulfill the above size, weight, and power consumption constraints. Working towards a MEMS-based micro-system solution requires dealing with a number of development and technology transfer issues that are pictorially represented in Figs. 1.34a and 1.34b.

1.5.2 Building Added-Value Lines

In the first place, it is not clear whether a detector combination consisting of a standard smoke detector and a MOX gas sensor array would indeed solve the

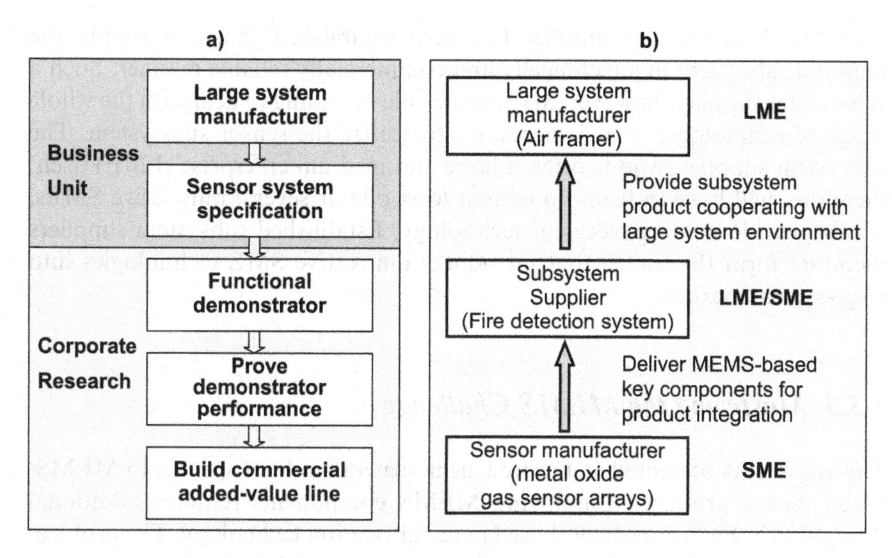

Fig. 1.34 (**a**) Workshare in sensor system development; (**b**) work share in the industrialization phase

problem of an airframe manufacturer. This large-system manufacturer is able to derive a sensor system specification from the safety requirements that are imposed on the aircraft and on its operation. The airframe manufacturer, however, will not be the actual sensor system manufacturer and thus needs to consult possible sensor system suppliers. In case a commercial product is not available or can be derived with a moderate effort from existing sensor system products, a sensor system manufacturer normally would not engage into the development of a prototype because the risk is very high that the prototype would not perform to satisfaction or that the problem could be solved by a competitor using some other technology. In such situations it normally happens that corporate research units become involved. Such internal units can be endowed with the task of building a demonstrator and proving that the demonstrator sensor system is actually able to perform in the specific application scenario. Very often such research is performed within the framework of public funded projects. Proceeding along this line allows specific know-how available in universities and research institutions to be used in the process of problem solving. Last but not the least, such consortia also need to involve other industrial partners, preferably small and medium enterprises (SME). Ideally such SMEs should be chosen in such a way that a complete industrial added value line can be formed, should the prototype turn out as a success. This process of research organization is schematically illustrated in Fig. 1.34a. In case of success, therefore, an added value line may develop in the end as shown is Fig. 1.34b. This latter figure presents the top-down point of view of the large-system manufacturer. From this point of view, the problem is solved, once a commercial sub-system supplier has been established that can supply the required subsystem in a technically and commercially reliable manner. Such a sub-system supplier, however, may not be able or willing to deal with the whole range of technologies that is necessary to realize the sensor sub-system. The subsystem supplier, who is often a large and medium enterprise (LME) itself, therefore, will have to team up with at least one or several innovative SMEs, which provide missing pieces of technology. Established subsystem suppliers therefore form the bridge that introduces innovative SME technologies into large-system markets.

1.5.3 Mastering the MEMS Challenge

The arguments presented so far are true in general and not specific to MEMS-based sensor products. Employing MEMS components requires additional thought as MEMS is a silicon-based mass fabrication technology. The problem that comes in at that level is that specialized sensor products such as aeronautic safety sensor systems seldom produce production lots that are sufficiently large to make production attractive for MEMS foundries. Moreover, MEMS foundries are not ideal sensor manufacturers as producing MOX gas sensors involves non-standard materials that are not part of and not normally allowed in silicon processing lines. Last but not the least, sensitive film production requires specialized knowledge in chemistry and thin-film deposition, special test equipment, and application know-how. This latter range of know-how is generally in the hands of established thick-film gas sensor manufacturers. In order to successfully introduce MEMS micro-heater technologies into the field of MOX gas sensors, a work share between silicon foundries and specialized sensor manufacturers needs to be set up that is profitable for both sides.

Profitability from the point of view of an MEMS foundry means that it can engage into large-volume silicon processing without the need of entering non-silicon technologies and unknown markets. Such a situation arises in case MEMS micro-heaters can be produced in large quantities for different market segments, some of them being potential mass drivers (Figs. 1.35 and 1.36). Such a situation is indeed the case considering the range of products that can be derived from the silicon micro-heaters described in this chapter. Here at least two applications – mass flow sensors [64, 67, 68] and air-bag igniters [64] – relate to automotive mass markets with potentially millions of units per year. Profitability from the point of view of a specialized sensor manufacturer means that the required relatively small number of MEMS heaters can be split from the large-volume production of the MEMS foundry and purchased for specialized processing in the SME sensor company. The added value generated in this latter company builds on specialized know-how in the field of sensitive layer deposition (thin-film, sol-gel, thick-film, etc.), sensor integration and software and specialized application know-how. Unlike the case of a silicon foundry, these

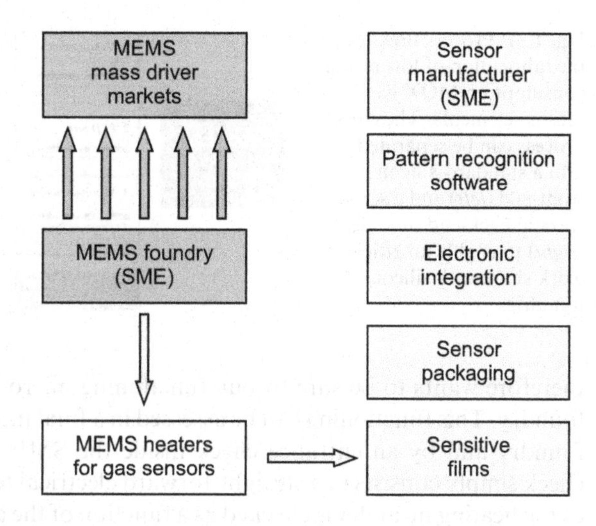

Fig. 1.35 Work share between silicon foundry and sensor system manufacturer

latter pieces of technology are much more based on specialized know-how than on heavy investment in semiconductor processing facilities.

1.5.4 Cooperation Across Technical and Economic Interfaces

In order to make the cooperation between a MEMS foundry and a sensor system manufacturer work, a process flow needs to be agreed on that produces a clear-cut technical and commercial interface between both partners. In order to achieve this, the sensor fabrication needs to be split into a MEMS front-end and a sensor-specific back-end as illustrated in Fig. 1.37. Here the front-end is concerned with the silicon processing that leads to a micro-heater that features the necessary heating meanders and sensing electrodes, but which does not yet contain the gas-sensitive layer itself. The know-how about these layers is proprietary knowledge of the SME sensor company. This latter company

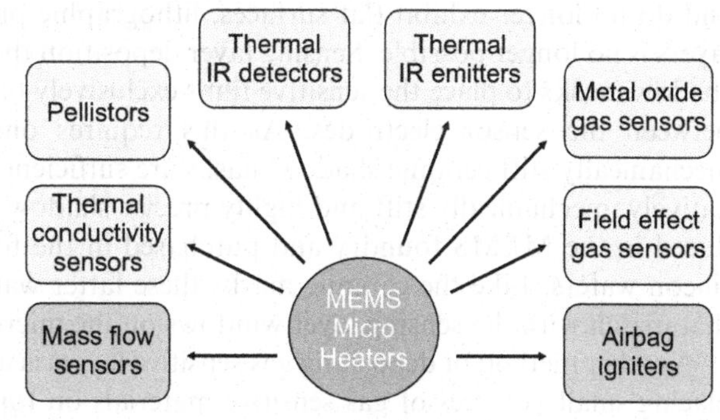

Fig. 1.36 Piling of related production lots in a MEMS foundry. The applications on the *left hand side* are best served by dielectric membrane-type devices; the *right hand side* is best served by SOI-type heaters. Automotive mass driver applications are indicated by *shading*

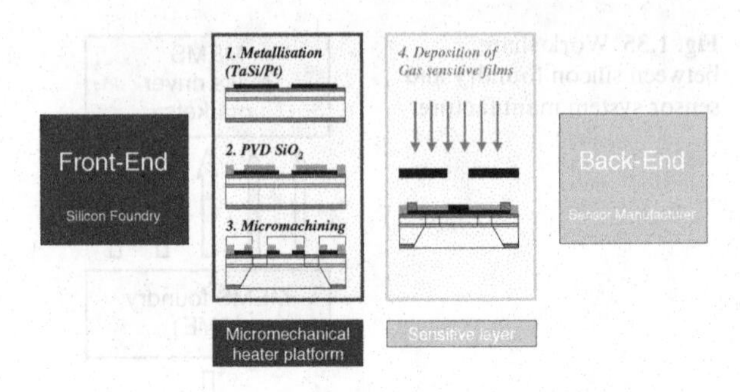

Fig. 1.37 Process flow for the fabrication of low power consumption MOX gas-sensing elements. The entire process can be separated into a standard-silicon front-end (*left*) and a sensor-specific back-end module (*right*) to enable an efficient work share with silicon foundries

therefore wants to be sure to buy functioning micro-heaters from the MEMS foundry. This functioning can be assessed in a final internal check in the MEMS foundry and by an entrance check inside the SME sensor company. Such a check simply consists in a straight-forward electrical test in which the resistance of the heating meander is checked as a function of the electrical input power and in which the light output of the micro-heater is recorded. Proceeding up to this level does not go beyond the technological capabilities of a MEMS foundry and passing this test also forms a clear commercial interface that leaves the responsibility for the sensor processing and all steps specific to gas sensor processing in the hands of the SME gas sensor manufacturer.

At this commercial interface, a point of interest is the cost of hot-plate chips. Assuming that the MEMS challenge has been met, i.e., a moderate number of gas sensor array chips can be derived from a large-volume production of hot-plate heaters, the production cost of such chips is very modest in the context of the much larger post-processing and system integration costs. From our experience with silicon foundries we estimate that silicon processing of a 3-sensor array chip (see Figs. 1.9 and 1.10) would amount to approximately 1.75 Euro per chip assuming 4-inch SOI wafers as starting materials. Due to the larger surface area of 6-inch wafers we estimate that this cost will decrease further to about 0.8 Euro per chip.

In Fig. 1.37 we have assumed that the sensing layer is a thin-film layer that is deposited by means of some classical PVD technology such as evaporation or sputtering. As the pre-fabricated micro-heaters already contain etch troughs and do no longer exhibit flat surfaces, lithographic processing of the sensing layers is no longer possible. Sensing layer deposition therefore needs to employ shadow masks to place the sensitive films exclusively onto the predefined spots between the sensor electrodes. As this requires only moderate precision, mechanically stiff ceramic shadow masks are sufficient for this purpose. Alternatively, mechanically stiff and highly precise shadow masks can also be produced in the MEMS foundry and purchased in the form of micro-machined silicon wafers. Like the ceramic masks these latter wafers contain open spots that match with the sensing layer windows on the micro-heaters.

Another method of depositing gas sensitive layers is using drop dispensers for placing small volumes of gas sensitive materials on top of the hot plates. The

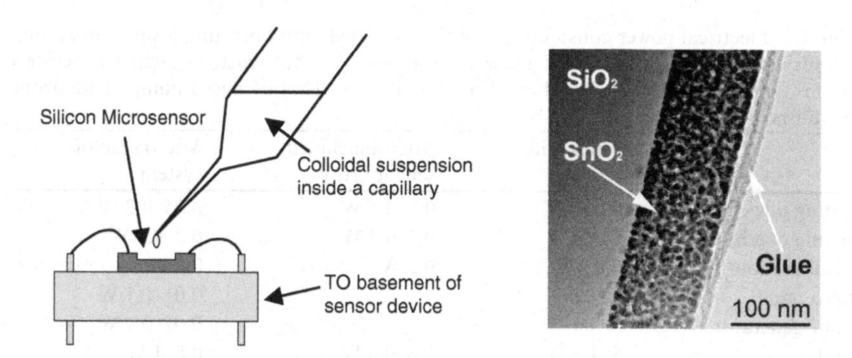

Fig. 1.38 (**a**) Deposition of a MOX sol onto a micro-heater using the drop dispenser technique; (**b**) cross-section transmission electron microscopy image of a SnO_2 thin film heat-treated at 500°C

method is schematically illustrated in Fig. 1.38a and a transmission electron microscopy image of a deposited gas sensitive layer is shown in Fig. 1.38b. In such a deposition process a sol is deposited with the help of a drop dispenser and the deposit is cured afterwards by thermal annealing to form a gel [75, 76, 77, 78].

Due to the good thermal insulation the center regions of the bridges or crosses can be heated to typical sensor operation temperatures (300–400°C) by electrical input powers of about 50 mW. The right-hand part in Fig. 1.39 shows the cross-shaped hot plate heated to a center temperature of about 1000°C by applying an increased electrical input power of about 200 mW. In this latter mode of operation local CVD can be performed. This method has been pioneered by the NIST group [79, 80, 81, 82]. In the future this may also prove to become useful for depositing nanorods and nanowires of MOX materials onto the center region [83, 84, 85, 86].

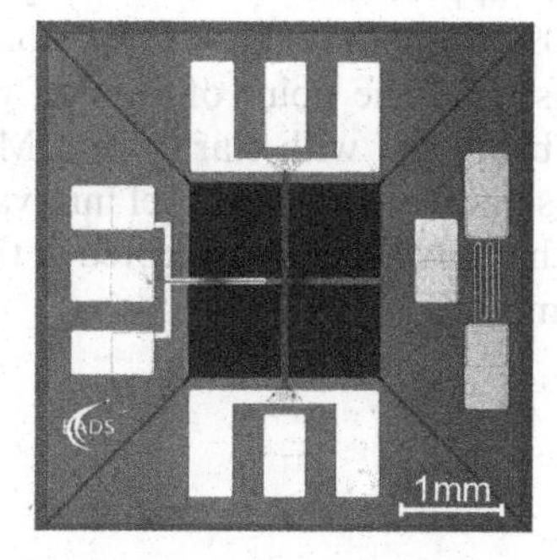

Fig. 1.39 SOI – based hot plate: (*left*) top view on a single gas sensor chip with a cross-bar silicon heater; (*right*) a cross-bar heater high-temperature heated and ready for micro-CVD deposition of sensitive layers

Table 1.2 Electrical power consumption levels associated with operating a single thick-film gas-sensing element, a miniaturized gas sensor array and a micro-reactor system with self-test features. In the latter case, low-duty-cycle operation (~1%) of micro-pumps and ozone generators was assumed

	Thick-flim sensor	Micromachined sensor array	Micro reactor system
Heating power	0.5–1 W	0.2–0.3 W	0.05–0.2 W
Heating overhead	0.5 W	0.2–0.3 W	0.2–0.3 W
Sensor readout	0.05 W	0.2 W	0.2 W
Micro pump	–	–	0.01–0.1 W
Ozone generator	–	–	0.01–0.1 W
Total	1–1.5 W	0.6–0.8 W	0.8–1 W

1.5.5 Creating Higher Added Value

In the sections above we have been considering the technical issues related to generating an industrial work share capable of producing miniaturized MOX gas sensors. With such sensors the heating power consumption of ceramic thick-film gas sensors can be reduced by roughly a factor of ten. The point we should like to make in this final chapter is that this reduced power consumption alone does not represent significant driving power for a technology change.]

A significant driving power only emerges, once higher added-value products are derived on the basis of the MEMS base technology. The basic idea to be pursued in such higher-added-value approaches is that the largely reduced heating power consumption of micro-machined gas-sensing elements is not so much interesting in itself but opens up novel possibilities towards assembling more complex gas-sensing systems. Within this concept the reduced power consumption is used to achieve enhanced selectivity, drift rejection and, eventually, introducing self-test features operating at the same time within the electrical power budget usually expended for a single commercial thick-film gas-sensing element. Such system-level benefits can be attained by using sensor array and micro-reactor concepts. Such approaches have already been described in some detail elsewhere [28, 109, 110]. In Table 1.2 we compare and contrast such system-level innovations from the point of view of power consumption. These latter considerations bring out with clarity that MEMS micro-miniaturization technologies enable significant system-level innovations within the size and power-level constraints that have become accepted in the use of single ceramic thick-film gas-sensing elements.

1.6 Conclusions and Outlook

In the pages above we have been considering gas-sensing microsystems that can be built up from MEMS micro-components. Proceeding through these arguments we have presented both a bottom-up and a top-down view onto this field.

Whereas the bottom-up view centers on scientific and technological issues, the top-down view focuses on aspects of technology transfer and industrial work-share. Both approaches complement each other and need to be matched early on during a technological development.

The task to be managed is a complex case of nano-, micro-, and macro-integration. On the level of nanotechnologies one has to deal with sensitive surfaces and sensitive films, on the level of micro-technologies MEMS processing and MEMS packaging technologies come in. Finally at the level of macro-technologies standard methods of system integration such as mechanical, electronic and software engineering must be considered. Building and commercializing gas sensor micro-systems therefore is heavily dependent on system engineering capabilities. Such capabilities rely on sound technical capabilities on all levels of technology stretching from the nano- to the macro-level. As the corresponding technical capabilities are normally distributed over different laboratories, the development of micro-gas-sensing systems is normally a highly cooperative effort. This is even truer when it comes to industrial exploitation. In this latter case it is essential to define early on clear-cut technical and commercial interfaces that allow different supplier SMEs and end-user LMEs to cooperate in a smooth and economically productive manner. This is a challenging task, but worthwhile considering the enormous potential of modern micro- and nanotechnologies.

Acknowledgments The authors acknowledge support by the EU under the projects NANOS4 (FP6-2002-NMP-1, No. 001528) and NetGas (FP5 IST – 2001 – 37802) and to the German Ministry of Education and Research BMBF under the contracts “MISSY”, “IESSICA” and “NACHOS”.

References

1. Westbrook MH, Turner JD. Automotive sensors. Institute of Physics Publishing: Bristol and Philadelphia; ISBN 0-7503-0293-3. 1994.
2. Moos R, Müller R, Plog C, Knezevic A, Leye H, Irion E, Braunc T, Marquardt KJ, Binder K. Selective ammonia exhaust gas sensor for automotive applications. Sensor Actuat. 2002;B83:181–189.
3. McGeehin P. Gas sensors for improved air quality in transportation. Sensor Rev. 2000;20:106–112.
4. Hunter GW, Chen LY, Neudeck PG, Knight D, Liu CC, Wu QH, Zhou HJ, Makel D, Liu M, Rauch WA. Chemical gas sensors for aeronautic and space applications. NASA/TM—1998-208504, http://gltrs.grc.nasa.gov/reports/1998/TM-1998-208504.pdf. 1998
5. Kallergis KM. New fire/smoke detection and fire extinguishing systems for aircraft applications. Space Eur.2001;3:197–200.
6. Kohl D, Kelleter J, Petig H. Detection of fires by gas sensors. Sens Update. 2001;9:161–223.
7. Helwig A, Schulz O, Spannhake J, Sayhan I, Krenkow A, Müller G. Aircraft applications of chemical sensors: from MEMS sensors to MEMS sensor systems, invited talk. 4th AIST International Workshop on Chemical Sensors, Nagoya, Japan; 30 Nov. 2006.
8. http://www.goodfood-project.org/

9. Wöllenstein J, Hartwig S, Hildenbrand J, Eberhardt A, Moreno M, Santander J, Rubio R, Fonollosa J, Fonseca L. A compact optical ethylene monitoring system. Microtechnologies for the new millenium 2007. Maspalomas, Gran Canaria, Spain; 2–4 May 2007.
10. Abad E, Zampolli S, Marco S, Scorzoni A, Mazzolai B, Juarros A, Gómez D, Elmi I, Cardinali GC, Gómez JM, Palacio F, Cicioni M, Mondini A, Becker T, Sayhan I. Flexible tag microlab development: gas sensors integration in RFID flexible tags for food logistic. Sensor Actuat B. 2007;127:2–7.
11. Gardner JW, Yinon J. Electronic noses and sensors for the detection of explosives. NATO Sci Ser II Math, Phys Chem.. Dordrecht: Kluwer; ISBN 1-4020-2317-0 (hardbound) and 1-4020-2318-9 (paperback). vol. 159. 2004.
12. Moseley PT, Tofield BC. Solid state gas sensors. Adam Hilger, Bristol; 1987.
13. http://www.figarosensor.com/
14. http://www.umweltsensortechnik.de/
15. Eranna G, Joshi BC, Runthala DP, Gupta RP. Oxide materials for development of integrated gas sensors – A comprehensive review. Crit. Rev Solid State Mater Sci. 2004;29:111–188.
16. Moseley PT, Norris J, Williams DE. Techniques and mechanisms in gas sensing. Adam Hilger, Bristol; 1991.
17. Sze SM. Semiconductor sensors. John Wiley & Sons; ISBN 0-471-54609-7. 1994.
18. Gardner JW, Bartlett PN. Electronic noses: principles and application. Oxford University Press: Oxford; ISBN 0-19-855955-0. 1999, p. 245.
19. Graf M, Gurlo A, Bârsan N, Weimar U, Hierlemann A. Microfabricated gas sensor systems with sensitive nanocrystalline metal-oxide films. J Nanopart Res. 2006;8:823–839.
20. Bourgeois W, Romain A-C, Nicolas J, Stuetz RM. The use of sensor arrays for environmental monitoring: interests and limitations. J Environ Monit. 2003;5:852–860.
21. Platt U, Stutz J. Differential optical absorption spectroscopy – principles and applications, Ser Phys Earth Space Environ. ISBN: 978-3-540-21193-8. 2008.
22. McNair HM, Miller JM. Basic gas chromatography. ISBN: 978-0-471-17261-1. 1997.
23. Eiceman GA, Karpas Z. Ion mobility spectrometry. 2nd ed. CRC Press, Taylor & Francis: Boca Raton; 2006.
24. Rubio R, Santander J, Fonseca L, Sabatér N, Gràcia I, Cané C, Udina S, Marco S. Non-selective NDIR array for gas detection. Sensor Actuat. 2007;B127:69–73.
25. Graf M, Barrettino D, Baltes HP, Hierlemann A. CMOS hotplate chemical microsensors. Ser.: Microtechnol MEMS. ISBN: 978-3-540-69561-5. 2007.
26. Hierlemann A. Integrated chemical microsensor systems in CMOS technology. Ser.: Microtechnol MEMS. ISBN: 978-3-540-23782-2. 2005.
27. Wapelhorst E, Hauschild JP, Müller J. Complex MEMS: a fully integrated TOF micro mass spectrometer. Sensor Actuat 2007;A138:22–27.
28. Müller G, Friedberger A, Kreisl P, Ahlers S, Schulz O, Becker T. A MEMS toolkit for metal-oxide-based gas-sensing systems, Thin Solid Films. 2003;436:34–45.
29. Sze SM. Semiconductor devices; physics and technology. John Wiley & Sons; ISBN-10: 0-471-33372-7; ISBN-13: 978-0-471-33372-2. 2001.
30. Waser R. Nanoelectronics and information technology. Wiley-VCH, ISBN 3527403639. 2003.
31. Heuberger A. Mikromechanik. Springer; ISBN 3-540-18721-9. 1989.
32. http://www.memsnet.org/mems/what-is.html
33. http://www.memx.com/
34. Bergveld, P. Thirty years of ISFETOLOGY: what happened in the past 30 years and what may happen in the next 30 years. Sensor Actuat. 2003;B88:1–20.
35. Lundström I, Shirvamavan MS, Svensson C. A hydrogen-sensitive MOS field-effect transistor. Appl Phys Lett. 1975;26:55–57.

36. Lundström I, Shirvamavan MS, Svensson C. A hydrogen sensitive Pd-gate MOS transistor. J Appl Phys. 1975;46:3876–3881.
37. Lundström I, Spetz A, Winquist F, Ackelid U, Sundgren H. Catalytic metals and field-effect devices – a useful combination. Sensor Actuat. 1990;B1:15–20.
38. Ahlers S, Müller G, Becker Th, Doll Th. Factors influencing the gas sensitivity of metal oxide materials. In: Grimes CA, Dickey EC, Pishko MV, editors. Encyclopedia of sensors. The Pennsylvania State University, University Park, USA; ISBN: 1-58883-056-X. 2005.
39. Fang Q, Chetwynd DG, Covington JA, Toh CS, Gardner JW. Micro-gas-sensor with conducting polymers. Sensor Actuat. 2002;B84:66–71.
40. Helwig A, Müller G, Sberveglieri G, Faglia G. Gas sensing properties of hydrogenated amorphous silicon films. IEEE Sens J. 2007;7:1506–1512.
41. http://www.sensorsmag.com/articles/0203/14/
42. Schjølberg-Henriksen K, Ferber A, Schulz O, Moe S, Wang DT, Lloyd MH, Legner W, Suphan KH, Bernstein RW, Rogne H, Müller G. Sensitive and selective photo acoustic gas sensor suitable for high-volume manufacturing. Proceedings of IEEE Sensors Conference, Daegu, Korea; October 22–25, 2006.
43. Tardy P, Coulon JR, Lucat C, Menil F. Dynamic thermal conductivity sensor for gas detection. Sensor Actuat. 2004;B98:63–68.
44. Miller JB. Catalytic sensors for monitoring explosive atmospheres. IEEE Sens J. 2001;1:88–93.
45. Sberveglieri G, Hellmich A, Müller G. Silicon hotplates for metal oxide gas sensor elements. Microsyst Technol. 1997;3:183–190.
46. Suehle JS, Cavicchi RE, Gaitan M, Semancik M. Tin oxide gas sensor fabricated using CMOS Micro hotplates and in-situ processing. IEEE Electr Device L. 1993;14: 118–120.
47. Semancik S, Cavicchi RE, Kreider KG, Suehle JS, Chaparla P. Selected area deposition of multiple active films for conductometric microsensor arrays. Proc. Transducers′95, EUROSENSORS IX, Stockholm, Sweden; 1995. pp. 831–834.
48. Semancik S, Cavicchi RE, Meier DC, Taylor CJ, Savage NO, Wheeler MC. Temperature-controlled MEMS chemical microsensors", Proc. 1st AIST International Workshop on chemical Sensors. Nagoya; March 13, 2003.
49. Friedberger A, Kreisl P, Rose E, Müller G, Kühner G, Wöllenstein J, Böttner H. Micromechanical fabrication of robust low-power metal-oxide gas sensors. Sensor Actuat. 2003;B93:345–349.
50. Spannhake J, Schulz O, Helwig A, Krenkow A, Müller G, Doll T. High-temperature MEMS heater platforms: long-term performance of metal and semiconductor heater materials. Sensors. 2006;6:405–419.
51. Zeitschel A, Friedberger A, Welser W, Müller G. Breaking the isotropy of porous silicon formation by current focussing. Sensor Actuat. 1999;74:113–117.
52. Düsco Cs, Vaszsonyi E, Adam M, Barsony I, Gardeniers JGE, van den Berg A. Porous silicon bulk micromachining for thermally isolated membrane formation. Proc. Eurosensors X, Leuven, Belgium; 1996. pp. 227–230.
53. Barrettino D, Graf M, Song WH, Kirstein KU, Hierlemann A, Baltes H. Hotplate-based monolithic CMOS microsystems for gas detection and material characterization for operating temperatures up to 500°C. IEEE J Solid-State Circ. 2004;39:1202–1207.
54. Kreisl P, Helwig A, Friedberger A, Müller G, Obermeier E, Sotier S. Detection of hydrocarbon species using silicon MOS capacitors operated in a non-stationary temperature pulse mode. Sensor Actuat. 2005;B106:489–497.
55. Kreisl P, Helwig A, Müller G, Obermeier E, Sotier S. Detection of hydrocarbon species using silicon MOS field effect transistors operated in a non-stationary temperature-pulse mode. Sensor Actuat. 2005;B106:442–449.

56. Müller G, Schalwig J, Kreisl P, Helwig A, Obermeier E, Weidemann O, Stutzmann M, Eickhoff M. High-temperature operated field-effect gas sensors. In: Grimes CA, Dickey EC, Pishko MV. Editors. Encyclopedia of sensors.The Pennsylvania State University, University Park, USA. ISBN: 1-58883-056-X. 2005.
57. Spannhake J, Helwig A, Müller G, Sberveglieri G, Faglia G, Wassner T, Eickhoff M. SnO_2:Sb – A new material for high-temperature MEMS heater applications – performance and limitations. Sensor Actuat. 2007;B124:421–428.
58. Schjølberg-Henriksen K, Wang DT, Rogne H, Ferber A, Vogl A, Moe S, Bernstein R, Lapadatu D, Sandven K. High-resolution pressure sensor for photoacoustic gas detection, EUROSENSORS XIX, Barcelona, Spain; 11–14 Sept. 2005.
59. Brida S, Beclin S, Metivet S, Martins P,Stojanovic O. High sensitivity piezoresistive silicon microphone for aerospace applications. 5th ESA MNT Round Table, Noordwijk, The Netherlands; 3–5 October 2005.
60. Brida S, Martins P, Beclin S, Metivet S, Stojanovic O, Malhaire C. Design of bossed silicon membranes for high sensitivity microphone applications, DTIP of MEMS MOEMS, Stresa, Italy, 26–28 April 2006.
61. Sekhar PK, Akellaa S, Bhansali S. A low loss flexural plate wave (FPW) device through enhanced properties of sol–gel PZT (52/48) thin film and stable TiN-Pt bottom electrode. Sensor Actuat. 2006;A132:376–384.
62. Neuberger R. PhD thesis, Department of Experimental Semiconductor Physics II, Technical University of Munich; 2003.
63. Schulz O, Müller G, Lloyd MH, Ferber A. Impact of environmental parameters on the emission intensity of micromachined infrared sources. Sensor Actuat. 2005;A121: 172–180.
64. Spannhake J, Helwig A, Friedberger A, Müller G, Hellmich W. Resistance heating. Wiley Encyclopedia of Electrical and Electronics Engineering, John Wiley & Sons, June, 2007, 10.1002/047134608X.W3223.pub2.
65. Meyer GCM, van Herwarden AW. Thermal sensors. Bristol and Philidelphia: Institute of Physics Publishing:; 1994. ISBN 0-7503-0220-8.
66. Spannhake J, Schulz O, Helwig A, Müller G, Doll T. Design, development and operational concept of an advanced MEMS IR source for miniaturized gas sensor systems. IEEE Sens. 2005;30:4. ISBN: 0-7803-9056-3/05.
67. http://aa.bosch.de/advastaboschaa/Category.jsp?ccat_id = 29&language = en-GB& publication = 1.
68. http://cms.hlplanar.de/index.php.
69. Kittel Ch, Kroemer H. Thermal physics. San Francisco: W.H. Freeman & Co Ltd; ISBN-10: 0716710889, ISBN-13: 978-0716710882.
70. http://www.thyracont.com/.
71. Puers R, Reyntjens S, De Bruyker D. The NanoPirani – an extremely miniaturized pressure sensor fabricated by focused ion beam rapid prototyping. Sensor Actuat. 2002;A97 & 98:208–214.
72. Zhang FT, Tang Z, Yu J, Jin RC. A micro-Pirani vacuum gauge based on micro-hotplate technology. Sensor Actuat. 2006;A126:300–305.
73. Ahlers S,Müller G, Doll Th. A rate equation approach to the gas sensitivity of thin-film SnO_2. Sensor Actuat. 2005;B107587–599.
74. Helwig A, Müller G, Sberveglieri G, Faglia G. Gas response times of nano-scale SnO_2 gas sensors as determined by the moving gas outlet technique. Sensor Actuat. 2007;B126:174–180.
75. Guidi V, Butturi MA, Carotta MC, Cavicchi B, Ferroni M, Malagù C, Martinelli G, Vincenzi D, Sacerdoti M, Zen M. Gas sensing through thick film technology. Sensor Actuat. 2002;B84:72–77.
76. Vincenzi D, Butturi MA, Stefancich M, Malagù C, Guidi V, Carotta MC, Martinelli G, Guarnieri V, Brida S, Margesin B, Giacomozzi F, Zen M, Vasiliev AA, Pisliakov AV.

Low-power thick-film gas sensor obtained by a combination of screen printing and micromachining techniques. Thin Solid Films. 2001;391:288–292.
77. Epifani M, Francioso L, Siciliano P, Helwig A, Mueller G, Díaz R, Arbiol J, Morante JR. SnO_2 thin films from metalorganic precursors: synthesis, characterization, microelectronic processing and gas-sensing properties. Sensor Actuat. 2007;B124:217–226.
78. Francioso L, Russo M, Taurino AM, Siciliano P. Micrometric patterning process of sol–gel SnO_2, In_2O_3 and WO_3 thin film for gas sensing applications: towards silicon technology integration. Sensor Actuat. 2006;B119:159–166.
79. Semancik S, Cavicchi RE, Kreider KG, Suehle JS, Chaparala P. Selected-area deposition of multiple active films for conductometric microsensor arrays. Proc. of Transducers 95/ Eurosensors IX. Norstedts Tryckeri AB, Stockholm, Sweden, 1995. pp. 831–834.
80. DiMeo F. Jr., Semancik S, Cavicchi RE, Suehle JS, Chaparala P, Tea NH. MOCVD of SnO_2 on silicon microhotplate arrays for use in gas sensing applications. MRS Proc. 1995;415:231–236.
81. Cavicchi RE, Suehle JS, Kreider KG, Shomaker BL, Small JA, Gaitan M, Chaparala P. Growth of SnO2 films on micromachined hotplates. Appl. Phys. Lett. 1995;66:812–814.
82. Semancik S, Cavicchi RE, Kreider KG, Suehle JS, Chaparala P. Selected-area deposition of multiple active films for conductometric microsensor arrays. Sensor Actuat. 1995;B34:209–212.
83. Comini E, Faglia G, Sberveglieri G, Pan ZW, Wang ZL. Stable and highly sensitive gas sensors based on semiconducting oxide nanobelts. Appl Phys Lett. 2002;81:1869–1871.
84. Choi YJ, Hwang IS, Choi KJ, Park JH, Park JG. Gas sensor based on the network of SnO_2 semiconducting nanowires. In: Li C, Zribi A, Nagahara L, Willander M, editors. Nanofunctional materials, nanostructures and novel devices for biological and chemical detection. Mater. Res. Soc. Symp. Proc. 951E, Warrendale, PA, 2007, pp. 08–03.
85. Meier DC, Semancik S, Button B, Strelcov E, Kolmakov A. Coupling nanowire chemiresistors with MEMS microhotplate gas sensing platforms. Appl Phys Lett. 2007;91:063118.
86. Baratto C, Comini E, Faglia G, Sberveglieri G, Zha M, Zappettini A. Metal oxide nanocrystals for gas sensing. Sensor Actuat. 2005;B109:2–6.
87. Hellmich W, Müller G, Doll T, Eisele I. Field-effect-induced gas sensitivity changes in metal oxides. Sensor Actuat. 1997;B43:132–139.
88. Ahlers S, Becker T, Hellmich W, Bosch-v.Braunmühl C, Müller G. Temperature- and field-effect-modulation techniques for thin-film metal oxide gas sensors. In: Doll T. editor. Advanced gas sensing: the electroadsorptive effect and related techniques. Kluwer Academic Publishers: Boston, London, Dordrecht; 2003.
89. Dalin J. Fabrication and characterisation of a novel MOSFET gas sensor. PhD thesis. Linköpings Institute of Technology, 2002.
90. Fan Z, Lu JG. Gate-refreshable nanowire chemical sensors. Appl Phys Lett. 2006;86:123510.
91. Zhang Y, Kolmakov A, Lilach Y, Moskovits M. Electronic control of chemistry and catalysis at the surface of an individual tin oxide nanowire. J Phys Chem. 2005;B109: 1923–1929.
92. Arbab A, Spetz A, ul Wahab Q, Willander M, Lundström I. Chemical sensors for high temperatures based on silicon carbide. Sens Mater. 1993;4:173–185.
93. Lloyd Spetz A, Baranzahi A, Tobias P, Lundström I. High temperature sensors based on metal-insulator-silicon carbide devices. Phys Stat Sol (a). 1997;162:493–511.
94. Lloyd Spetz A, Unéus L, Svenningstorp H, Tobias P, Ekedahl L-G, Larsson O, Göras A, Savage S, Harris C, Mårtensson P, Wigren R, Salomonsson P, Häggendahl B, Ljung P, Mattsson M, Lundström I. SiC based field effect gas sensors for industrial applications. Phys Stat Sol (a). 2001;185:15–25.

95. Käpplinger I, Brode W, Krenkow A, Spannhake J, Müller G. High temperature silicon-on-insulator based hotplates: long term performance of platinum heater materials. Proc. AMA 2007. Nürnberg, Germany, 22- 4 May 2007.
96. Nachos: http://www.micro-hybrid.de
97. Schulz O, Legner W, Müller G, Schjølberg-Henriksen K, Ferber A, Moe S, Lloyd MH, Suphan K-H. Photoacoustic gas sensing microsystems. Proc. AMA 2007. Nürnberg, May 2007.
98. Helwig A, Schulz O, Sayhan I, Müller G. "Multi-criteria fire detectors for aeronautic applications. Proc. TRANSFAC, San Sebastian, Spain, September 2006. Invited talk at Global Symposium on Innovative Solutions for the Advancement of the Transport Industry/Transfac'06, San Sebastian, Spain, October 4–6, 2006.
99. Grosshandler W.L. editor. Nuisance alarms in aircraft cargo areas and critical telecommunication systems. Proc. Third NIST Fire Detector Workshop, December 4–5, 1997, NISTIR 6146, National Institute of Standards and Technology, Gaithersburg, MD, March 1998.
100. Airbus Internal Directive ABD0100, Equipment-Design General Requirements for Suppliers; Source: Airbus Documentation Office, Toulouse, Blagnac, France.
101. Millenium Sensor Systems (MISSY), http://www.ipc.uni-tuebingen.de/weimar/research/researchprojects/missy/info_missy.pdf.
102. IMOS (EU FP6); http://imos.fhnon.de.
103. Sayhan I, Helwig A, Becker Th, Müller G, Elmi I, Zampolli S, Cardinali GC, Padilla M, Marco S. Discontinuously operated metal oxide gas sensors for flexible tag microlab applications. IEEE Sens J. 2008;8:176–181.
104. Bell AG. On the production and reproduction of sound. Am J Sci. 1880;20:305.
105. Kreuzer LB. Ultra-low gas concentration absorption spectroscopy. J Appl Phys. 1971;42:2934–2943.
106. Ohlckers P, Ferber AM, Dmitriev VK, Kirpilenko G. A photo-acoustic gas sensing silicon microsystem. Transducers 2001, Germany, June 2001, pp. 780–783.
107. Schulz O. PhD thesis. Technical University of Ilmenau, 2007.
108. http://www.multimems.com
109. Becker T, Mühlberger S, Bosch-von Braunmühl C, Müller G, Meckes A, Benecke W. Microreactors and microfluidic systems: an innovative approach to gas sensing using tin-oxide-based gas sensors. Sensor Actuat. 2001;B77:48–54.
110. Becker T, Mühlberger S, Bosch - von Braunmühl C, Müller G, Ziemann T, Hechtenberg KV. Air pollution monitoring using tin-oxide based micro-reactor systems. Sensor Actuat. 2000;B69:108–119.

Chapter 2
Electrical-Based Gas Sensing

Elisabetta Comini, Guido Faglia, and Giorgio Sberveglieri

2.1 Introduction

Metal oxides represent an assorted and appealing class of materials whose properties cover the entire range from metals to semiconductors and insulators and almost all aspects of material science and physics including superconductivity and magnetism. In the field of chemical sensing, for more than five decades it has been known that the electrical conductivity of semiconductor varies with the composition of the gas atmosphere surrounding them [17].

The first generation of commercial devices was prepared in the 1960s by Taguchi in Japan. They were made of SnO_2 prepared by thick-film technology and employed in the warning of possible leaks of explosive gases. This was and is of major importance in densely built-up Japanese cities characterized by wooden houses and widespread gas furnaces. Taguchi established Figaro Engineering Inc., which remains today the major manufacturer of gas sensors.

Since then, the need of cheap, small, low-power-consuming and reliable solid-state gas sensors has grown over the years and, recently, the development of information technology have triggered a large amount of research worldwide to overcome metal oxide sensor drawbacks, summed up in improving the well-known "3S": sensitivity, selectivity, and stability. A variety of devices have been developed mainly by an empirical approach, and a lot of basic theoretical research and spectroscopy studies have been carried out.

The sensing properties of semiconductor metal oxide other than SnO_2, like TiO_2, WO_3, ZnO, Fe_2O_3, and In_2O_3, have been studied as well as the benefits from the addition of noble metals – Pd, Pt, Au, Ag – in improving selectivity and stability. In 1991 Yamazoe [171] showed that reduction of crystallite size resulted in a huge improvement in sensor performance. The challenge became to prepare materials with small crystallize size which were stable when operated at high temperature for long periods. Since the incorporation of a second oxide

E. Comini
Sensor Laboratory, CNR-INFM and Brescia University, Via Valotti, 9, 25133 Brescia, Italy
e-mail: comini@ing.unibs.it

E. Comini et al. (eds.), *Solid State Gas Sensing*,
DOI: 10.1007/978-0-387-09665-0_2,

phase even in small quantity changes the growth of the base oxide and prevents grain coalescence at high temperature, material research shifted towards mixed oxide systems.

From the preparation side, first-generation devices were prepared by thick-film technology starting from powders, quite an irreproducible technique. Since sensor performance depends on percolation through intergranular resistance, a parameter varying with small details of preparation, each sensor differed slightly in its initial characteristics. Therefore, the materials fabrication processes have been improved in second-generation thin-film technology, a more automated production method that offers higher reproducibility and compatibility with Si technology, by physical (mostly evaporation and sputtering) and chemical vapour deposition. However, the technological improvement went along with a reduction of sensing performances due to lower porosity of the prepared devices. The introduction of sol-gel methods of preparation offered a new technique to prepare low-cost porous nanocrystallite thin-film devices.

The effort towards miniaturization and reduction of the heating power necessary to keep the device at the operating temperature (600–900 K) prompted the development of Si "micro-hot-plates" substrates with shrinked overall dimensions and reduced thermal mass.

New detection principles and techniques, other than simple conductance monitoring, have been explored. Noteworthy are the development of suspended-gate Field Effect Transistors (FET), which transduces the metal oxide work function, a truly surface effect that encloses contribution from both physisorbed dipoles and ionosorbed species, and the development of photo-assisted detection. To improve selectivity, devices with partially overlapping sensitivities were introduced in sensor arrays and became, matched to a suitable data analysis, the core of Artificial Olfactive Systems (AOS).

Basic research has dealt with application of imaging and spectroscopies to study the surface structure of metal oxide semiconductors and understand adsorption processes at atomistic level. Besides, a great effort has been made to develop self-sustained models which could relate the response of the sensors to the concentration of target species in the gas phase. However, progresses have been slow since the involved physicochemical systems are very complex, enclosing gas diffusion and adsorption, and surface reactions, and their influence on electrical properties of hardly known microstructures.

An unexpected breakthrough, yet to fulfill all its promises, has been the successfully synthesize stable single-crystal quasi-one-dimensional semiconducting oxides nanostructures (so-called nanobelts, nanowires, or nanoribbons) by simply evaporating the desired commercial metal oxide powders at high temperatures [120, 31]. Due to their peculiar characteristics and size effects, these materials are of great interest both for fundamental study and for potential nanodevice applications, i.e., the third generation of metal oxide gas sensors.

The subject treatment will start with an overview of metal semiconductor surfaces, in terms of physical and electrical properties of surfaces and how they affect the conduction in mono- and polycrystalline sensors, and with the physicochemical grounds of the sensing process, i.e., the role of surface reactions in modifying electrical properties. After presenting the preparation techniques and their historical development and improvement, the production steps in the fabrication of a conductometric sensor will be detailed. The monograph will close with the description of how the chemical sensing step is transduced in a physical measurable quantity.

2.2 Metal Oxide Semiconductor Surfaces

Metal oxide semiconductors are normally high-band-gap metal oxides that are insulating in which the semiconducting behavior arises from a deviation of stoichiometry [158]. They should always be regarded as compensated semiconductors: cation vacancies are acceptors, yielding holes and negatively charged vacancies, as in p-type semiconductors like Cu_2O. In contrast, it is in the case of SnO_2 [135] that shallow states made up of oxygen vacancies act as n-type donors, since the bonding electrons on the adjacent Sn cations are easily removed and donated to the conduction band:

$$O_o^X \rightleftarrows \frac{1}{2}O_2 + V_O^{n+} + ne^- \tag{2.1}$$

Subscript O means an oxygen lattice position.

Therefore, bulk reduction and oxidation originates from surface lattice vacations reaction with gaseous oxygen. Oxygen vacancies can be singly ($n=1$) or doubly ($n=2$) ionized, depending on the temperature; the presence of Sn interstitials can also contribute to formation of the donor states [83].

2.2.1 Geometric Structures

Studying the geometric structures of metal oxide semiconductor surface is a key factor in the understanding of the mechanism of gas adsorption and in the design of innovative gas sensors. An extensive review of metal oxide surface has been written by Henrich and Cox [74].

In a metal oxide lattice the bonding forces are those that operate between the positively charged small metal cations and the negatively charged large oxygen anions, so that coordination of the metal ion is, together with stoichiometry, the most important parameter to characterize a structure. Many structures appear to be based on a close-packed or nearly-close-packed array of oxide ions, with metal cations occupying "interstitial" sites.

Two of the metal oxide semiconductors whose surface properties have been most thoroughly studied, TiO_2 and SnO_2, have the tetragonal rutile lattice [12, 36]. The most stable crystal face of rutile is the (110) surface: rows of "bridging" O ions lie above the main surface plane, which contains equal numbers of five- and sixfold coordinated metal cations; symmetric rows of O ions lie an equal distance below the surface plane, resulting in a non-polar surface. Surface lattice vacancies are produced by removal of bridging or in-plane oxygen.

What makes SnO_2 highly surface reactive is the fact that bridging and in-plane O ions can be easily removed by heating (heating to 800 K removes all of them) or replaced by annealing [33]. When surface O_2^- ions are removed the two electrons left convert Sn^{4+} to Sn^{2+}. In contrast, on the TiO_2 (110) surface, the rows of bridging ions are tightly bound.

2.2.2 *Electronic Structures*

As for the electronic structure, metal oxide are normally subdivided in non-transition metal oxides, in which the cation valence orbitals are of s or p type and transition metal oxides where the cation valence orbitals have d symmetry.

Among the transition metal oxide, post-transition SnO_2 and ZnO are semiconductor.

Measurements of the surface electrical conductivity of SnO_2 (110) as a function of density of O removed showed that the Sn^{2+} ions formed still possessed a localized electronic structure [33] and that increasing the temperature conductivity of the surface did not increase significantly until after the bridging O ions have been removed. At a temperature above 700 K, removal of in-plane oxygen produces a large change of conductivity. Ultraviolet photoemission spectroscopy (UPS) showed that the increased surface conductivity goes together with occupation of surface states in the bulk band gap near E_F.

Many of the properties of the electronic structure of transition metal oxides, like the existence of variable oxidation states, are due to the different bonding properties associated with d orbitals. On the TiO_2 (110) surface, the rows of bridging ions are tightly bound and removal of ions changes strongly the electronic structures. The bridging O vacancy role in gaseous oxygen adsorption has been demonstrated recently for TiO_2 [73].

2.3 Electrical Properties of Metal Oxide Semiconductor Surfaces

2.3.1 *Semiconductor Statistics*

This section aims to provide a brief introduction to the electrical properties of metal oxide semiconductor surfaces by laying a solid foundation of definitions, concepts, and equations for the subsequent sections. Sources are the excellent books from [154] and [169].

According to the Fermi–Dirac statistic, which electrons and holes obey, in equilibrium

$$f_n(E) = \frac{1}{1+e^{\frac{E-E_f}{kT}}} \quad f_p(E) = \frac{1}{1+e^{\frac{E_f-E}{kT}}} \quad f_n(E)+f_p(E) = 1 \tag{2.2}$$

where E_f is the electrochemical potential, $f_n(E)$ and $f_p(E)$ the probability that a state of energy E is occupied by an electron or by a hole. In a non-degenerate semiconductor (SC), as long as the electrochemical potential lies deep under the conduction band and high above the valence band,

$$n = N_c e^{\frac{E_C - E_f}{kT}} \quad p = N_v e^{\frac{E_f - E_V}{kT}} \tag{2.3}$$

Here E_C is the bottom of the conduction band and E_V the top of the valence band. N_C and N_V are the effective density of states of the two bands. In the simplest case of an s band we have

$$N_c = 2\left(\frac{m_n kT}{2\pi h^2}\right)^{3/2} \quad N_v = 2\left(\frac{m_n kT}{2\pi h^2}\right)^{3/2} \tag{2.4}$$

where m_n and m_p are the effective masses of the electrons at the bottom of the conduction band and of the holes at the top of the valence band.

The conduction band is considered as an N_C degenerate discrete level of enery E_C (bottom of the CB) and the valence band as an N_V degenerate discrete level of energy E_V (top of the VB).

For an intrinsic SC, $n = p = n_i$

$$np = n_i^2(T) = N_c N_v^{\frac{E_C - E_V}{kT}} \tag{2.5}$$

And the mass action law for a SC – intrinsic or doped,

$$np = f(T) = n_i^2(T) \tag{2.6}$$

As regard the population of local levels by electrons and holes, in the energy spectrum these levels represent lattice imperfections (defects) that can produce donor levels or acceptor levels. A donor level may be either in the neutral state – occupied by an electron – or (freeing itself by an electron) in the positively charged state – occupied by a hole, while an acceptor level may be either in the neutral state – occupied by a hole – or (accepting an electron) in the negatively charged state – occupied by an electron.

When the SC is doped by N_d donor dopants of which N_d^0 are neutral and N_d^+ are charged,

$$N_d = N_d^0 + N_d^+ \tag{2.7}$$

The density of charged dopants N_d^+ of energy E_d is

$$N_d^+ = \frac{N_d}{1 + e^{\frac{E_f - E_d}{kT}}} \tag{2.8}$$

Level degeneration is enclosed in the energy E_d. Generally $E_d - E_f \gg kT$ so that $N_d \approx N_d^+$.

When the SC is p-doped by N_a acceptor dopants of which N_a^0 are neutral and N_a^- are charged,

$$N_a = N_a^0 + N_a^- \tag{2.9}$$

The density of charged dopants N_a^- of energy E_a is

$$N_a^- = \frac{N_a}{1 + e^{\frac{E_a - E_f}{kT}}} \tag{2.10}$$

Level degeneration is enclosed in the energy E_a. Generally $E_f - E_a \gg kT$ so that. $N_a \approx N_a^-$

The position of the electrochemical potential is determined by imposing the conditions of charge neutrality

$$n + N_a^- = p + N_d^+ \tag{2.11}$$

The bulk conductivity of a semiconductor is given by

$$\sigma = qn\mu_n + qp\mu_p \tag{2.12}$$

where q is the absolute value of the electronic charge, n and p are densities of electrons and holes and μ_n and μ_p their mobility. One of the two terms normally dominates over the other.

2.3.2 Surface States

A surface is defined as a boundary of media with different physical properties. For example, the surface between a semiconductor and a gas is referred to as a "free surface" or just "surface." The surface between a semiconductor and another solid is usually referred to as an "interface."

An ideal termination of the semiconductor eliminates the lattice symmetry in the direction perpendicular to the surface. Therefore, the unit cells next to the surface are, in general, not equivalent to those in the bulk, and states which are localized in the vicinity of the surface may arise.

The termination of the periodic structure of a semiconductor at its free surface may form surface-localized electronic states within the semiconductor band gap and/or a double layer of charge, known as a surface dipole. The formation of surface-localized states typically involves additional, more complex phenomena which make the surface unit cells not equivalent to the bulk cells. These include the following: "dangling bonds," i.e., the formation of surface atoms with no upper atom to bind to; surface reconstruction or relaxation, i.e., a change in the position and/or chemical bonding configuration of surface atoms, which minimizes the surface energy; steps and kinks at the surface; impurity atoms adsorbed on the surface; etc.

There have been two dominant models used in describing the chemical and electronic behavior of a surface. One is the atomistic model or the "surface molecule" model; the other is the band model. The atomistic model has been preferred in general in discussion of the chemical processes at a solid surface; the rigid-band model is preferred in the discussion of electron exchange between the solid and a surface group. The atomistic model describes the solid in terms of surface sites, atoms or group of atoms at the surface, essentially ignoring the band structure of the solid. The band model describes the surface in terms of surface states, localized electronic energy levels available at the surface, to a great extent ignoring the microscopic details of atom–atom interaction between the surface species and its neighboring substrate atom. The two models are complementary; nevertheless, in the following we will deal with the band model since it is most useful when dealing with SC and can explain their transport properties.

Surface states can be of donor or acceptor type. A donor defect level may be either in the neutral state – occupied by an electron – or (freeing itself by an electron) in the positively charged state – occupied by a hole, while an acceptor level may be either in the neutral state – occupied by a hole – or (accepting an electron) in the negatively charged state – occupied by an electron.

$$N_{td} = N_{td}^0 + N_{td}^+ \tag{2.13}$$

$$N_{td}^+ = \frac{N_{td}}{1 + e^{\frac{E_f - E_{td}}{kT}}} \tag{2.14}$$

$$N_{ta} = N_{ta}^0 + N_{ta}^- \tag{2.15}$$

$$N_{ta}^- = \frac{N_{ta}}{1 + e^{\frac{E_{ta} - E_f}{kT}}} \tag{2.16}$$

The appearance of surface-localized states induces charge transfer between bulk and surface in order to establish thermal equilibrium between the two. The charge transfer results in a non-neutral region (with a non-zero electric field) in the semiconductor bulk, usually referred to as the surface space charge region (SCR). This region may extend quite deeply into the bulk. Similar considerations apply to a semiconductor interface.

The position of the electrochemical potential is determined by imposing the conditions of charge neutrality between bulk and surface.

2.3.3 Surface Space Charge Region

The charge found in surface states is clearly supplied by the underlying bulk. We therefore expect the carrier density in the vicinity of the surface to deviate from its equilibrium value and result in a surface space charge region (SCR). The surface may be found in three different configurations:

- accumulation, where the majority carrier concentration at the surface is larger than its bulk value,
- depletion, where the majority carrier concentration at the surface is smaller than its equilibrium value, but larger than the minority carrier concentration at the surface, and
- inversion, where the majority carrier concentration at the surface is smaller than the minority carrier concentration at the surface.

The mathematical description used from here afterwards will deal with an n-type SC with acceptor surface states, as usual for metal oxide semiconductor gas sensors, producing a depletion SCR. The results for p-type SC can be easily obtained by analogy. Besides, we will assume that lateral dimensions are much larger than the vertical ones, thus adopting a one-dimensional analysis that goes through the solution of the Poisson equation:

$$\frac{d}{dx}\left(\varepsilon\frac{dV(x)}{dx}\right) = -\rho(x) \tag{2.17}$$

where x is the coordinate, V(x) is the electric potential, ρ(x) is the charge density in the SCR, and ε is the dielectric permittivity of the semiconductor.

Figure 2.1 reports the band showing the formation of a SCR in an n-type SC in presence of surface acceptor states, starting from an ideal surface before charge is transferred to the surface (a) and after the formation of the SCR (b); the potential of the electron in vacuum is used as a constant reference.

According to the Poisson equation (2.17), a non-equilibrium carrier density implies a non-zero electric field and potential. Therefore, even under equilibrium conditions the surface potential, denoted as V_s, is different from the electric potential far away in the bulk. This means that the semiconductor bands are bent in the vicinity of the surface.

For a given set of semiconductor bulk and surface properties, the value of V_s is dictated by the charge conservation rule:

$$Q_{ss} = -Q_{sc} \tag{2.18}$$

where Q_{ss} is the net surface charge and Q_{sc} is the net charge in the SCR (both per unit area). This is because the underlying crystal is the sole supplier of the surface charge.

For calculating V_s, we must know how Q_{ss} and Q_{sc} depend on it. From (2.16):

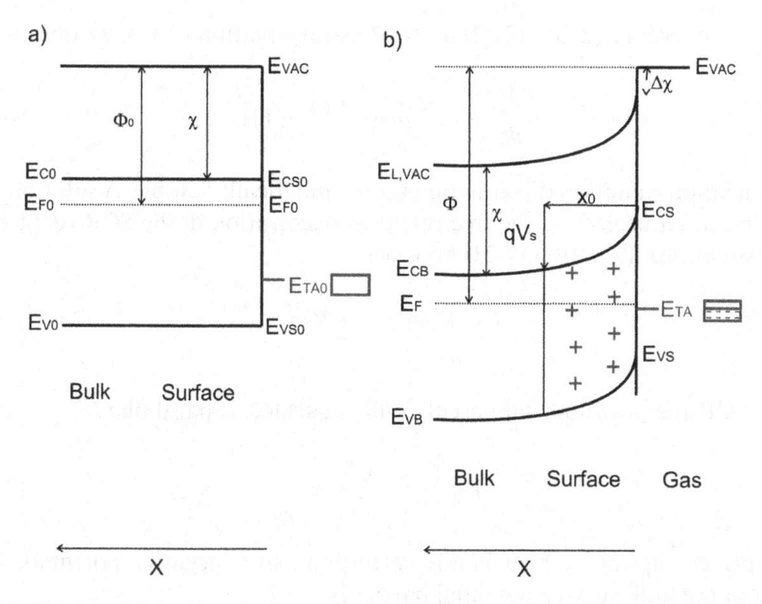

Fig. 2.1 Band model showing the formation of a SCR in an n-type SC in presence of surface acceptor states before (**a**) and after (**b**) adsorption. The potential of the electron in vacuum E_{vac} is used as a constant reference

$$Q_{ss} = -q\frac{N_{ta}}{1+e^{\frac{E_{ta}-E_f}{kT}}} = -q\frac{N_{ta}}{1+e^{\frac{E_{ta0}-E_{f0}-qV_s}{kT}}} \tag{2.19}$$

where E_{ta0}–E_{f0} is the interval between the state energy and electrochemical potential in the absence of bending. If multiple states are present, the total charge may be found by summation and integration.

We now consider the dependence of Q_{sc} on V_s. The total (static and dynamic) charge density may be expressed, ignoring minority carrier contribution, as:

$$\rho(x) = q(N_d - n(x)) \tag{2.20}$$

where N_d is the donor density, assumed to be uniform and due to fixed oxygen vacancies. If the semiconductor is non-degenerate, we obtain the relation of the electron in the SCR n(x) to their densities in the quasi-neutral bulk n_b from the Boltzmann relations:

$$n(x) = n_b e^{\frac{qV(x)}{kT}} \tag{2.21}$$

Moreover, in the quasi-neutral region the net charge density is zero so that

$$N_d = n_b \tag{2.22}$$

Placing expressions (2.20)–(2.22) in the Poisson equation (2.17), we obtain

$$\frac{d^2V(x)}{dx^2} = \frac{q}{\varepsilon}\left[n_b\left(e^{qV/kT} - 1\right)\right] \tag{2.23}$$

Without approximation this equation is not analytically soluble. A solution can be found when neglecting the free carrier concentration in the SCR (depletion approximation). Equation (2.23) becomes

$$\frac{d^2V(x)}{dx^2} = -\frac{qN_d}{\varepsilon} \tag{2.24}$$

In the SCR the potential, supposedly null at surface, is parabolic

$$V(x) = -\frac{qN_d}{2\varepsilon}x^2 \tag{2.25}$$

As expected, upwardly bent bands correspond to a negative potential. The height of the bulk surface potential barrier is

$$V_s = \frac{qN_d}{2\varepsilon}x_0^2 \tag{2.26}$$

where x_0 is the width of the depletion layer. By solving respect to x_0 one obtains

$$x_0 = \sqrt{\frac{2\varepsilon V_s}{qN_d}} \tag{2.27}$$

For reasonable values , e.g., $V_s = 0.4$ V, $\varepsilon = 11.8$, ε_0 and $N_d = 5\ 10^{15}$ cm^{-3}, we obtain $x_0 = 300$ nm. This results reveal the striking influence of surface states on the semiconductor bulk: states localized over no more than several monolayers affect the semiconductor electrically even thousand monolayers away from the surface.

Since from (2.18) it follows that $N_{ta}^- = N_d x_0$, Eq. (2.26) can be written as

$$V_s \frac{qN_{ta}^{-2}}{2\varepsilon N_d} \tag{2.28}$$

V_s cannot increase too much even in the presence of a high density of states N_{ta}, i.e., a small surface charge can be accomodated and limited to about 10^{-3}–10^{-2} monolayers of equilibrium ionosorption [165]. Weisz first pointed out this limitation on surface coverage, which in term of bands is known as "pinning" of the electrochemical potential to a surface state energy.

A simple explanation can be obtained from Fig. 2.1. The following relation holds

$$E_{Cs} = E_{ta} = E_{Cs0} - E_{ta0} = qV_s + (E_{Cb} - E_F) + (E_F - E_{ta}) \quad (2.29)$$

together with (2.28) and (2.16)

When the state density N_{ta} is low, at equilibrium the surface states are completely filled with a slight shift of the electrochemical potential from equilibrium value. The surface/bulk potential barrier V_s is low. When the state density N_{ta} is high, E_f–E_{ta} ranges between $\pm kT$ (the electrochemical potential is pinned) and the potential barrier V_s is limited to values less than 1 V.

The (very low) density of electrons at the surface n_s is, by using (2.21) and (2.28),

$$n_s = N_d e^{\frac{q^2 N_{ta}^{-2}}{2\varepsilon kTN_d}} \quad (2.30)$$

The depletion approximation assumes that free carrier density changes abruptly from N_d in the bulk to zero in the SCR. It could be shown that it is questionable only within a few Debye lengths L_D from the edges of the SCR at x_0

$$L_D = \sqrt{\frac{\varepsilon kT}{q^2 n_b}} = \sqrt{\frac{\varepsilon kT}{q^2 N_d}} \quad (2.31)$$

Typical values of L_D for tin oxide ranges from 130 to 10 nm when temperature changes from 400 to 700 K.

The width of the depletion layer as a function of L_D is given through (2.27) and (2.31) by the relation

$$x_0 = L_D 2\sqrt{\frac{qV_S}{kT}} \quad (2.32)$$

2.3.4 Surface Dipoles

In addition to surface states, another important phenomenon associated with a semiconductor surface is the surface dipole [90]. An adsorbate layer may result in a surface dipole, the magnitude of which depends on the ionicity of the adsorbate–substrate bond. The ionicity is related to the difference in electronegativity between the adsorbate species and the semiconductor substrate. Partial charge transfer between the adsorbate and a gap state may likewise cause a surface dipole and is also governed by the degree of ionicity of the bond.

As shown in Fig. 2.1, the absolute vacuum level is defined as the reference energy of an electron at rest which is situated very far from the semiconductor so that it is "unaware" of the existence of the semiconductor. At the local vacuum level E_{vac}^{l}, the electron at rest is free from the crystal microscopic potentials caused by the atomic forces, but not from macroscopic potentials due to SCR and dipoles electric field.

The surface dipole manifests itself as a step in the local vacuum level ($\Delta\chi$ in Fig. 2.1), i.e., in the electric potential at the surface because the potential changes abruptly over several monolayers. This is in contrast to the macroscopic dipole created by the surface states and surface SCR.

The work function at the semiconductor surface in equilibrium Φ, defined as the energy separation between the Fermi level and the local vacuum level at the surface, may be expressed in the form:

$$\phi = E_{cb} - E_F - qV_s + \chi + \Delta\chi \tag{2.33}$$

where $\chi = E_{vac} - E_{c0}$ is the electron affinity that changes by $\Delta\chi$ due to surface dipoles. If surface dipoles can be considered as an ordered dipole layer of moment p and density N oriented at an angle θ respect to the surface, $\Delta\chi$ rough estimation is [99]

$$\Delta\chi = \frac{Np\cos\theta}{\varepsilon} \tag{2.34}$$

2.4 Conduction Models of Metal Oxides Semiconductor

Metal oxide gas sensors are generally operated in air in the temperature range between 500 and 800 K, where conduction is electronic and oxygen vacancies are doubly ionized and fixed. At higher temperatures, oxygen vacancies mobility become appreciable and the mechanism of conduction turns into a mixed ionic–electronic. From the point of view of semiconducting sensing, this is a highly undesirable result that leads to mobile donors and produces a slow and irreversible changes in the resistance of the sensor. To avoid long-term changes, oxide sensors should be operated at a temperature either low enough so that appreciable bulk variation never occurs (very low diffusion constant of oxygen vacancies: semiconductor gas sensors) or high enough so that bulk variation occurs in a time on the order of the desired response time or less (very high diffusion constant of oxygen vacancies: metal oxide ionic oxygen sensors).

When operated in the semiconducting temperature range, the overall resistance of the sensor element is determined by the charge transfer process produced by surface reactions and by the transport mechanism from one electrode to the other through the sensing layer. Therefore, the microstructure of the sensing layer plays a key role for the development of an effective gas sensor. Rather that in compact and porous, the description will be subdivided in poly- and mono-crystalline materials, since the active geometrical surface exposed to gaseous species always corresponds to crystallite grain boundaries, which account for the overall resistance of any polycrystalline layer. In fact, an ideal compact polycrystalline layer would be completely inert to gases. Indeed, the gases must be able to diffuse through the layer in order to adsorb at grains surface; i.e., layer porosity is an essential requisite to develop effective metal oxide gas sensors. Greater porosity is accounted, on an experimental basis, as the main explanation of compressed-powders thick films' higher response compared to more compact physical or chemical deposited thin films.

In polycrystalline gas sensors, a high-resistance SCR is created around the surface of each grain and in particular at the intergrain contact as shown in Fig. 2.2a: the carriers must overcome the barrier qV_s in order to cross from one grain to the neighbour. As shown by Yamazoe in a fundamental experimental article [171], metal oxide polycrystalline gas sensors response increases abruptly when the grain size is reduced. In a low-grain-size device, almost all the carriers are trapped in surface states and only a few thermal activated carriers are available for conduction as shown in Fig. 2.2b. In this configuration the transition from activated to strongly not activated carrier density, produced by target gases species, has a great effect on sensor conductance.

In monocrystalline gas sensors – like the recently developed quasi-mono-dimensional structure – the current flows parallel to the surface and is modulated by the surface reactions like the channel of an FET by the gate voltage as shown in Fig. 2.3. When the channel is fully depleted, carriers thermally activated from surface states are responsible for conduction.

The metal semiconductor junction that forms at the interface between the layer and the contacts can play a role in gas detection, enhanced by the fact that

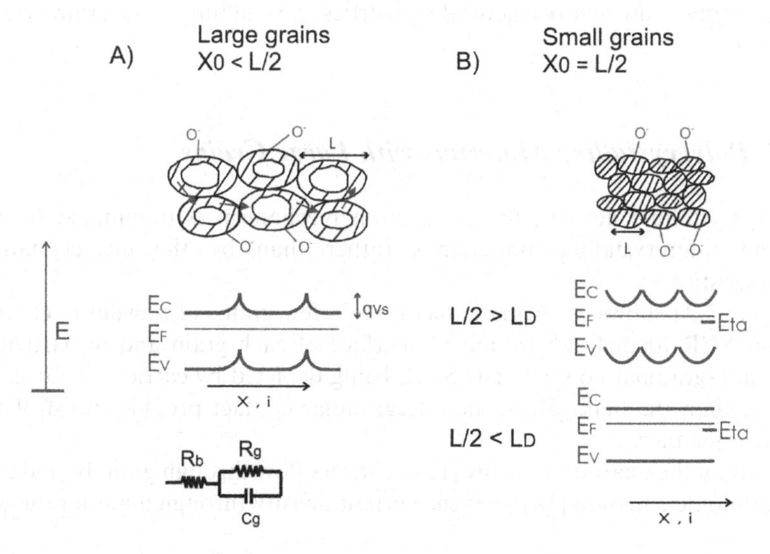

Fig. 2.2 Conduction mechanisms and electronic bands for a polycrystalline layer when grains are (**a**) partially or (**b**) fully depleted

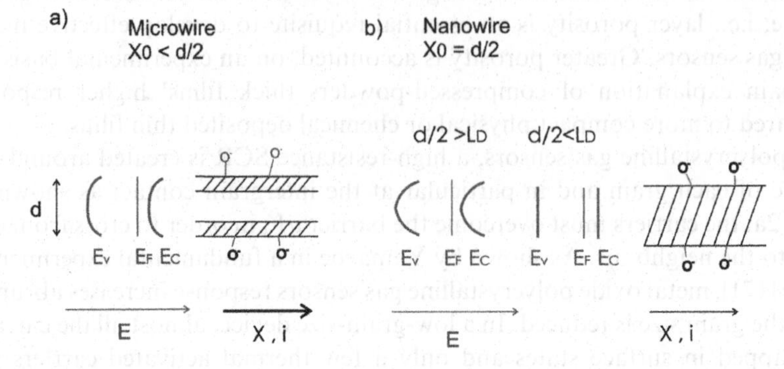

Fig. 2.3 Conduction mechanisms and electronic bands for a monocrystalline layer when grains are (**a**) partially or (**b**) fully depleted

the metal used for the contact acts also as a catalyst. The contact resistance is more important for mono-crystalline layers since it is in series with the semiconductor resistance for polycrystalline ones where it is connected to a large number of resistances. Indeed, when considering nanowires bundles, the conduction mechanism is dominated by the inherent intercrystalline boundaries at nanowires connections – like in polycrystalline samples – rather than by the intracrystalline characteristics; the intergranular contact provides most of the sample resistance.

In the following a more detailed description of conduction as a function of most important microstructural properties, crystallinity, and grain size, is reported.

2.4.1 Polycrystalline Materials with Large Grains

In polycrystalline devices, the conduction mechanism is dominated by the inherent intercrystalline boundaries rather than by the intracrystalline characteristics.

Figure 2.2a shows a schematic picture of a few grains of powder in contact and the SCR (depletion) around the surface of each grain and in particular at the intergranular contact. The SCR, being depleted by carriers, is far more resistive than the bulk. Thus, the intergranular contact provides most of the sample resistance.

Based on the analysis made by [123], carriers flow through grain boundaries by thermionic emission [154] and the current density through a grain results

$$J = Mn_b e^{\frac{qV_s}{kT}}\left(e^{\frac{qV_b}{kT}} - 1\right) \tag{2.35}$$

where V_b is the voltage drop across the barriers and M a factor that is barrier dependent.

A similar result is obtained if one make use of the diffusion theory [154] with an additional dependence as square root of V_S. Instead, free carriers tunneling through the barrier is usually disregarded due to the long extension x_0 of the SCR.

Since the film has many barriers, the voltage drop across any one is small compared to kT/q and

$$J = Mn_b e^{\frac{qV_s}{kT}} \frac{qV_b}{kT} \tag{2.36}$$

if V is the total voltage drop across the film and there are n_c crystallites or grains for unity length along the film of length L, then

$$J = \left[qn_b \mu_0 e^{\frac{-qV_s}{kT}} \right] E \tag{2.37}$$

where $\mu_0 = M/n_c kT$, E is the electric field and the quantity in brackets is the conductivity that has a thermally activated mobility term $\mu = \mu_0 e^{\frac{-qV_s}{kT}}$.

The conductance of a polycrystalline gas sensor can therefore be expressed as

$$G = G_0 e^{\frac{-qV_s}{kT}} = G_0 e^{\frac{-q^2 N_{ta}^{-2}}{2\varepsilon N_d kT}} \tag{2.38}$$

where G_0 is a pre-exponentiál term independent in first approximation from the surface adsorption and temperature.

2.4.2 *Polycrystalline Materials with Small Grains*

When the size l of the grain is small, almost all available electrons are trapped in surface states $N_{ta}^- = N_d l$ size of the grain/2 and the grains are depleted from free carriers.

Seto [143] and Orton and Powell [116, 115] have developed a model for electrical transport through polycrystalline thin films that takes into account the possibility that the grains are completely depleted of carriers. When grains are fully depleted of carriers and Debye length $L_D < l$ size of the grain/2, the depletion approximation is still valid but in a small region of radius L_D from the center of the grain. The depletion layer extends through all the grain and bands are parabolic as in Fig. 2.2b up, both mobility and average density of carrier are thermally activated, the latter from electrons trapped in surface states:

$$n \propto e^{\frac{(E_{cs}-E_{ta})-qV_s}{kT}}, \quad \mu \propto e^{\frac{qV_s}{kT}}, \quad G \propto e^{\frac{(E_{cs}-E_{ta})}{kT}} \tag{2.39}$$

When $L_D > l$ size of the grain/2 the depletion approximation cannot be applied, calculation of the band profiles shows they are flat through the grain as in Fig. 2.2b, mobility is constant and average density of carrier is thermally activated from surface states:

$$n \propto e^{\frac{(E_c - E_{ta})}{kT}}, \quad \mu = \mu_0, \quad G \propto e^{\frac{(E_c - E_{ta})}{kT}} \tag{2.40}$$

It follows that if the intergrains charge density is changed by an external agent, this may result in quite large variations of carrier concentrations produced by the transition, from not activated to strongly activated carrier density.

In the above treatment, we have derived expressions for the height of the potential energy barrier between two particles. In a porous film or body of a polycrystalline SC, there inevitably are a great number of barriers between different particles. For thick films made of compressed powders, necks between grains should also be taken into account, which could be opened or closed in relation to dimensions and size of the depletion layer [10, 163]. Therefore, the inevitable spread in grain sizes will smear the transition region and percolation effects will become important.

Thus, we can envisage a situation in which a particular change in N_{ta}^- results in almost complete depletion of some grains while having only minimal effects on others. Some grains will be switched from conducting to non-conducting states while others remain unchanged. As a result, not only will the overall resistance be increased but the distribution of current paths will be appreciably modified, and this can only be treated in terms of percolation theory. This makes detailed calculation much more complex, but the general conclusion that n may be very sensitive to changes in N_{ta}^- remains valid.

Obviously, the overall height of the resulting barrier, or the measurable activation energy, must depend on the height of individual barriers and on their distribution. Sinkkonen [149] has derived, on the basis of the effective medium theory, equations for the dc conductance of different random barrier networks. The following results are applicable in the ohmic region. When the distribution of the height of barriers is a delta function, the activation energy of conductance is the value that all of the barriers have. When there is a uniform distribution of barriers between qV_{s1} and qV_{s2}, the activation energy E_{act} at practical temperatures in the expression of conductance (2.38) is given by

$$E_{act} = q(V_{s2} - (1 - 2/z)\Delta V) \tag{2.41}$$

where $\Delta V = V_{s2} - V_{s1}$ and z is the coordination number of the network.

When the distribution is broad, i.e., the height of barriers extends from zero to a maximum value qV_{s2}, the activation energy becomes

$$E_{act} = \frac{4}{z^2} qV_{s2} \tag{2.42}$$

In both cases, the dependence of E_{act} on z is very strong if ΔV is large.

In metal oxide SCs, the high temperature required for the surface reactions to take place can induce grain growth by coalescence and prevent the achievement of a stable distribution of barriers. The consequent lack of long term stability of conductance has until today prevented a wide range application of this type of sensors. A strong effort has been therefore made to develop stable nanocrystalline metal oxide, like mixed oxides in which stabilization of the active sensing material is obtained by the presence of the second oxide, that inhibit the cations interdiffusion and crystallite growth despite the surface energy stored at the interface.

2.4.3 Mono-crystalline Materials

The interest in mono-crystalline metal oxides has arisen after innovative low cost preparation techniques of quasi-mono-dimensional structures have been established. In the case of mono-crystalline materials, the electrical conduction takes place in the direction parallel to the surface band bending; in the presence of a high-resistance surface SCR, the conductivity is modulated as in the channel of an FET as shown in Fig. 2.3. We will consider the case of a mono-dimensional belt along the main axis; the treatment for a nanowire is straightforward using the Poisson equation in cylindrical coordinates.

The conductance before adsorption of a n-type nanobelt of length L and thickness d, is given by

$$G = \sigma \frac{d}{L} = q n \mu_n \frac{d}{L} \tag{2.43}$$

When dealing with a microbelt (see Fig. 2.3a), gas adsorption creates a surface depletion layer, the thickness of the conducting channel is reduced and sheet conductance becomes – by using (2.27) –

$$G = \frac{\sigma}{L}(d - x_0) = \frac{\sigma}{L}(d - \frac{N_{ta}^-}{N_d}) = \frac{\sigma}{L}\left(d - \sqrt{\frac{2\varepsilon V_s}{q N_d}}\right) \tag{2.44}$$

With $\sigma = q\mu_n n_b$ since the dependence of mobility on surface coverage can be neglected because electron diffusion length (about 1 nm) is much shorter than diameter (tens of nm). The dependence of G on the density of occupied surface states is linear, which is very weak compared to the case of polycrystalline materials (2.38).

Electrical transport changes completely when the thickness of the belt is small like in a nanowire; the SCR extends through all the cross-section of the belt as depicted in Fig. 2.3b and all electrons are trapped in the surface states. The shape of the bands can be parabolic ($L_D<d/2$) or flat ($L_D>d/2$) as for small-grain polycrystalline materials, with the difference that current flows parallel instead that perpendicular to the surface like in a pinched-off FET channel. Density of carriers is thermally activated from surface states into the conduction band and the overall conductance can be calculated by integration over the belt section:

$$G \propto e^{\frac{(E_{cs}-E_{ta})-qV_s}{kT}} \quad L_D < d/2 \tag{2.45}$$

$$G \propto e^{\frac{(E_c-E_{ta})}{kT}} \quad L_D > d/2 \tag{2.46}$$

When charge density is reduced due to surface reactions with ionosorbed oxygen, quite large variations of carrier concentrations – and therefore of conductivity – are produced by the transition from strongly activated to not activated carrier density.

In addition to the better stochiometry and greater level of crystallinity compared to polycrystalline oxides, which reduce instability associated with hopping and coalescence, quasi-mono-dimensional single-crystal structures can be configured as a single nanowire transistor (SNT) as in Fig. 2.4.

For n-type MOX nanowires-based SNTs, application of a negative gate bias V_{GS} depletes the channel of electrons and shuts the device off; a positive V_{GS} greater than threshold voltage creates a majority carrier accumulation conducting channel as in n-channel thin-film transistors (TFT). At low V_{DS}, the SNT shows typical transistor behavior as I_{DS} increases linearly with V_{DS}. Current saturation is observed at high V_{DS} as the accumulation layer is pinched off at the drain electrode. The configuration allows estimation of mobility μ_n in the linear regime by calculating transconductance g_m in the linear zone [104].

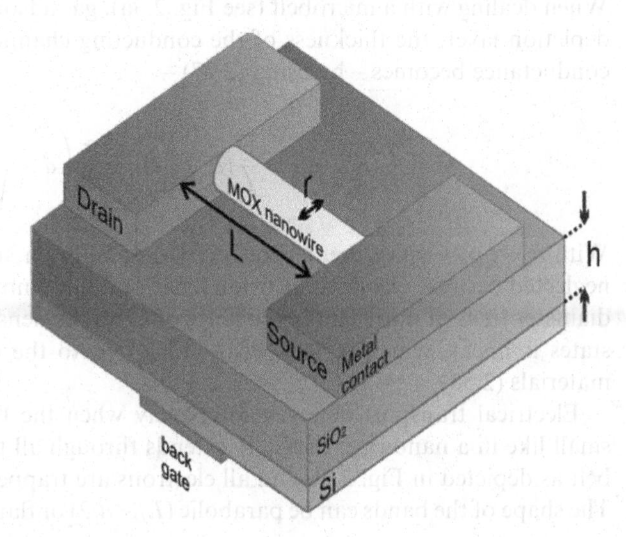

Fig. 2.4 Layout of a single nanowire transistor with a metal oxide nanowire acting as channel

When the gate potential is floating, the surface space charge region (SCR) extends through all the nanowire nanometer size section, bands are flat (Debye length L_D for tin oxide ranges from 130 to 10 nm when temperature changes from 400 to 700 K), and carriers thermally activated from surface states are responsible for conduction. By applying a gate potential, the position of the electrochemical potential and the availability of electrons for surface adsorption and desorption reactions and the sensing and recovering performance of the device can be modulated [174, 176, 44], exploiting the well-known electro-adsorbitive effect [16].

2.5 Adsorption over Metal Oxide Semiconductor Surfaces

2.5.1 Physical and Chemical Adsorption

Figure 2.5 shows the Lennard-Jones representation of gas adsorption over a surface when adsorption is accompanied by decomposition of the molecule into separate atoms. Adsorption proceeds from physisorption (a) – weak adsorption associated with dipole interactions and van der Waals forces – to chemisorption (b) – strong bonding, charge transfer between adsorbent and adsorbate, since less energy E_a is required to provide the total energy of dissociating the molecule E_{dis}. Therefore, physisorption is the first step of association of gas species with a

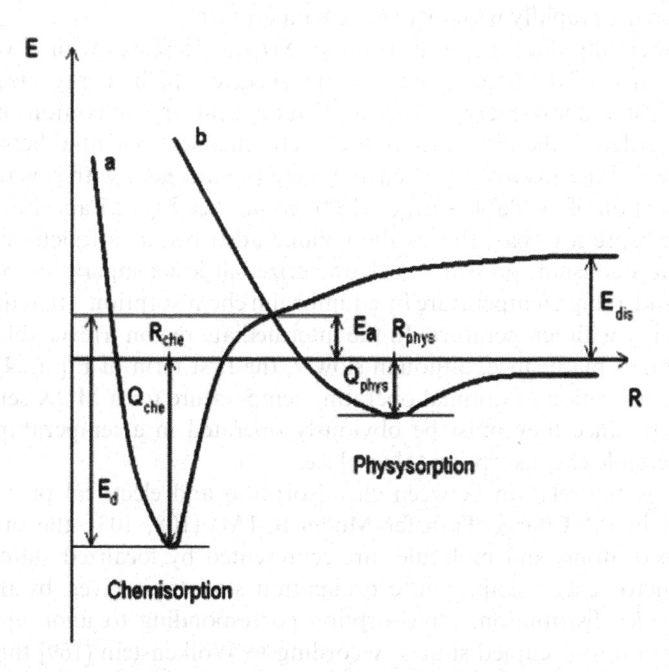

Fig. 2.5 Lennard-Jones model for physisorption (**a**) and chemisorption (**b**)

solid surface, afterwards the physisorbed species can be chemisorbed if they exchange electrons with the semiconductor surface.

Physisorption is a slightly exothermic process (ΔH_{phys}) characterized by high coverage θ at low temperature and a low coverage at high temperature. If the partial pressure is very low, Henry's Law applies and the amount physisorbed is simply proportional to the partial pressure.

Desorption, like chemisorption, requires an energy $E_d = Q_{chem} + E_a$. So, in contrast of physisorption, that is a slightly exothermic, inactivated process, chemisorption and desorption are activated processes. The activation energies can be supplied either thermally or by a non-equilibrium process such as illumination.

The net rate of chemisorption is given by [103]

$$\frac{d\theta}{dt} = k_{ads} \exp\left(-\frac{E_a}{kT}\right) - k_{des}\theta \exp\left(-\frac{E_a + \Delta H_{chem}}{kT}\right) \quad (2.47)$$

where θ is the fraction of chemisorbed available surface sites, and k_{ads} e k_{des} are the rate constants for adsorption and desorption.

At steady state the equilibrium coverage results

$$\vartheta = \frac{k_{ads}}{k_{des}} \exp\left(\frac{\Delta H_{chem}}{kT}\right) \quad (2.48)$$

And θ decreases rapidly with increasing temperature

Experimentally the heat of adsorption ΔH_{chem} decreases with coverage θ, mainly because of the heterogeneity of the surface – high energy sites will be occupied first and low-energy ones later. On the contrary, activation energy E_a can be regarded as the difference in the electrochemical potential between the SC surface and the adsorbed oxygen and therefore increases with coverage, due to the reduction of available surface electrons n_s – see Fig. 2.1 and Eq. (2.30).

The adsorption isobar, that is the volume adsorbed as a function of temperature at a constant pressure, is characterized at low temperature by physisorption and at high temperature by equilibrium chemisorption – that decreases exponentially with temperature. In the intermediate region irreversible chemisorption takes place since, although slowly, the first term of Eq. (2.47) dominates. Identification of optimal operating temperature for a MOX sensor is a crucial step, since they must be obviously operated in a temperature range where reversible chemisorption takes place.

As far as the relation between chemisorption and electrical properties is concerned in the Charge Transfer Model (CTM) [108, 103], the physi and chemisorbed atoms and molecules are represented by localized states in the semiconductor energy gap, whose occupation statistic is given by the same Fermi– Dirac distribution, physisorption corresponding to unoccupied and chemisorption to occupied states. According to Wolkenstein [169] this model is quite inaccurate since physisorbed and chemisorbed states represent electron states of different stable lattice–ion configurations of radius R_{phys} and R_{chem}. In the Volkenstein Model (VM), physical and chemical adsorption therefore belong to different adsorption curves and there are three forms of adsorbed species: physisorbed neutral, weak chemisorbed species with oriented dipole momenta, and charged chemisorped species. The VM model can be applied for explaining work function measurements: in the CTM the $\Delta\chi$ contributions are only due to the dipole moments of charged chemisorbed species, while in the VM there are also neutral weak chemisorbed species [54, 55].

2.5.2 Surface Reactions Towards Electrical Properties

When the adsorbate acts as a surface state capturing an electron or an hole, chemisorption is often called ionosorption.

An important ubiquitous species that ionosorbs over MOX surfaces is water [10]. The chemisorption of water onto oxide from air can be very strong, forming an "hydroxylated surface," where the OH^- ion is bounded to the cation and the H^+ ion to the oxide anion. The overall effect of water vapor is to increase the surface conductance, although the surface reactions are still debated and many mechanisms have been proposed [71, 103, 74].

The study of oxygen ionosorption is of particular importance for MOX gas sensors which are operated in an atmospheric environment. The case study of SnO_2 will be examined here, since it is the most studied material in the field and can be considered as a prototype metal oxide.

In the temperature range between 400 and 800 K oxygen ionosorbs over SnO_2 in a molecular (O_2^-) and atomic form (O^-) [10]:

$$O_{2,gas} \rightleftarrows O_{2,ads} \tag{2.49}$$

$$e^- + O_{2,ads} \rightleftarrows O^-_{2,ads} \tag{2.50}$$

$$e^- + \frac{1}{2} O_{2,ads} \rightleftarrows O^-_{ads} \tag{2.51}$$

Since O_2^- has a lower activation energy it is dominating up to about 500 K, at higher temperature the O^- form dominates.

By considering only Eq. (2.51) (T>400 K) and taking into account that concentration of available electrons at surface n_s is given by (2.30) and concentration of ionosorbed oxygen relative to the available surface states, represented by physisorbed oxygen, by (2.16), the mass action law gives

$$k_{ads}(N_{ta} - N^-_{ta}) n_s p^{1/2}_{o_2} = k_{des} N^-_{ta} \tag{2.52}$$

The activation energies for adsorption and desorption are included in the reaction constants k_{ads} and k_{des}

Equations (2.52), (2.16), (2.30), and (2.38) give a frame in which a relation between conductivity and oxygen partial pressure can be estimated. In practical applications the surface will normally be saturated with oxygen at any pressure approaching atmospheric and N_{ta}^{-}, n_s, and G are independent from oxygen pressure [103].

In many cases the process of gas detection is intimately related to the reactions between the species to be detected and ionosorbed surface oxygen. When a reducing gas like CO comes into contact with the surface, the following equations that consumes ionosorbed oxygen adds to (2.51):

$$CO_{gas} \rightarrow CO_{ads} \tag{2.53}$$

$$CO_{ads} + O_{ads}^{-} \rightarrow CO_{2,gas} + e^{-} \tag{2.54}$$

The mass action law becomes, by considering the concentration of CO reacting at the surface proportional to its pressure, valid at ppm level:

$$k_{ads}(N_{ta} - N_{ta}^{-})n_s p_{o_2}^{1/2} = (k_{des} + k_{react} p_{CO})N_{ta}^{-} \tag{2.55}$$

The overall effect is a shrinking at equilibrium of the density of ionosorbed oxygen N_{ta}^{-} that is detected as an increase of sensor conductance – see Eq. (2.38).

By combining Eqs. (2.55), (2.16), (2.30), and (2.38) and making a few assumptions, a square law dependence ($\alpha = 1/2$) of conductance versus CO concentration has been determined by Madou and Morrison [103] for large-grain polycrystalline materials

$$G \propto p_{CO}^{\alpha} \tag{2.56}$$

Barsan and Weimar [10] have determined that α changes in the range 0.3–2.4 by considering the other ionosorbed oxygen forms and the effect of microstructure on conductance.

The extreme variability of predictions and the great difficulty to determine the relation between measured overall conductance and available carrier density n_s have prevented a widespread use of Eq. (2.56) in the design of innovative gas sensors. Besides, surface reactions are still debated: CO sensing, for example, could take place through reaction with hydroxyl groups, producing atomic hydrogen that recombines with oxygen lattice and releases a free electron [11] and even by direct adsorption as CO^{+} [67].

Direct adsorption is also proposed for the gaseous species – like strongly electronegative NO_2^{-} , whose effect is to decrease sensor conductance:

$$NO_{2,gas} \rightleftarrows NO_{2,ads} \tag{2.57}$$

$$e^{-} + NO_{2,ads} \rightleftarrows NO_{2,ads}^{-} \tag{2.58}$$

The occupation of surface states, which are much deeper in the band gap than oxygen's, increases the surface potential and reduces the overall sensor conductance.

A few theories of gas diffusion reaction also take into account the influence of the gas transport diffusion through the porous sensing layer [52, 166, 134, 3]. As a matter of fact the target gas, to react with ionosorbed oxygen, diffuses in the sensing layer where it is consumed gradually. In steady-state conditions, the gas concentration inside the sensing layer decreases with depth and results in a profile that depends on the rates of diffusion and surface reactions. Film thickness, pore-size and microstructure, temperature, and electrodes geometry are the most important parameters of gas diffusion-reaction theories. Unfortunately, due the lack of experimental data on well-defined structures and to intrinsic complexity of gas diffusion theories, their benefits have been so far limited.

2.5.3 Catalysts and Promoters

Semiconductor oxides are widely employed in heterogeneous catalysis, which take place at the interface between solid and gaseous phases, including oxidation and hydrogenation. Of course, in catalysis research, one is interested in obtaining a particular product (product selectivity) while in the sensor field selectivity is toward the reactants (reactants selectivity). Anyway, catalytic properties of a semiconductor are very closely connected with the electronic processes taking place inside and at the surface.

The oxide could also be the active support for other catalysts. Addition of a small amount of noble metals such as Au, Pd, Pt, and Ag speeds up surface reactions and improves selectivity towards target gas species. The activation of species involved in a surface reaction may be the dissociation of a molecule, the ionization of the species or some other intermediate reaction. From the energetic point of view, the effect of catalysis is to provide a more favorable reaction path.

Instead promoters are additives that show no strong catalytic activity in themselves but modify the semiconductor in a favorable way, e.g., by stabilizing the surface area.

Following Yamazoe [171], noble metals supported over a semiconductor improve response and selectivity by chemical or electronic sensitization.

In chemical sensitization – well known for oxygen and hydrogen [103] – the catalyst activates the gaseous species which are "spilled over" the surface of the support. Much used are catalyst – like Pd – that favor molecular oxygen adsorption and dissociation in ionic form O^-, made available at lower temperature and greater concentration over the support surface. The catalyst can be another metal oxide which perform the receptor function (being chemically active) while the first oxide acts as transducer. Catalysts should be dispersed as small crystallites over the surface of the oxide in order to be active near the grain boundaries where carrier transport takes place.

Electronic sensitization derives from a direct electronic interaction between the catalyst and the support, in which a SCR is produced. When the oxidation state of the catalyst changes due to interactions with gaseous species, the SCR is modified accordingly. Electronic sensitization is also called Fermi energy control [103] since the Fermi energy of the semiconductor is pinned to the catalyst one. Beside that well dispersed, catalyst crystallites to be effective must be very thin, in order for the SCR to reach the supporting oxide.

Poisoning and mainly coalescence by surface migration of the metal particles at high temperature could deactivate the supported catalyst and should be carefully avoided.

2.6 Deposition Techniques

In the last years deposition techniques of oxides were engineered, improved for both process technology and the performances of the materials deposited, furthermore new techniques were developed. At the beginning there were only a limited number of applications of coating processes, while nowadays the industrial applications cover a wide range that spans from resistive films, protective layers, sensors, and catalyzers.

We will distinguish the different deposition technique as a function of the nanostructures obtained:

- Three-dimensional nanostructures (thick films)
- Two-dimensional nanostructures (thin films)
- One-dimensional materials (nanorods, nanowires, nanobelts, etc.)

2.6.1 Three-Dimensional Nanostructures

Thick-film technology was introduced about 40 years ago to produce hybrid circuits, such as semiconductor, sensing, and other devices [126]. A key factor distinguishing a thick film is the method of film deposition, screen-printing, e.g., that is one of the oldest forms of graphic art reproduction.

Thick-film deposition procedure is similar to the one used for screen-printing of objects. Of course, the precision and the sophistication of a printing system for thick film is not comparable to the one used for printing objects; furthermore, screen materials are different for microelectronic circuit production requirements. A typical thick-film screen will be made from a finely woven mesh of stainless steel, polyester, or nylon. This is mounted under tension on a metal frame and coated with an ultraviolet-sensitive emulsion upon which the circuit pattern can be formed photographically. The finished screen has open-mesh areas through which the desired pattern can be printed. The screen is held in position within a screen-printing machine at an adequate distance from the surface of the substrate. The thick-film paste is transferred on the top surface of the stencil and a squeegee traverses the screen under pressure. This brings the screen into contact with the substrate and forces the paste through the open areas. The required pattern deposited onto the substrate has to be dried; the pastes contain organic solvents that are needed in order to produce the correct viscosity for screen-printing and to remove these parts the paste must be dried. This step can be performed drying the film in an infrared belt drier (or a conventional box oven) at a temperature of around 150°C. After drying, the adhesion of the films on the substrate is enhanced.

In general further annealing of the thick films is performed at high temperatures. During this step, called firing, the glass melts, the fine powders sinter and the overall film becomes a solid composite material. The function of the glass is to bond the film to the substrate and also to bind the active particles together. The firing is performed in a furnace at temperatures of about 600–800°C. The temperature ramps of the furnace have to be controlled. After firing, the film is firmly attached to the substrate. The thickness of the deposited layer depends on deposition parameters such as the viscosity of paste and size of the apertures in the net.

Most thick-film gas sensors are fabricated from a printable paste of semiconducting metal oxide powder, inorganic additives, and organic binders. The paste is printed over an alumina substrate equipped with metal film electrodes and a heating resistor on the back of the substrate; the paste is then fired in an infrared (IR) or thermal belt furnace.

Thick-film technology is an easy and quite simple technique that allows the fabrication of low-cost devices; crucial steps are the powder selection and the firing step. As described in the previous sections grain size is a key factor influencing the sensing properties, and thus nanostructured powder has to be used. The firing process stabilizes the material, but induces grain growth for some oxides if the use of dopant or additive is not foreseen.

Possible technique for powder preparations for screen-printing are, for example, precipitation, pyrolysis, or thermal decomposition of a solid precursor. The powders can be produced also by the sol-gel technique in order to control the grain size down to the nanometer scale. Another useful technique proposed recently for the production of controlled grain size is flame spray pyrolysis [101, 102, 133].

Table 2.1 reports some references of metal oxide prepared by thick-film technology as semiconductor gas sensor, the most studied material, of course, is tin oxide (it is also one of the few that has been commercialized), and other oxides are titanium oxide, tungsten oxide, zinc oxide, indium oxide, and iron oxide.

2.6.2 Two-Dimensional Nanostructures

This paragraph presents a survey of deposition processes like physical or chemical vapor deposition (PVD and CVD) used to obtain two-dimensional nanostructured films.

Table 2.1 Examples of metal oxide semiconductor thick-film gas sensor

Material	Gas	Reference
$MnWO_4$	Humidity	[127]
$BaTiO_3$	CO_2	[66]
SnO_2	CO	[167]
Ga_2O_3	Reducing gases	[48]
SnO_2	H_2	[8]
Fe_2O_3	Ethanol	[155]
ZrO_2	Oxygen	[155]
SnO_2	Propanal, NO_2	[133]
$LaFeO_3$	H_2S, CH_3SH, $(CH_3)(2)S$	[26]
$Cr:TiO_2$	NO_2	[131]
SnO_2	HC and VOC	[124]
WO_3	NO_2	[25]
WO_3	NO_2	[27]
SnO_2	CO, NO_2, CH_4	[63]
ZnO	NH_3	[130]
In:ZnO	VOCs, benzene, acetone, ethyl alcohol, toluene, xylene	[177]

PVD processes consist in a phase transformation of the material to be deposited, the starting point is the solid phase of the source material, an intermediate vapor phase and finally a solid phase on the substrate. The starting material has to be vaporized and then condensed on a substrate in form of a thin film. Depositions technique can be distinguished with the different process used for the vaporization of the source material. For example, thermal and electron beam evaporation induce the phase transformation by heat transfer, while in sputtering the vaporization is obtained by bombardment of energetic ions.

The oldest PVD process is the evaporation technique, the source material is heated to the point of vaporization and the vapor phase condenses on the substrate in form of thin film. In order to control the composition of the deposited material and to allow a lower operation temperature the process is performed under vacuum conditions. The basic deposition system consists of a vacuum chamber, a mechanical roughing pump, a high-vacuum pump, a heated crucible, and a substrates holder.

To describe the deposition process we can use the ideal gas law:

$$PV = NkT \tag{2.59}$$

where P is the pressure, V the volume, T the temperature, and N the Avogadro's and k is the Boltzmann constant.

Form this equation we can derive the concentration of the gas molecules $n = N/V = P/kT$.

The deposition rate can be determined from the number of molecules hitting the substrate surface.

$$\varphi = \frac{P}{\left(\sqrt{2\pi mkT}\right)} = a\frac{P}{\sqrt{MT}} \tag{2.60}$$

where m is the mass of the molecule, M is the molecule weight and a is a constant. The deposition rate of course depends by the sticking coefficient of the impinging molecules.

There are two important parameters that influence the characteristics of the growing film: temperature and pressure. Varying temperature and pressure, the mean free path of the molecules (the average distance the molecule travels before it collides with another molecule) changes according to $\lambda = \frac{kT}{\sqrt{2}\pi Pd^2}$, where d is the diameter of the molecule. As the pressure increases the mean free path decreases and reaching the substrates becomes less probable, the molecules suffer from multiple collisions and can interact with the atmosphere. This is one of the reasons why, if we want to obtain a pure layer, the pressure has to be maintained as low as possible. Furthermore, if the pressure is decreased, the evaporation of the source material can be obtained at a lower temperature, easing the choice of the crucible material and the heating system.

The easiest evaporation system is the thermal one. A crucible, where the source material is placed, is heated by Joule effect. The crucible can have different shapes depending on the source material; there are, e.g., basket shape, filament, and box shape crucibles. The source material can be in forms of drops, slabs, rods, or powders; in the last case the crucible is a closed box with holes on the top surface in order to let only the vapor come out and not the powder itself. Depending on the material, the solid source can sublimate or just fuse and then evaporate, e.g., Cr, Ti, and Mo sublimate. Although thermal evaporation is the simplest system, unintentional doping in the film can be high due to the high temperature of vaporization that has to be reached by the source material in order to deposit the film. Particular attention must be made to the material that composes the crucible and to the possible alloys that it can form with the source material at the evaporation temperature that it should reach. That is another reason why the deposition is preferably made in high-vacuum condition: first to reduce the evaporation temperature and furthermore to improve the purity of the film (decreasing possible inclusion in the film due to the surrounding atmosphere).

Table 2.2 reports some reference of sensitive metal oxide layers deposited by thermal evaporation and used as gas sensors.

Table 2.2 Examples of metal oxide semiconductor thin film gas sensor deposited by evaporation technique

Metal oxide	Gas	Reference
SnO_2	O_2 NO_2	[126, 59]
WO_3	H_2S	[150, 151]

Another kind of evaporation technique is the electron beam evaporation (EBE): in such systems, the vaporization of the source material is obtained with heat transfer from an electron beam accelerated toward the material. A high-energy beam of electron is focused on the target with the material that we want to deposit. The target stops the electrons beam, upon impingement most of the kinetics energy is converted into heat and temperatures exceeding 300°C can be achieved, causing a local melt in the region of the target where the beam is focused. Since the energy is imparted by charged particles, it can be concentrated and a particular portion of the target can be heated reducing the interaction between source material and support materials.

The most used for industrial application among PVD techniques is sputtering. The impact of an atom or ion on a surface produces sputtering from the surface as a result of momentum transfer from the incoming particle. Unlike many other vapor-phase techniques, there is no melting of the material.

This technique was discovered by W.R. Grove when in 1852, he studied the discharge in a gas containing tube and noticed a deposition on the tube. Although its usage in industry and research started in the last decades.

The process occurs by bombarding the surface of the sputtering target with gaseous ions under high-voltage acceleration. As these ions collide with the target, atoms or occasionally entire molecules of the target material are ejected and can reach the substrate, where they form a very tight bond.

The basic setup is shown in Fig. 2.6 and consists in a vacuum chamber where a gas is introduced, a voltage is applied to the target material and the ionized atoms are accelerated to the target. Sputtering has proven to be a successful method of coating a variety of substrates with thin films of electrically conductive or non-conductive materials. One of the most striking characteristics of sputtering is its universality. Since the coating material is passed into the vapor phase by mechanical rather than chemical or thermal processes, virtually any material can be deposited. Direct current (DC) sputtering is used to sputter

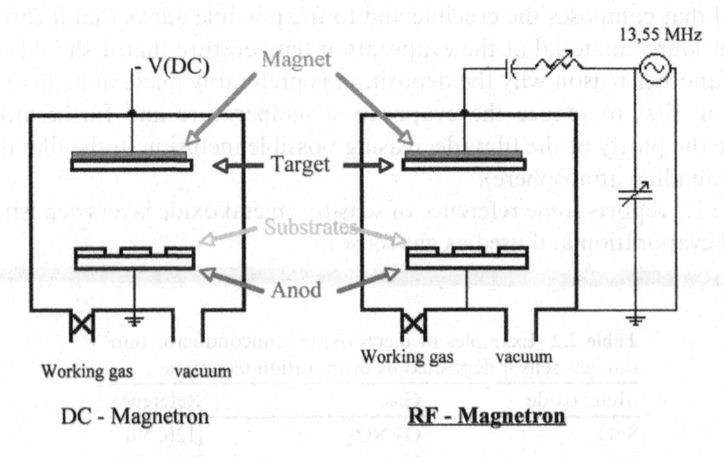

Fig. 2.6 Experimental set-up for a sputtering plant

conductive materials, while radio frequency (RF) sputtering is used for non-conductive materials in order to avoid charging of the non-conductive target due to the bombarding ions impinging on it. During part of each half cycle, the potential accelerate ions to the surface with energy high enough to cause sputtering, while on the other half cycle, electrons are accelerated towards the target and prevent any charge build-up. The frequency of the cycle has to be high enough to be followed by electrons and not by ions that have a lower mobility. The frequency 13.56 MHz has been assigned so as to avoid interference with commercial transmissions in the RF range. Another particular configuration that can be used for DC and RF sputtering is the magnetron sputtering. An appropriate magnetic field is added in the target region, in order to deflect and confine electrons near the target. This high flux of electrons creates higher density plasma, allowing the maintenance of the glow discharge at lower working pressure.

The sputtering yield (ratio of atoms sputtered to the number of incident particles) depends on mass and energy of the bombarding ion; at very high energies it decreases because ions lose their energy far from the surface and due to possible implantation phenomena.

The energy of the molecules arriving on the substrate is higher with respect to TE and EBE and this improves the quality of the layer in terms of crystallinity and adhesion to the substrate.

The target is cooled in sputtering systems to minimize the heat radiation and prevent the melting of the target material and diffusion processes. Since sputtering is usually from a large area the angular distribution of the depositing flux at a point on the substrate is large compared to thermal evaporation where the angular distribution is typically small.

Due to its versatility and reliability, there is a lot of research activity on metal oxide gas sensors prepared by sputtering, tin oxide, titanium oxide, molybdenum oxide, vanadium oxide, niobium oxide, and many other oxides as reported in Table 2.3.

Table 2.3 Examples of metal oxide semiconductor thin-film gas sensor deposited by sputtering technique

Metal oxide	Gas	Reference
SnO_2	O_3, NO_2	[13, 1, 2, 45]
	H_2S	
	C_2H_5OH, H_2, NH_3, CO	
WO_3	NO_2	[83]
MoO_3	CO, NO_2	[30, 82]
	NH_3	
Nb_2O_5	NH_3	[22]
TiO_2	NO_2	[47]
ZnO	NH_3, H_2, CH_4, C_4H_{10}	[111, 91]
	O_2	

Pulsed laser deposition (PLD) is a promising technique for the deposition of films. This method is based on rapid evaporation by means of high power laser shot and on following condensation of the vaporized material on a substrate. The PLD is relatively fast, simple and highly versatile technique for deposition of thin films of a variety of materials. Since 1987, when YBaCuO superconducting thin films were prepared by PLD, this technique has emerged for growth of stoichiometric and epitaxial thin-films. The PLD technology can be used also for the deposition of the complicated multicomponent materials, and single-crystals nanostructures.

Laser ablation is a macroscopic process that is difficult to approximate theoretically, mainly because of the extreme conditions that are typically involved. The high photon flux provides essentially instantaneous heating of the material, sometimes to temperatures near the critical point. The material reacts to this temperature increase by evaporation. The particles in the dense emitted gas can also interact with the laser photons, which heats the gas.

The setup of the experimental technique is represented in Fig. 2.7. A laser beam is focused onto a target material placed in a high-vacuum chamber or a low pressure of background gas. The photon density on the target surface causes the ejection of neutral and ionized material via thermal and/or photo-chemical ejection mechanisms. The ejected material from a target arising after laser irradiation is called an ablation plume. The flux of material then impinges on a substrate, on which deposition takes place. The plume is a plasma-like substance consisting of molecular fragments, neutral particles, free electrons and ions, and chemical reaction products.

If the removal is by vaporization, special attention must be given to the plume. The plume will be responsible for optical absorption and scattering of the incident beam. Ablation depth is determined by the absorption depth of the material and the heat of vaporization of the material. The depth is also a function of beam energy density, the laser pulse duration, and the laser wavelength. There are several key parameters to consider for PLD technique. The

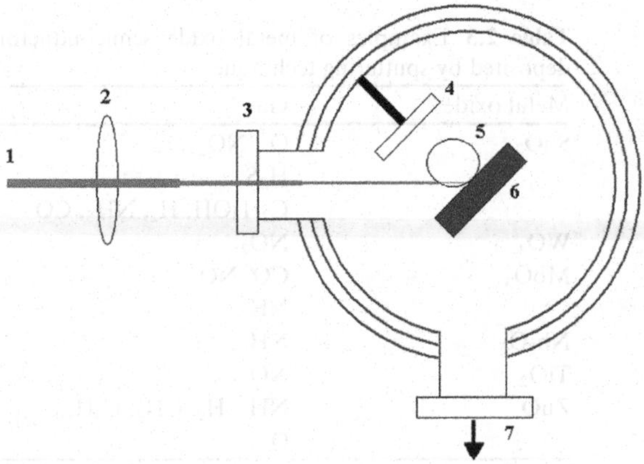

Fig. 2.7 Schematic diagram of an apparatus for PLD of a solid target with deposition onto a substrate: (1) laser beam, (2) focusing lens, (3) transparent window, (4) substrates holder, (5) plume, (6) target, (7) gas pumping system

first is selection of laser wavelength, intensity, and pulse duration. The temperature of the plasma depends on the laser intensity and the pulse duration. In some cases it is necessary to use a gas to obtain the correct composition in the film. The introduction of a background gas in the deposition chamber leads to a complicated interaction between the laser ablation plume and the gas. Langmuir probes and optical spectroscopy are used to control the plasma conditions and optical reflectometry is used to monitor the film growth. For example, the stoichiometry of the target surface will differ from that of the bulk, if one or more of these surface collisions have a component-dependent efficiency. The modified surface stoichiometry that results will be reflected in the plume generated in any subsequent ablation events and in turn in the composition of the deposited film. Such behavior provides one explanation for the frequent observation that films grown by PLD from multicomponent target, in vacuum, are enriched in the less volatile component. In the case of oxide films, at least, it is possible to compensate for such non-stoichiometry by performing the PLD in background pressure of oxygen.

Different oxides have been deposited by this technique, such as ZrO_2, Al_2O_3, MoO_3, ZnO, SnO_2, and TiO_2 [125, 98, 114], but only few works present the gas-sensing properties of these layers [168, 38, 39, 85, 113, 152] and some report the gas-sensing properties of thick film prepared with PLD deposited powders.

Chemical vapor deposition process is based on a chemical reaction that transform the gaseous molecules, called precursors, into a solid material in form of thin films or powders onto the surface of the substrate. There are different kinds of CVD:

- Atmospheric pressure chemical vapor deposition (APCVD)
- Low pressure chemical vapor deposition (LPCVD)
- Plasma enhanced chemical vapor deposition (PECVD)

Furthermore PECVD can be additionally divided in:

- Photochemical vapor deposition (PCVD)
- Laser chemical vapor deposition (LCVD)
- Metal-organic chemical vapor deposition (MOCVD)

CVD is a widely used method for depositing thin films of a large variety of materials. In typical CVD process reactant gases (in general diluted in carrier gases) enter at room temperature in the reaction chamber. The gas is then heated while approaching the deposition surface. The reactant gases may undergo homogeneous chemical reactions in vapor phase before striking the surface, depending on the operation conditions. Near the surface the gas reactants slow down due to viscous drag and the chemical composition changes. Heterogeneous reactions of the source gases or reactive intermediate species (formed from homogeneous pyrolysis) occur at the deposition surface forming the deposited material. Gaseous reaction by-products are then transported out of the reaction chamber.

Nevertheless, in each case CVD process have to provide a volatile precursor containing the elements that composes the deposited film, transport the precursor towards the substrate surface, enhance or reduce reactions in the gas phase, and provide surface reaction needed to form the film in a reproducible, fast, and reliable manner.

Figure 2.8 reports the fundamental steps during the CVD deposition process: the first is vaporization and transport of the precursor molecules into reactor, the second is diffusion of the precursor molecules towards the surface. Then, if we have do not have a homogeneous chemical reaction before hitting the substrate, the third step is the adsorption on the surface. Finally there will be the decomposition and incorporation into solid films with desorption into the gas phase of by-products.

A generic CVD system includes: gas and vapor delivery lines a reactor main chamber – hot wall, cold wall, substrate loading and unloading assembly, energy source, vacuum systems (LPCVD), exhaust system (by-products removal), process control, measurement gauges (gas flow rate, pressure, temperature, deposition time), and safety equipments. Hazardous vapor-phase reactants and products are frequently used and produced by chemical reactions then ventilation, sensors, and alarms are needed for safety reasons.

The core of a CVD system is the reactor; it has to be designed with great accuracy in order to force the reaction on the substrates and not everywhere else. Undesired reactions results in particles that can fall onto the substrate affecting the purity of the grown film. Temperature, time, pressure and surface treatments have to be taken in account and finely controlled to obtain this selective deposition.

The energy necessary to have the desired chemical reaction can be assigned as thermal energy (resistive heating, radiant heating, inductive heating), photon energy, or glow discharge plasma.

The advantages of CVD are good adhesion to the substrate, the possibility to have good step coverage, and high versatility regarding substrates and materials

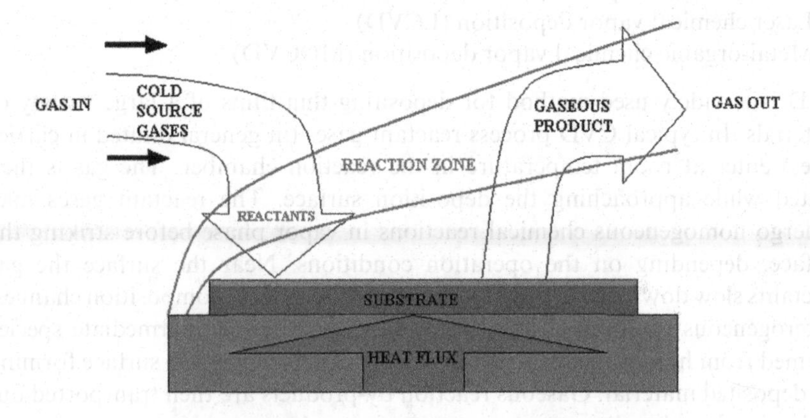

Fig. 2.8 Fundamental steps during the CVD deposition process

Table 2.4 Metal oxide gas sensors prepared by chemical vapor deposition process

Metal oxide	Gas	Reference
SnO_2	CO ethanol H_2 CO, C_2H_5OH, CH_4	[118, 107]
Fe_2O_3	Alcohol	[96, 21]

to deposit. The disadvantages are the formation of hazardous and corrosive by-products and the high temperature required. Some of the metal oxide gas sensor prepared by CVD are reported in Table 2.4.

Other thin film growth techniques from liquid and solution phase are sol-gel, spray coating, spin coating, electroplating, liquid phase epitaxy.

We will describe the sol-gel process since it is a well-known and used method for the deposition of metal oxide for gas sensors. Furthermore, it is used also for powder preparation for thick-film production. It is a versatile solution process for making ceramic and glass materials.

In general, the sol-gel process involves the transition of a system from a liquid sol into a solid gel phase. By applying the sol-gel process it is possible to fabricate ceramic or glass materials in a wide variety of forms: ultra-fine or spherical-shaped powders, thin film coatings, ceramic fibers, microporous inorganic membranes, monolithic ceramics and glasses, or extremely porous aerogel materials. An overview of the sol-gel process is presented in Fig. 2.9.

The starting materials used in the preparation of the sol are usually inorganic metal salts or metal organic compounds such as metal alkoxides. In a typical sol-gel process, the precursor is subjected to a series of hydrolysis and

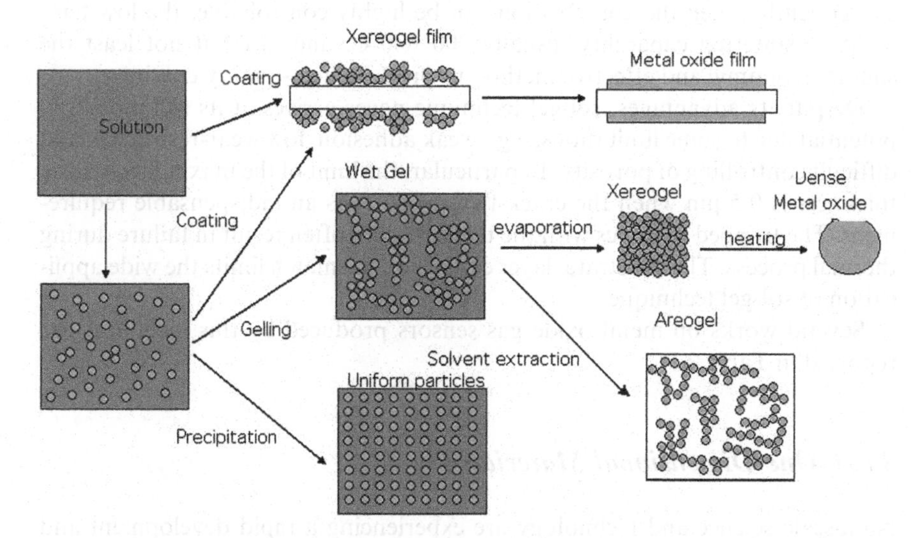

Fig. 2.9 Schematic representation of sol-gel technology and possible products

Table 2.5 Metal oxide gas sensors prepared by the sol-gel technique

Metal oxide	Gas	Reference
SnO_2	C_4H_{10}, CO, CH_4, H_2	[119]
MoO_3	CO, NO_2, O_2, O_3	[51, 94]
WO_3	CO, NO_2, O_2, O_3	[51]
TiO_2	O_2, H_2, NH_3	[86, 131, 53, 51]
	CO, humidity, ethanol	
	ethanol	
	CO, NO_2, O_2, O_3	
Ga_2O_3	H_2	[157]
In_2O_3	NO_2	[15]

polymeration reactions to form a colloidal suspension, or a sol. Further processing of the sol enables one to make ceramic materials in different forms.

A sol is a dispersion of the solid particles in a liquid where only the Brownian motions suspend the particles. A gel is a state where both liquid and solid are dispersed in each other, which presents a solid network containing liquid components. The sol-gel coating process usually consists of four steps: the desired colloidal particles are dispersed in a liquid to form a sol, the deposition of sol solution produces the coatings on the substrates by spraying, dipping, or spinning, the particles in sol are polymerized through the removal of the stabilizing components and produce a gel in a state of a continuous network, and finally the heat treatments pyrolyze the remaining organic or inorganic components and form an amorphous or crystalline coating.

The advantages of sol-gel technique are the ability to produce high-purity metal oxide because the organo-metallic precursor of the desired ceramic oxides can be mixed, dissolved in a specified solvent and hydrolyzed into a sol, and subsequently a gel; the composition can be highly controllable, the low temperature sintering capability, usually 200–600°C; and last but not least the simple, economic and effective method to produce high-quality coatings.

Despite its advantages, sol-gel technique never arrives at its full industrial potential due to some limitations, e.g., weak adhesion, low wear-resistance, and difficult controlling of porosity. In particular, the limit of the maximum coating thickness is 0.5 μm when the crack-free property is an indispensable requirement. The trapped organics with the thick coating often result in failure during thermal process. The substrate–layer expansion mismatch limits the wide application of sol-gel technique.

Several works on metal oxide gas sensors produced by this technique are reported in Table 2.5.

2.6.3 One-Dimensional Materials

Nanoscale science and technology are experiencing a rapid development and they will have a deep impact on every field of research in the first decades of the twenty-first century. Due to their peculiar characteristics and size effects, these materials often exhibit novel physical properties that are different from those of the bulk, and are of great interest both for fundamental study and for potential nanodevice applications. Carrier transport takes place in a regime in which the Boltzmann equation is invalid. Mesoscopic devices are structures large compared to the atomic scale but small compared to the macroscopic scale, where the Boltzmann equation holds. Carriers do not exhibit fully wavelike behavior but are sufficiently de-localized to exhibit also some particle-like properties. The finite size of the metal oxide particles confines the electrons wave functions, leading to quantized energy levels and a complete modification of the transport and optical properties of the material. The hugely enhanced surface–volume ratio augments the role of surface states in the sensor response. Sensing mechanisms controlled at the nanoscale level will therefore bring many benefits to the three "S" of sensor technology (sensitivity, selectivity, and stability). Interest in nanowires of semiconducting oxides has grown exponentially in the last years, due to their attracting potential applications in electronic, optical, and sensor field. Above all, the possibility to control the dimensionality of these materials is crucial to build devices with specific physical properties.

Numerous quasi-one-dimensional oxide nanostructures with useful properties, compositions, and morphologies have recently been fabricated using so-called bottom-up synthetic routes. Some of these structures could not have been created easily or economically using top-down technologies. A few classes of these new nanostructures with potential as sensing devices are summarized schematically in Fig. 2.10.

A nomenclature for this peculiar structures has not been well established. In the literature a lot of different names have been used, like whiskers, fibers, fibrils, nanotubules, nanocable, etc. The definition of these one-dimensional nanostructures is not well established. In the recent literature the terms nanowires/nanorods were used for structures with dimensions not exceeding few hundreds of nanometers.

In the past years, the number of synthesis techniques has growing exponentially.

In order to obtain one-dimensional structures there have to be a preferential growth direction with a faster growth rate in a particular direction. The evaporation condensation method will be extensively described since it is one of the most explored in the recent papers. The experimental setup consists of a

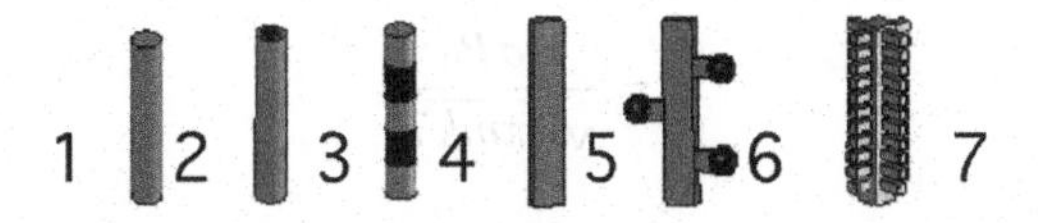

Fig. 2.10 Different kinds of one-dimensional metal oxide (1) nanowires-nanorods; (2) core-shell structures; (3) nanotubules/nanopipes and hollow nanorods; (4) heterostructures; (5) nanobelts/nanoribbons; (6) dendrites; (7) hierarchical nanostructures

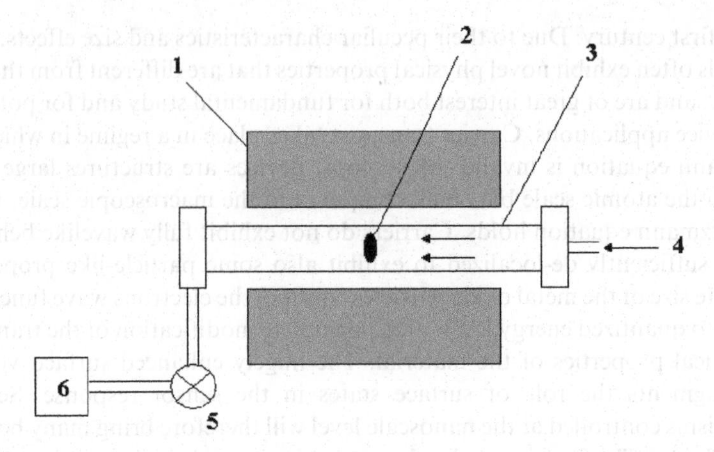

Fig. 2.11 Experimental set-up for evaporation-condensation deposition of one-dimensional structures: (1) heating furnace, (2) source material (metal, metal oxide), (3) vacuum sealed tube (quarzt, alumina), (4) inlet for gas carrier (Ar, O_2, H_2) (5) valve (6) vacuum pump

vacuum-sealed tubular furnace with a inlet for the carrier gas and an output connected to a vacuum pump as reported in Fig. 2.11. The growth chamber has to be designed in order to obtain the proper temperature gradient for one-dimensional nanostructures preparation. The metal oxide powder is placed in the higher temperature region and the gas carrier transports the evaporated oxide towards the substrates.

The growth process can be summarize in some steps:

- diffusion of the growth species towards the surface of the substrate,
- adsorption/desorption mechanism of the molecules on the substrate surface,
- diffusion of the adsorbed species on the substrate surface since a growth site is reached,
- aggregation of adsorbed species to form the crystal,
- diffusion and evacuation of possible by-products

Some of these growth steps can limit the crystal growth rate. If the concentration of the growth species is low then the limiting step is the adsorption mechanism, the growth rate is controlled by the condensation rate, which is proportional to the vapor pressure of the growth species.

As described before:

$$J = \frac{\alpha\sigma P_0}{\sqrt{2\pi mkT}} \tag{2.61}$$

where α is the accommodation coefficient, P_o is the equilibrium vapor pressure of the crystal at a temperature T, and sigma is the super-saturation of the growth species in the gas phase.

When the growth species concentration is high the limiting factor is the surface growth step.

As the vapor pressure increases, the defect formation probability increases and furthermore possible secondary nucleation site can happen losing the mono-crystallinity.

Some growth mechanism can be distinguished: vapor-solid (VS), vapor-liquid-solid (VLS), and solution liquid-solid (SLS). The VS growth stand for vapor-solid growth and it is used to describe all the growth in which the nanowire crystallization originates from the direct condensation from the vapor phase without the use of a catalyzer. At the beginnings the growth was attributed to the presence of lattice defects, but when defects-free nanowires where observed this explanation had to be abandoned. Another peculiar effect registered was a nanowire growth rate higher than the condensation rate from the vapor phase calculated the equation for a flat surface.

A possible explanation proposed is that also the lateral faces of the nanowire adsorb the molecules that diffuse on the surface of the wire.

The other possible growth mechanism is the VLS in which we have the contribution of liquid and solid phase; in general, the liquid phase is the catalyzer that drives and controls the nanowire growth. This is an old growth mechanism proposed in 1964 by Wagner. In order to obtain this process, there are some requirements to fulfill, such as the following: the catalyzer has to allow the formation of a liquid solution with the material that we want to deposit in the deposition conditions, the vapor pressure of the catalyzer has to be low, and the catalyzer has to be chemical inert. For metal oxide nanowires the metal can be a catalyzer itself. A control in the growth direction can be obtained with the choice of a mono-crystalline substrate with the right orientation. The growth species in the vapor phase diffuses in the liquid drop of catalyzer when the concentration in the drop is too high the growth species precipitate and form the nanowire. The liquid drop of the catalyzer is a preferential growth site because all the adsorbed atoms are not able to desorb. This induces a higher growth rate of the VLS with respect to the VS growth process. Another advantage is an enhanced control in the nanowire diameter induced by the control in the dimension of the catalyzer dispersion and dimension. In general, high temperatures are required for VLS growth of nanowires, an alternative method is the SLS. This method is similar to VLS, but the source material is in solution and the produced nanowires are in general polycrystalline.

Among these studies, several investigations have been devoted to tin oxide (SnO_2) nanowires/nanobelts and different approaches to their synthesis have been developed [120, 89, 79, 23, 35].

At the present, vapor deposition processes seem to be the most promising ones, due to their simplicity and low costs. Some works drew the attention on the VLS (vapor-liquid-solid) mechanism, through which thermal evaporation techniques can favor a fast growth below 1000°C (crystal growth by thermal evaporation of SnO_2 powders usually occurs at temperatures higher than 1300°C). Many studies have been performed on the gas sensing properties of these quasi one dimensional nanostructures we report just some reference [31, 92, 93, 174, 175, 176], for an updated list of references the reader can refer to these review papers [28, 88, 95, 97, 162].

2.7 Conductometric Sensor Fabrication

2.7.1 Substrate and Heater

The choice of the sensing film is important for the receptor function (the reactions between the gas and the metal oxide surface), but the choice of the transducer and the operation mode can affect the response of the complete device. Furthermore, not only the proper design of the transducer but also a proper choice of bonding and packaging is needed to provide a final sensor device for practical applications.

The material used as substrate can be chosen; alumina- or silicon-based substrates, e.g, are the most used for gas sensors. If the bulk substrate is conductive then an additional insulating layer must be provided.

The transducer has to provide at least two electrical contacts on the metal oxide to measure conductance changes and a heating system in order to maintain the metal oxide at the suitable operation temperature that is of the order of hundreds degrees centigrade. The first example of semiconductor gas sensor presented on the market is a Taguchi-type sensor and the heater is embedded in an alumina ceramic tube and the metal oxide is mounted on the tube with two gold electrodes.

The simplest transducer is reported in Fig. 2.12: a bulk substrate with electrical contacts on one side of the substrate and a heater on the backside. The working temperature of the metal oxide gas sensors is achieved applying a constant voltage to the heating meander: the current, flowing through it, heats the substrate by Joule effect. The temperature control can be made with a feedback circuit to keep either resistance or power consumption constant. Since the power consumption can depend also on changes in the gas flow or outside temperature, the first method is preferred. The temperature of the substrate can be read by a thermocouple or directly on the heating meander, using the metal

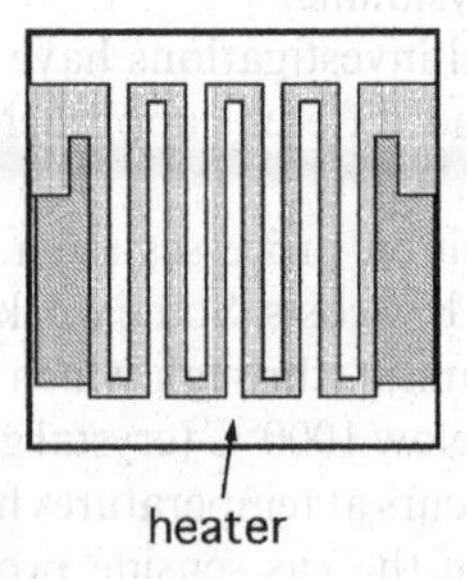

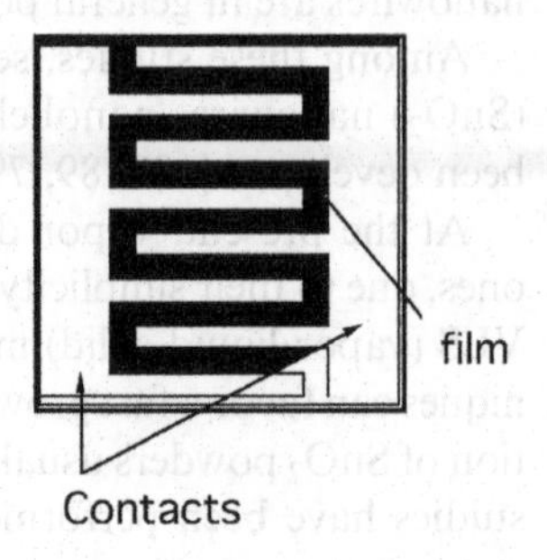

Fig. 2.12 Possible configuration of bulk transducer for conductometric gas sensors

used for the meander deposition (platinum, for example) as a thermometer. The resistance of the meander changes as its temperature changes and this relation can be used to predict the real working temperature of the metal oxide layer.

These transducers can be fabricated with thin or thick film technologies and are very robust, the temperature uniformity on the sensing layer side is good, but of course the power consumption of these bulky transducer is quite high, about 400–1000 W for 300–400°C, depending on the substrate dimensions.

A way to reduce power consumption is the use of micromachined silicon substrate. The principal characteristics that a low power consumption substrate has to provide are the following: controlled temperature profile across the sensing layer and mechanical stability. The first point is crucial because the gas-sensing properties of metal oxide are temperature dependent and different temperatures can produce different changes in electrical conductance in the same gas environment. The second point is needed for all the steps of the production of the final device and in particular during deposition and annealing of the sensing layer.

The typical process flow for the production of micromachined substrates is reported in Fig. 2.13. The starting point is a silicon bulk substrate with thickness of hundreds of micrometers followed by the deposition of a dielectric layer, the deposition of a metal layer, and the pattering process in order to form heater and electrodes, the passivation of the heater, and finally the backside etching in order to reduce the thermal mass of the heated part. There are two possible structures: a closed membrane and a suspended membrane as reported in Figs. 2.14 and 2.15.

Heater and thermometer can be integrated in two different ways: in the same layer of the electrodes (horizontal approach) or in different layers (vertical approach). The first approach has the advantages of a single deposition step but the disadvantage is the temperature uniformity. The second approach allows different configurations and a well-defined temperature distribution. Heater and

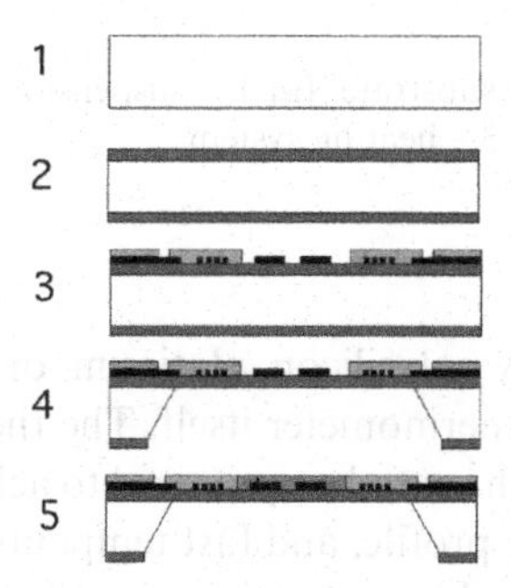

Fig. 2.13 The typical process flow for the production of micromachined substrates. First the silicon substrate (1), then the insulating layer deposition (2), metallization and passivation steps to form heating systems and metal contacts (3), backside etching (4), and finally the metal oxide-sensing layer deposition (5)

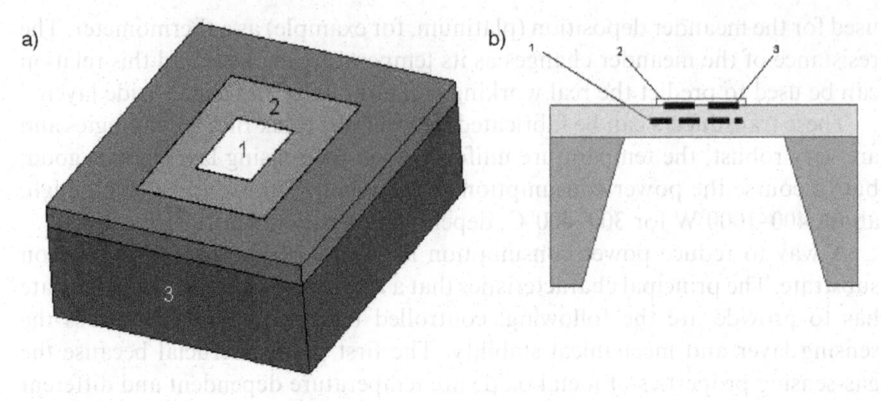

Fig. 2.14 Closed-membrane substrate: (a) 1 – active area, 2 – membrane, 3 – bulk silicon substrate; (b) 1 – heater, 2 – metal oxide, 3 – contacts

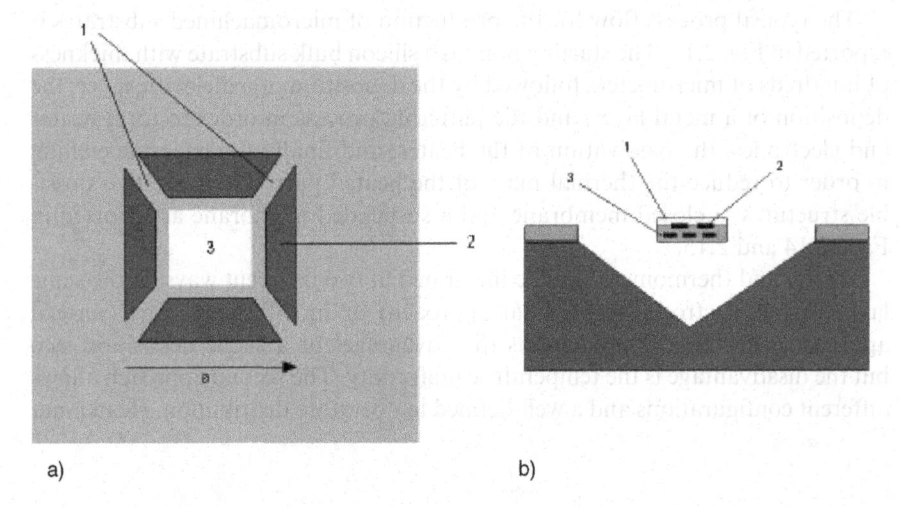

Fig. 2.15 Suspended membrane substrate: (a) 1 – suspensions, 2 – empty, 3 – active area; (b) 1 – metal oxide, 2 – contacts, 3 – heating system

thermometer can be made by polysilicon, platinum, or p^{++} doped structures and the heater can be used as a thermometer itself. The thermal characteristic of the micromachined transducers have to be optimized to achieve low power consumption, controlled temperature profile, and fast temperature transient response for at the possibility of fast pulsed temperature operation.

The total power consumption can be reduced constructing thin membranes of material with a low thermal conductivity, using suspensions with high length-to-width ratio, decreasing the heated area (Table 2.6).

Table 2.6 Power consumption and thermal time constant for some substrates used for gas sensor development

Type	Membrane dimension	Heater dimension	Temperature [°C]	Power consumption [mW]	Thermal time constant	Reference
Si_3N_4/SiO_2	1.7	900 micron	500	130	10 ms	[137]
Nitrided PS	1.5		400	170	56 ms	[100]
Alumina 254 micron thick	2	2 mm	500	650	3s	

Heating the active area with constant power per area is not sufficient for a homogeneous temperature profile; to solve this problem, an heater and a ring heater have been used or a double spiral, with variable width [64].

There are also substrates in which the gradient of temperature is intentionally created such as the one developed at the Karlsruhe Research Center Forschungszentrum Karlsruhe (FZK), it is a compact array of sensor based on a unique and highly integrated gradient micro-array chip. It allows the characterization of the layers at different temperatures simultaneously. On the back of the chip four heating meanders maintain a temperature difference of about 50°C across the micro-array. The top of the substrate is coated with a metal oxide film. By simple partitioning of the metal oxide film with parallel electrode strips, an array of individual gas sensor segments is formed. As every part of the metal oxide is kept at a different temperature, the gas responses will be different for each one. Moreover, in this particular micro-array, the metal oxide layer is coated with a gas-permeable membrane of which the thickness varies across the micro-array. Thus, each sensor segment is heated to a slightly different temperature and further differs in its membrane thickness from its neighboring sensor segments. Consequently, the individual sensor segments have a different sensitivity spectrum [153].

Variations in the electrode configuration produce modifications in the sensing properties of the final device as reported in [75, 76, 77].

Furthermore, an electrical field can be applied perpendicularly to the sensing layer in order to change the surface reactions. Hellmich et al. [72] used a micromachined sensor, and by application and variation of the bias voltage they achieve a modification of the sensor signal.

Another way to modify the sensor signal is the temperature modulation with duty cycle of the orders of ms or minutes. In the case of ceramic sensors only slow modulation can be used while in the case of micromachined substrate also fast modulation can be used. In such a way the concentration of adsorbed oxygen is modulated and influences the surface reactions with the target gas.

For example in Ref. [42] thin SnO_2 sensing layer for CO monitoring, deposited upon a Si micromachined substrate characterized by a low thermal mass, has been operated in a very fast temperature mode to detect CO. It has been demonstrated that with this fast-pulsed temperature operation mode, it is possible to drastically reduce the power consumption down to W level without compromising the sensing layer performances. Stability over 1 week is satisfactory and interference due to water vapor, benzene, and nitrogen dioxide does not increase with respect to isothermal characterization.

2.7.2 Electrical Contacts

Another crucial step, other than the sensing layer and the transducer design, is the choice of the metal use for making electrical contacts on the sensing layer in order to measure the electrical properties as a function of the environment. First of all the contact has to be good: i.e., it must provide as low sheet resistance as possible in order to minimize the voltage drops along the interconnections (that's why metals are used in general) and the contact should be ohmic. Platinum is one of the most used metallization since it have a good ohmic contact with most of the metal oxide, it does not oxides at high temperatures, it has a low diffusivity and it is resistive to corrosive gases. Gold, instead, diffuses quickly and it may introduce an unintentional doping of the metal oxide.

To develop stable and sensitive semiconductor gas sensors, it is important to understand deeply the mechanism of gaseous species detection by change of material resistance. This variation is due to modulation not only of the intergrain potential barrier but also of the metal semiconductor barrier at the contacts. The dependence of the contact resistance on the surrounding atmosphere can be due to electrical contribution of the MOX–metal interface to the overall sensor resistance or to possible chemical interactions; for the catalytic nature of the electrode material (the materials used are often platinum and gold), it can increase surface reactions; surface species (which can be more easily adsorbed on the metal) may diffuse fast and interact with MOX surface.

Usually semiconductor gas sensor resistance is calculated by two probe measurements without separating the two contributions to the sensing properties. Different metals used can produce a variation in the sensing performances of the sensing layer. A way to distinguish this two contribution is the semiconductor four-probe array analysis either in square or in collinear configuration. It is possible to determine, besides to the semiconductor resistivity, the contribution to the overall resistance due to the metal-semiconductor junction. In Ref. [43], e.g., Pt and Au were used as metal contacts and their effect was studied in CO and CH_4 detection. The technique has been applied to carbon monoxide and methane detection by tin dioxide thin films. The main result is that the contact contribution is very important for CO detection, while the material contributes to CH_4 detection. In the last case, the sensitivity may be related, as above described, to the intergrain conduction processes.

2.7.3 *Heating Treatments*

The structure of metal oxide can be greatly influenced by thermal treatments, change in morphology, grain dimension and oxide phase are measured after thermal treatments in reactive or inert atmospheres. Most of the metal oxides suffer from grain growth when treated at high temperatures, and this lead to an undesired decrease in the sensing performances. Nevertheless, a thermal treatment is in general needed, for screen-printed or sol-gel films, e.g., due to the deposition technology or just to stabilize the metal oxide that has to work at high temperature. A compromise has to be considered to obtain a sensitive and stable material.

Moreover, annealing treatments can also be used as a powerful tool to change the material properties, the chemical composition, and porosity. As an example we can cite the selective sublimation process (SSP) that has been extensively described in [65] and the rheotaxial growth and thermal oxidation [134, 139].

The SSP consists of the co-deposition of two poorly miscible oxides, one of which featuring a relatively low sublimation temperature. Annealing at suitable temperature causes the sublimation of the most volatile compound, leaving a nanostructured layer with adjustable composition and properties. The method proved to be effective for the synthesis of nanostructured gas-sensing films. This methodology allows fine-tuning of experimental parameters and in turn of their performance as gas sensors. The technical advantages of SSP are manifold. First, grain-coalescence inhibition is expected to take place. As was extensively studied in the past, addition of a foreign element or a second phase often results in coalescence inhibition of the grains during annealing. Second, the relative proportion of the oxides can be modulated as sublimation starts being effective. Third, an effect on film porosity takes place, depending on the extent of oxide segregation from the nanosized film. The method has been worked out extensively, such as for WO_3-TiO_2 [46] MoO_3-TiO_2 [62], MoO_3-WO_3 [160] with very good performance both in terms of gas response and stability of the layers.

Another peculiar technique, developed by Sberveglieri et al. [139], is the Rheotaxial Growth and Thermal Oxidation (RGTO) technique. It consists of two processes: first the deposition of a metallic thin metallic layer on a substrate, kept at a temperature higher than the melting point of the metal, then a thermal oxidation in order to get a stoichiometric composition for the metal oxide. The deposition of the metal on a "hot" substrate (400°C) produce a rough film. Just after the deposition, there are not interconnected spheres, hence the film is insulating.

The metal-semiconductor phase transformation (thermal oxidation) is achieved by keeping the thin film in a furnace with a humid synthetic air flow and making a thermal cycle of 4 hours at 250°C and 30 hours at 600°C. During the first step at low temperature a little layer of oxide grows on the metal surface avoiding the melting during the second step at 600°C that perform a complete oxidation.

The oxygen incorporation in the lattice increases the volume of about the 30%, causing interconnection of the agglomerates and creating percolating paths for the current flow. The surface of the semiconductor film after thermal oxidation is made of porous agglomerates and a porous surface is well suited for gas absorption. Furthermore this surface is perfectly suited for catalyst dispersion. The porous surface of the RGTO semiconductor metal oxide prevent the formation of a continuous metal layer.

2.7.4 Dopings, Catalysts and Filters

A useful way to tailor the sensing layer properties is the use of dopants, catalyzers or filters. Dopants are added to the metal oxide in the deposition step in order to change the electrical properties of the material itself, as in the case of classical doping in silicon technology, but dopant concentration in this case is fractions of a percent. Catalyzers instead are in general finely dispersed on the surface of the metal oxide; it is a substance that accelerates the rate of a particular chemical reaction as we have seen in Section 2.5.3. Differently a filter can alter the molecule diffusion and let only some species arrive on the metal oxide surface. Of course, there is no sharp separation between these three functionalizers; example.g., dopants and filter can have also a catalyzer effect.

The use of catalyzers is of course very effective in the gas sensor world [87, 24, 136, 159, 138, 80, 37, 140, 112, 9, 50], moreover the materials used for gas sensing are often chosen among the oxides used in catalysis field. Furthermore dispersions of fine particles of catalyzers on the top of the metal oxide surface enhance a particular reduction or oxidation reaction and can decrease the working temperature of the sensor, increase the response to a particular gas, improve the reaction kinetics.

For example, a low-temperature CO metal oxide sensor can be obtained with the use of catalysts [81]. Low-temperature CO oxidation catalysts has become an important research topic during the last 10 years due to the many potential areas of application not only for the sensor field. The several catalysts, prepared and tested for low-temperature CO oxidation activity, are composed of a metal (Au or a Pt group metal PGM) supported on oxides or mixed oxides. The promising technical performance of gold with respect to the Platinum Group Metals (PGMs) is shown, e.g., in the preferential oxidation (PROX) of carbon monoxide in hydrogen-rich reformer gas and in the water gas shift reaction used for producing hydrogen from carbon monoxide and water. For the former, a gold catalyst (Au/α-Fe_2O_3) is significantly more active at lower temperatures than the commercial PROX catalyst, i.e., Pt/γ-Al_2O_3, currently used for this purpose, and for the latter, gold supported on iron oxide (Au/α-Fe_2O_3) catalyst is more active at lower temperatures than both the α-Fe_2O_3 support and the mixed copper/zinc oxide (CuO/ZnO) catalyst currently used commercially [78, 122, 20].

Gold, like other noble metals, is not a reactive material in the bulk form, but when it is deposited on metal oxides as ultra-fine particles, its chemistry changes [58, 68, 70]. It is also reactive for CO oxidation at 0°C [81], the reactivity depends on the electrical charge and atomic number of gold clusters. Effective supports for CO oxidation with gold catalyst are, e.g, titanium oxide, iron oxide, cobalt oxide, and nichel oxide, and as reported in [69, 70], they are active also at 70°C.

All these studies in the catalysis field has driven researcher in the sensing field in the use of these catalyzers on metal oxides already used for CO detection.

Many papers show that the use of other additives can considerably affect the sensing properties of the metal oxide. Also metal oxides themselves are used as additives and they can provide faster response or recovery, enhancement in the sensing performances such as sensitivity, selectivity, and stability.

Nanocrystalline mixed metal oxides (nanocomposites) are attracting a considerable attention with regard to their applications in chemical gas sensors and heterogeneous catalysis [74, 60, 56, 57].The nanocomposites comprise at least two metal oxides MOX_1/MOX_2, one of which is the nanocrystalline matrix and the other is distributed between the bulk and the surface of the MOX_1 grains.

Several synergetic effects can derive from these nanocomposite systems, i.e.:

- The presence of one oxide phase can stabilize the second oxide that can act as sensing material, ensuring for example the maintenance of its nanosized character [32]. The grain boundaries between the two oxides inhibit interdiffusion and crystallite growth despite the surface energy stored at the interface. Furthermore the presence of a second phase can maintain the grain sizes and the interface layers thickness within dimensions comparable with the Debye Length (5–50 nm), giving to these systems new physical properties [148, 19].
- One of the two phases could act as a filter and the other one as a sensing material. Interfering gases could be consumed in the catalytic active superficial layer.
- One of the two phases could perform the receptor function (being chemically active) while the other one could perform the transducer (in electrical signal) function (this is more effective in the case of the metallic phases added).

Finally the use of filters can diminish the cross interferences of specific molecules, charcoal filters are the most often used [109, 156]. Commercially available metal oxide gas sensors for CO detection (see, e.g., Figaro Gas Sensor Figaro Engineering Inc., Products Catalogue, Gas Sensors FiS Inc., FiS products review) are equipped with charcoal filters to absorb NOx, alcohols, and longer-chain hydrocarbons in order to fulfill standard regulations such as UL2034 in the US.

The direct coupling of filters to the sensor layer had also been realized with zeolite filters [110] and porous SiO_2 layers [142, 104, 61, 41], with the latter prepared by TEOS (tetra-ethyl ortho silicate) CVD or spin casting processes, and Al_2O_3 layers [5, 6, 4, 49].

In addition, catalytically active layers such as Pt and Pd layers on top of Al_2O_3 or SiO_2 membranes utilize catalytic activity in the conversion of certain gases as a pretreatment. Reactive gases are burned in the catalytic layer, the less reactive ones reaching and interacting with the SnO_2 layer [40, 121].

Schweizer-Berberich et al. [141] report on the use charcoal filters with a micromachined SnO_2 sensor operated in a pulsed mode to fulfill UL2034 or BS7860 standards. Small reducing molecules were distinguished from CO by mode of operation and a suitable multicomponent analysis.

2.8 Transduction Principles and Related Novel Devices

2.8.1 DC Resistance

The change of density of trapped carriers on surface states, which is related to concentrations in the gaseous phase, can be transduced as a change of resistance, impedance, or work function.

The easiest measurable physical quantity that can be transduced is the sensor conductance in DC conditions. In laboratory tests it is normally measured by a voltamperometric technique at constant bias, while in commercial gas sensors the layer is usually inserted inside a voltage divider.

The kinetic response of DC conductance to a concentration step is reported in Fig. 2.16. When a reducing species like CO at time τ_0 is introduced, sensor

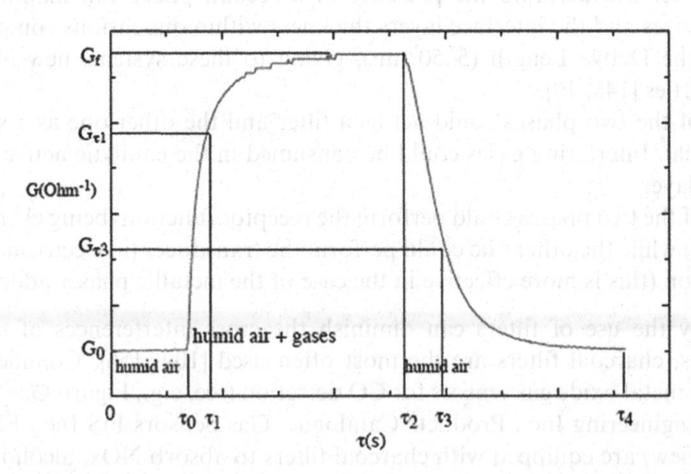

Fig. 2.16 Kinetic response of the sensor conductance produced by a step in a reducing gas concentration

conductance G_0 increases slowly towards the new equilibrium value G_f, due to the time taken for surface reactions to reach the new thermodynamic equilibrium.

In some cases due to irreversible chemisorption or to material instability, steady state is not reached; therefore, linear regression or other techniques should be applied to estimate that the response is in steady state.

Response time is the term $\tau_1 - \tau_0$ necessary for the conductance to reach a threshold value (normally 90%) of the difference between the steady state and the air value $G_f - G_0$. Recovery time is the term $\tau_3 - \tau_2$ necessary for the conductance to recover to a band expressed as a percentage fraction (normally 90%) of $G_f - G_0$. Of course, flow conditions and test chamber size strongly affect characteristics times.

Since for gas sensors response, the linearity hypothesis is not verified; i.e., when a gas mixture is present, the output is not equal to the sum of the individual response, the superimposition principle cannot be applied and dynamic response has a restricted importance; i.e., it is not possible to determine a sensor transfer function.

The sensor response towards a target gas concentration for metal oxide semiconductor sensors is usually defined as the (relative) change of conductance, with the obvious relation

$$S_G = \frac{G_S}{G_0} = \frac{G_S - G_0}{G_0} + 1 = S_G^* + 1 \tag{2.62}$$

In case of an oxidizing target species, resistance increases following gas introduction and the sensor response is defined as the (relative) change of resistance

$$S_R = \frac{R_S}{R_0} = \frac{R_S - R_0}{R_0} + 1 = S_R^* + 1 \tag{2.63}$$

Starting from the sensor response it is possible to derive the sensor response curve, that is the representation of the steady-state output as a function of the input concentration [34]. The sensor response curve is frequently called, erroneously, sensitivity curve.

Since the relation between concentration and conductance follows a power law as in Eq. (2.56), the sensor response curve is normally linearized by reporting input and output in a bilogaritmic scale. Beside representation of response is often given by reporting the conductance normalized to its value at a fixed gas concentration [164] as reported in Fig. 2.17. Sensor response at each concentration can be easily obtained as the distance – in logarithmic scale – between the curve and the normalized value of air level:

$$S_G = \frac{G_f}{G_0} = \frac{G_f\, G_S}{G_S\, G_0} \Rightarrow \log S_G = \log\frac{G_f}{G_S} - \log\frac{G_0}{G_S} \tag{2.64}$$

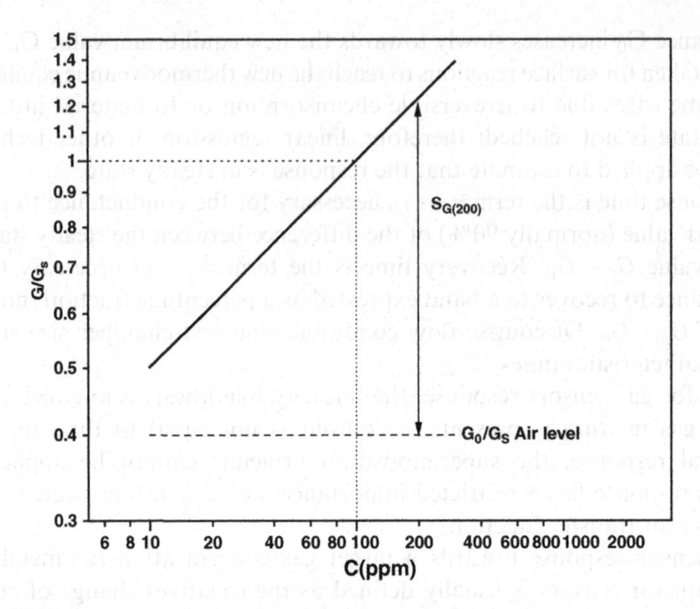

Fig. 2.17 Calibration curve for a reducing species

On each sensor response curve is normally reported the separate response towards each gases (target and interfering) to which the device is sensitive and the effect of humidity on the sensor response curve.

2.8.2 AC Impedance

Since every SCR of size x_0, related through Eq. (2.27) to the surface potential barrier, has associated an equivalent capacitance per unit area [154] expressed by

$$C = \frac{\varepsilon}{x_0} \tag{2.65}$$

A metal semiconductor layer can be considered as a network of parallel capacitance C_g/resistance R_g produced by the double Schottky layers at grain boundaries, facing each other, in series to the grain bulk resistance R_b as shown in Fig. 2.2. Besides, the electrode contact can also be represented by an RC element, so that the sensor equivalent circuit is usually made of two (or more) series RC.

Change of impedance can be measured by spectroscopic analyzer or by LCR bridges and is a very useful tool for identification of grain, bulk, and contact contribution to the overall sensor response. Despite this, due to higher cost and failure in improving selectivity and sensitivity, no commercial devices based on impedance measurements are available in the market yet.

2.8.3 Response Photoactivation

The necessity of high operation temperatures of available commercial metal oxide gas sensors affects the potential applications of these devices. For example, monitoring gas composition in an environment containing an explosive species such as methane cannot be safe with MOX-based sensors, because the high temperature could cause explosions. Recently, new approaches have been explored in order to overcome these drawbacks; among these there is the use of light to activate surface reactions.

A sensor that can work at room temperature does not need a heater and this can also lead to a complete integration on Silicon transducers with conventional IC techniques to provide low cost devices.

As described in the previous paragraphs, when the surface of a semiconductor is in contact with a gas, the adsorption's phenomenon takes place. Illumination can influence the adsorption rate and its time dependence, with a positive effect (increasing it) or with a negative effect (decreasing it). Moreover, also the activation energy may be changed by illumination. Most of the studies of heterogeneous photocatalysis were made using visible or near-ultraviolet radiation ($E_A \sim 5$ eV) [144, 145, 146].

Two typical adsorbents, which are effective for photocatalysis, are zinc oxide (ZnO) and titanium oxide (TiO_2), both with energy gap around 3 eV [14]. There are different sources of radiation suitable to produce photons at these energies (discharge lamps or lasers). There are some particular phenomena that the light can produce, such as "after-effect" (adsorption proceeds for some time after the light is turned off) or "memory effect" (adsorptive properties of the surface in the dark are modified also when the light is turned off). The sign and magnitude of these effects depend not only on the experimental conditions, such as pressure, temperature and light frequency, but also on the previous history of the sample.

When we want to induce a photo-activation the light has to be absorbed by the semiconductor itself, otherwise the observed photodesorption may only be an apparent effect due to heating of the transducer.

When the metal oxide semiconductor absorbs light, there is a modification of the electrical properties of the semiconductor itself: free carriers are created by either intrinsic or extrinsic optical absorption. There is therefore a close relation between the optical absorption spectrum and the photoconductivity spectral response $\Delta\sigma$ vs ω. The general shape of photoconductivity curves is given in Fig. 2.18. The photoconductivity is controlled by the surface lifetime in the first region. Whereas in second region, there is still a strong absorption and the photoconductivity is controlled by the bulk lifetime, with a maximum occurring when the absorption constant is approximately equal to the reciprocal of the sample thickness. Finally, in the third region the photoconductivity, controlled by the bulk lifetime, decreases with increasing wavelength as the absorption decreases.

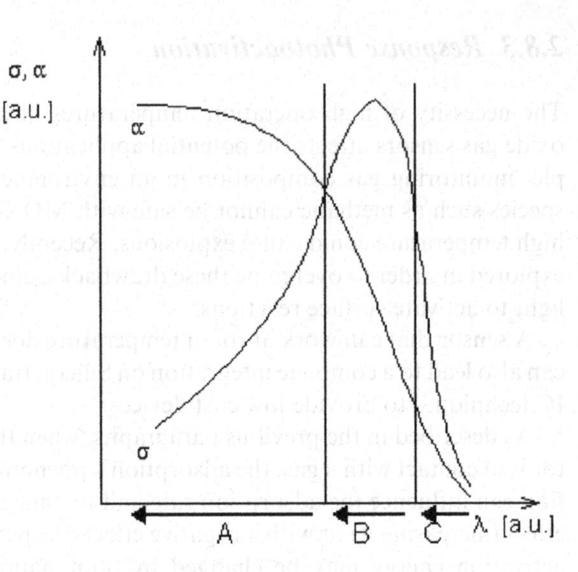

Fig. 2.18 Variation of the absorption coefficient and of the photoconductivity as a function of the wavelength

In the case of metal oxide semiconductor gas sensors we generally deal with polycrystalline material, made up of grains with different orientations.

Photoexcitation can affect transport across the grain boundaries by:

- increasing the density of free carriers throughout the material,
- decreasing the intergrain barrier height by changing the intergrain states charge,
- increasing the probability of tunneling through the intergrain barriers by decreasing the depletion layer widths in the adjacent grains.

As introduced before, the conductivity of a polycrystalline layer is given by Eq. (2.37)

$$\sigma = qn_b\mu_0 e^{\frac{-qV_s}{kT}} \tag{2.66}$$

And changes due to illumination

$$\Delta\sigma = qn_b\mu_0 e^{\frac{-qV_s}{kT}}\Delta n_b - \frac{q^2}{kT}n_b\mu_0 e^{\frac{-qV_s}{kT}}\Delta V_s \tag{2.67}$$

Illumination, changing the occupancy of the defects by electrons and holes, changes the concentration of the adsorption centers of each given type and the capacity of adsorption on the surface of the semiconductor.

When the radiation does not change the chemical composition of the semiconductor, but produces structural (radiation) defects, i.e., introduces additional disorder into the lattice, the semiconductor remembers illumination for some time and then gradually "forgets" the event. Radiation-induced disorder gradually relaxes itself after the illumination has been turned off; the higher the temperature, the greater the relaxation rate. Therefore, heating proves to be a factor helping to "erase" the memory of illumination. The "impurities" introduced into the sample by the effect of illumination gradually disappear.

Both the magnitude and the sign of the photoadsorption effect are determined not only by the experimental conditions, but also by the previous history of the sample: they depend on the treatment the sample was subjected to before illumination.

As a consequence of the chemisorption of a reactive gas at the surface of the solid, the position of the Fermi level (E_F) changes relative to the positions of the conduction band edge (E_C), and the valence band edge (E_V). In situations in which the adsorption of an adsorbate, such as di-oxygen (O_2), takes place the Fermi level is depressed by an amount q φ_b, where φ_b is the surface potential created through the adsorption process. The presence of an electric field restricts the access of electrons and favors the access of holes to the surface. This provides a displacement of the position of the thermal equilibrium by desorbing molecules of adsorbed species according to the following equation: $h^{+}{}_{vb} + O^{-}{}_{2}(ads) \rightarrow O_2(ads)$, $O_2(ads) \rightarrow O_2(g)$.

As the Fermi level is raised by desorption process until a new equilibrium position is established, the rate of arrival of electrons and holes at the surface again becomes equal.

On the contrary, if the transfer of electronic charge takes place from the adsorbate to the adsorbent, the Fermi level will be raised and the resulting electric field in the space charge layer will encourage the net transfer of electrons to the surface via downwards bending of the bands if hole–electron pairs are generated within the solid by the absorption of photons of appropriate energy. In such circumstances one can anticipate also desorption of a neutral species as a consequence of the interaction of a conduction band electron with a positively charged adsorbed species.

In each of the situations that have been described, if the photon flux is interrupted, the system relaxes towards the thermal equilibrium position produced before the system was irradiated. Under these circumstances a reversible change can be induced, i.e., adsorption under irradiation (photoadsorption) and desorption in the dark to re-establish the original position of thermal equilibrium or desorption under irradiation (photodesorption) and re-adsorption in the dark.

This behavior has been measures for zinc oxide; both types of reversible behavior have been observed as a result of different thermal pretreatments of the zinc oxide powdered specimen prior to the establishment of the initial thermally activated equilibrium condition. Zinc oxide is an n-type oxide but the position of the Fermi level relative to the upper edge of the valence band (E_V) and the lower edge of the conduction band (E_C) is dependent upon the extent of the deviation of its composition from the (1:1) stoichiometric value. Thermal conditioning of the oxide in vacuum, or in a reducing atmosphere such as in hydrogen gas, results in an excess of the metallic component which raises the Fermi level closer to the conduction band edge, i.e., the difference in energy $E_C - E_F$ is less for the reduced solid than it is for the oxidized form.

Photoelectronic studies of solids and surface species are one of the most difficult aspects of studies of photoadsorption and photodesorption. The absorption of photons can produce an electrically excited state of the solid, or can produce subsequently highly reactive surface radical species. In the former situation, there is a transient increase in the number of charge carriers, which will be rapidly restored to the initial condition via charge-carrier recombination if the source of photons is extinguished.

Dealing with photoadsorption phenomena on semiconductor films requires a great care to avoid "memory effect." Every time the irradiation is stopped, the semiconductor has to be heated at high temperature so that the "impurities" introduced into the sample by illumination gradually disappear. In this way measurements become reproducible, an essential feature for gas-sensing applications. There is also an increase of the response and a decrease in the response time and memory effects as the incident power increase. All these problems then could be solved with a high power density on the film.

Comini et al. [29] report that the employment of light in gas sensors development is quite promising for the detection of nitrogen dioxide, carbon monoxide, and methane also at room temperature; the measurements were performed with UV light activation on tin dioxide RGTO layers.

Anothainart et al. [7] studied NO_2 gas-sensing behavior of SnO_2 at room temperature under illumination. Two main effects were observed: the conductance increased and the time constant for NO_2 desorption was drastically reduced. Furthermore, the gas induced baseline shift was reduced.

Shukla et al. [147] report the effect of UV radiation exposure on the room-temperature hydrogen sensitivity of indium oxide-doped tin oxide thin-film gas sensor. The H_2 response was reduced from 65000 to 200 under UV radiation for 900 ppm of H_2. The observed phenomenon was attributed to the reduced surface coverage by the chemisorbed oxygen under UV radiation. As we have already stress previously the positive or negative effect is strongly dependent on the experimental conditions used such as wavelength, intensity, pressure of the measurement.

Photoactivation measurements have been performed also on quasi-one-dimensional nanostructures; Zhang et al. [172] report photoconducting properties of In_2O_3 nanowires. The tested devices is based on individual In_2O_3 nanowires and it showed a substantial increase in conductance of up to four orders of magnitude upon exposure to UV light. It exhibited short response times and significant shifts in the threshold gate voltage. The sensitivity to UV of different wavelengths was studied and compared. The use of UV light was used to enhance the response for In_2O_3 nanowire chemical sensors, leading to a recovery time as short as 80s.

All these results are interesting for the production of sensors that can be used in dangerous atmosphere, for example when there are gases with concentrations higher than the lower explosive limit (LEL). The application on semiconductor

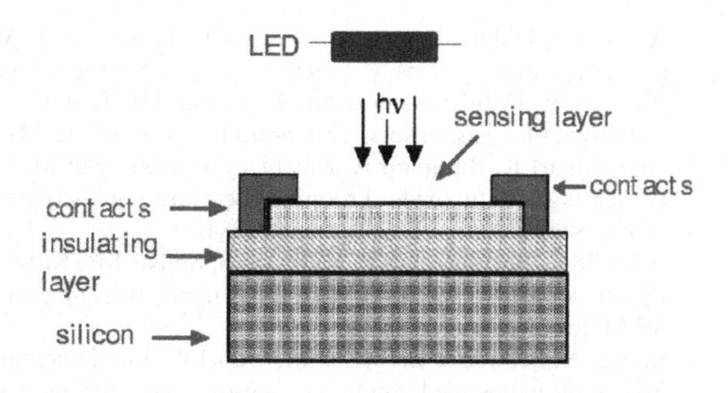

Fig. 2.19 Schematic rapresentation of a possible complete gas sensor device using light activation

thin films with lower energy-gap could permit employment of higher wavelength sources like light-emitting diods (LEDs) and so a complete device by IC technology could be implemented as is sketched in Fig. 2.19.

2.9 Conclusions and Outlook

A survey of properties and performances of semiconductor gas sensors has been presented. Considerable efforts have been made in many aspects of the development and testing of new semiconducting metal oxides; however, modern application of sensors still face problems of selectivity and long term drift produced by stoichiometry changes and coalescence of crystallites. What seems achievable nowadays is the development of sensors tailored for more specific and focused applications, instead of preparing multipurpose devices.

The improvement of computer calculus power and of imaging and spectroscopical surface techniques will increasingly provide a powerful tool for better understanding of sensing mechanism and sensor design. On this subject, the greater surface to volume ratio, the better stochiometry and greater level of crystallinity compared to polycrystalline oxides, which reduce instability associated with hopping and coalescence, make newly developed quasi-monodimensional semiconducting oxide very promising for the better understanding of sensing principles and the development of a new generation of sensors.

References

1. Advani GN, Komen Y, Hasenkopf J, Jordan AG. Improved performance of SnO_2 thin-film gas sensors due to gold diffusion. Sensor Actuat B. 1981;2:139–147.
2. Advani GN, Nanis N. Effects of humidity on hydrogen-sulfide detection by SnO_2 solid-state gas sensors. Sensor Actuat B. 1981;2:201–206.
3. Ahlers S, Müller G, Doll T. A rate equation approach to the gas sensitivity of thin film metal oxide materials. Sensor Actuat B. 2005;107:587–599.
4. Althainz P. Multisensor microsystem for contaminants in air. Sensor Actuat B. 1996;33:72–76.

5. Althainz P, Dahlke A, Frietsch-Klarhof M, Goschnick J, Ache HJ. Reception tuning of gas-sensor microsystems by selective coatings. Sensor Actuat B. 1995;24:366.
6. Althainz P, Dahlke A, Goschnick J, Ache HJ. Low-temperature deposition of glass membranes for gas sensors. Thin Solid Films. 1994;241:344–347.
7. Anothainart K, Burgmair A, Karthigeyan A, Zimmer M, Eisele I. Light enhanced NO_2 gas sensing with tin oxide at room temperature: conductance and work function measurements. Sensor Actuat B. 2003;93:580–584.
8. Ansari SG, Boroojerdian P, Kulkarni S, Sinkar SR, Karekar RN, Aiyer RC. Effect of thickness on H_2 gas sensitivity of SnO_2 nanoparticle-based thick film resistors. J Mater Sci Mater Electron. 1996;7:267–270.
9. Barsan N, Schweizer-Berberich M, Gopel W. Fundamental and practical aspects in the design of nanoscaled SnO_2 gas sensors: a status report. Fresenius J Anal Chem. 1999;365:287–304.
10. Barsan N, Weimar U. Conduction model of metal oxide gas sensors. J Electroceram. 2001;7:143–167.
11. Barsan N, Weimar U. Understanding the fundamental principles of metal oxide based gas sensors; the example of CO sensing with SnO_2 sensors in the presence of humidity. J Phys-Condens Mat. 2003;15:R813–R839.
12. Batzill M, Diebold U. The surface and materials science of tin oxide. Prog Surf Sci 2005;79:47–154.
13. Becker T, Ahlers S, Bosch-von Braunmuhl C, Muller G, Kiesewetter O. Gas sensing properties of thin- and thick-film tin-oxide materials. Sensor Actuat B. 2001;77: 55–61.
14. Bickley RI. Photoadsorption and photodesorption at gas solid interface. John Wiley & Sons; 1997.
15. Bogdanov P, Ivanovskaya M, Comini E, Faglia G, Sberveglieri G. Effect of nickel ions on sensitivity of In_2O_3 thin film sensors to NO_2. Sensor Actuators B. 1999;57:153–158.
16. Bogner M, Fuchs A, Scharnagl K, Winter R, Doll T, Eisele I. Electrical field impact on the gas adsorptivity of thin metal oxide films. Appl Phys Lett. 1998;17:2524–2526.
17. Brattain WH, Bardeen J. Surface properties of germanium. Bell Syst Tech J. 1953;32:1–41.
18. Bube RH. Photoelectronic properties of semiconductors. Cambridge University Press; 1992.
19. Bush MB. Modelling of nanophase materials. Mater. Sci. Eng. 1993;A161:127–134.
20. Cameron D, Holliday R, Thompson D. Gold's future role in fuel cell systems. J Power Sources. 2003;118:298–303.
21. Chai CC, Peng J, Yan BP. Preparation and gas-sensing properties of alpha-Fe_2O_3 thin-films. J Electron Mater. 1995;24:799–804.
22. Chambon L, Maleysson C, Pauly A, Germain JP, Demarne V, Grisel A. Investigation for NH_3 gas sensing applications, of the Nb_2O_5 semiconducting oxide in the presence of interferent species such as oxygen and humidity. Sensor Actuat B. 1997;45:107–114.
23. Chen Y, Cui X, Zhang K, Pan D, Zhang S, Wang B, How JG. Bulk-quantity synthesis and self-catalytic VLS growth of SnO_2 nanowires by lower-temperature evaporation. Chem Phys Lett. 2003;369:16–20.
24. Cheong H, Choi J, Kim HP, Kim J, Churn G. The role of additives in tin dioxide-based gas sensors. Sensor Actuat B. 1991;9:227–231.
25. Choi YG, Sakai G, Shimanoe K, Miura N, Yamazoe N. Wet process-prepared thick films of WO_3 for NO_2 sensing. Sensor Actuat B. 2003;95:258–265.
26. Chu XF, Siciliano P. CH_3SH-sensing characteristics of $LaFeO_3$ thick-film prepared by co-precipitation method. Sensor Actuat B. 2003;94:197–200.
27. Chung YK, Kim MH, Hum WS, Lee HS, Song JK, Choi SC, Yi KM, Lee MJ, Chung KW. Gas sensing properties of WO_3 thick film for NO_2 gas dependent on process condition. Sensor Actuat B. 1999;60:267–249.

28. Comini E, Baratto C, Fagila, G, Ferroni M, Vomiero A, Sberveglier, G. Quasi-one dimensional metal oxide semiconductor: preparation, characterization and application as chemical sensors. Progress in Materials Science. 2008, In Press, Accepted Manuscript, Available online 10 July 2008 doi: 10.1016/j.pmatsci.2008.06.003.
29. Comini E, Faglia G, Sberveglieri G. UV light activation of tin oxide thin films for NO_2 sensing at low temperatures. Sensor Actuat B. 2001;78:73–77.
30. Comini E, Faglia G, Sberveglieri G, Cantalini C, Passacantando M, Santucci S, Li Y, Wlodarski W, Qu W. Carbon monoxide response of molybdenum oxide thin films deposited by different techniques. Sensor Actuat B. 2000;68:168–174.
31. Comini E, Faglia G, Sberveglieri G, Pan ZW, Wang ZL. Stable and highly sensitive gas sensors based on semiconducting oxide nanobelts. Appl Phys Lett 2002;81: 1869–1871.
32. Comini E, Sberveglieri G, Ferroni M, Guidi V, Frigeri C, Boscarino D. Production and characterization of titanium and iron oxide nano-sized thin films. J Mater Res. 2001;16:1559–1564.
33. Cox DF, Fryberger TB, Semancik S. Oxygen vacancies and defect electronic states on the $SnO_2(110)$ 1 × 1 surface. Phys Rev B. 1988;38:2072–2083.
34. D'Amico A, Di Natale C. A contribution on some basic definition of sensors properties. IEEE Sens J. 2001;1:183–190.
35. Dang HY, Wang J, Fan SS. The synthesis of metal oxide nanowires by directly heating metal samples in appropriate oxygen atmospheres. Nanotechnology. 2003;14:738–741.
36. Diebold U. The surface science of titanium dioxide. Surf Sci Rep. 2003;48:53.
37. Dieguez A, Vila A, Cabot A, Romano-Rodriguez A, Morante JR, Kappler J, Barsan N, Weimar U, Gopel W. Influence on the gas sensor performances of the metal chemical states introduced by impregnation of calcinated SnO_2 sol-gel nanocrystals. Sensor Actuat B. 2000;68:94–99.
38. Dolbec R, El Khakani MA, Serventi AM, Trudeau M, Saint-Jacques RG. Microstructure and physical properties of nanostructured tin oxide thin films grown by means of pulsed laser deposition. Thin Solid Films. 2002;419:230–236.
39. Dominguez JE, Pan XQ, Fu L, Van Rompay PA, Zhang Z, Nees JA, Pronko PP. Epitaxial SnO_2 thin films grown on ((1)/012) sapphire by femtosecond pulsed laser deposition. J Appl Phys. 2002;91:1060–1065.
40. Dutronc P, Lucat C, Menil F, Loesch M, Horrillo MC, Sayago I, Gutierrez J, de Agapito JA. A potentially selective methane sensor-based on the differential conductivity responses of Pd-doped and Pt-doped tin oxide thick layers. Sensor Actuat B. 1993;16:384–389.
41. Ehrmann S. Automated cooking and frying control using gas sensor microarray. Sensor Actuat B. 1999;66:43–45.
42. Faglia G, Comini E, Cristalli A, Sberveglieri G, Dori L. Very low power consumption micromachined CO sensors. Sensor Actuat B. 1999;55:140–146.
43. Faglia G, Comini E, Sberveglieri G, Rella R, Siciliano P, Vasanelli L. Square and collinear four probe array and Hall measurements on metal oxide thin film gas sensors. Sensor Actuat B. 1998;53:69–75.
44. Fan Z, Jia GL. Gate-refreshable nanowires chemical sensors. Appl Phys Lett. 2005;86:123510.
45. Fang YK, Lee JJ. A tin oxide thin-film sensor with high ethanol sensitivity. Thin Solid Films. 1989;169:51–56.
46. Ferroni M, Guidi V, Martinelli G, Comini E, Sberveglieri G, Vomiero A, Della Mea G. Selective sublimation processing of a molydbelum-tungsten mixed oxide thin film. J Vac Sci Technol B. 2003;21:1442–1448.
47. Ferroni M, Guidi V, Martinelli G, Faglia G, Nelli P, Sberveglieri G. Characterization of a nanosized TiO_2 gas sensor. Nanostruct Mater. 1996;7:709–718.

48. Frank J, Fleischer M, Meixner H. Electrical doping of gas-sensitive, semiconducting Ga_2O_3 thin films. Sensor Actuat B. 1996;34:373–377.
49. Frietsch M. CuO catalytic membrane as selectivity trimmer for metal oxide gas sensors. Sensor Actuat B. 2000;65:379–381.
50. Gaidi M, Chenevier B, Labeau M. Electrical properties evolution under reducing gaseous mixtures (H_2, H_2S, CO) of SnO_2 thin films doped with Pd/Pt aggregates and used as polluting gas sensors. Sensor Actuat B. 2000;62:43–48.
51. Galatsis K, Li YX, Wlodarski W, Comini E, Sberveglieri G, Cantalini C, Santucci S, Passacantando M. Comparison of single and binary oxide MoO_3, TiO_2 and WO_3 sol-gel gas sensors. Sensor Actuat B. 2002;83:276–280.
52. Gardner JW. A diffusion–reaction model of electrical conduction in tin oxide gas sensors. Semicond Sci Tech 1989;4:345–350.
53. Garzella C, Comini E, Bontempi E, Depero LE, Frigeri C, Sberveglieri G. Sol-gel TiO_2 and W/TiO_2 nanostructured thin films for control of drunken driving. Sensor Actuat B. 2002;83:230–237.
54. Geistlinger H. Electron theory of thin-film gas sensors. Sensor Actuat B. 1993;17:47–60.
55. Geistlinger H, Eisele I, Flietner B, Winter R. Dipole- and charge transfer contributions to the work function change of semiconducting thin films: experiment and theory. Sensor Actuat B. 1996;34:499–505.
56. Gleiter H. Materials with ultrafine microstructures: retrospective and perspectives. Nanostruct Mater. 1992;1:1–19.
57. Gleiter H. Nanostructured materials: basic concepts and microstructure. Acta Mater. 2000;48:1–29.
58. Gluhoi AC, Lin SD, Nieuwenhuys BE. The beneficial effect of the addition of base metal oxides to gold catalysts on reactions relevant to air pollution abatement. Catal Today. 2004;90:175–181.
59. Gopel W, Shierbaum KD, Shmeisser D, Wiemhofer HD. Prototype chemical sensors for the selective detection of O_2 and NO_2 in gases. Sensor Actuat B. 1989;17:377.
60. Göpel W. Ultimate limits in the miniaturization of chemical sensors. Sensor Actuat A. 1996;56:83–102.
61. Goschnick J, Natzeck C, Sommer M, Zudock F. Depth profiling of non-conductive oxidic multilayers with plasma-based SNMS in HF-mode. Thin Solid Films. 1998;332:215–219.
62. Guidi V, Boscarino D, Casarotto L, Comini E, Ferroni M, Martinelli G, Sberveglieri G. Nanosized Ti-doped MoO_3 thin films for gas-sensing application. Sensor Actuat B. 2001;77:555–560.
63. Guidi V, Butturi MA, Carotta MC, Cavicchi B, Ferroni M, Malagu C, Martinelli G, Vincenzi D, Sacerdoti M, Zen M. Gas sensing through thick film technology. Sensor Actuat B. 2002;84:72–77.
64. Guidi V, Cardinali G, Dori L, Faglia G, Ferroni M, Martinelli G, Nelli P, Sberveglieri G. Thin film gas sensor implemented on low-power consumption micromachined silicon structure. Sensor Actuat B.1998;49:88–92.
65. Guidi V, Martinelli G, Schiffrer G, Vomiero A, Scian C, Della Mea G, Comini E, Ferroni M, Sberveglieri G. (2005) Selective sublimation processing of thin films for gas sensing. Sensor Actuat B. 1998;108:15–20.
66. Haeusler A, Meyer JU. A novel thick film conductive type CO_2 sensor. Sensor Actuat B. 1996;34:388–395.
67. Hahn SH, Barsan N, Weimar U, Ejakov SG, Visser JH, Soltis RE. CO sensing with SnO_2 thick film sensors: role of oxygen and water vapour. Thin Solid Films. 2003;436:17–24.
68. Haruta M, Souma Y. Copper, silver and gold in catalysis. Catal Today 1997;36:153–166.
69. Haruta M, Tsubota S, Kobayashi T, Kageyama H, Genet MJ, Delmon B. Low-temperature oxidation of CO over gold supported on TiO_2, alpha-Fe_2O_3, and Co_3O_4. J Catal 1993;144:175–192.

70. Haruta M, Yamada N, Kobayashi T, Iijima S. Gold catalysts prepared by coprecipitation for low-temperature oxidation of hydrogen and of carbon-monoxide. J Catal 1989;115:301–309.
71. Heiland G, Kohl D. Physical and chemical aspects of oxidic semiconductor gas sensors. In: Seiyama T, editors. Chemical sensor technology. Tokyo: Kodansha; 1988. vol. 1.
72. Hellmich W, Müller G, Bosch-v.Braunmühl C, Doll T, Eisele I. Field-effect-induced gas sensitivity changes in metal oxides. Sensor Actuat B. 1997;43:132–139.
73. Henderson MA, Epling WS, Perkins CL, Peden CHF, Diebold U. Interaction of molecular oxygen with the vacuum annealed $TiO_2(110)$ Surface: molecular and dissociative channels. J Phys Chem B. 1999;103(25):5328–5337.
74. Henrich VE, Cox PA. The surface science of metal oxides. Cambridge: University Press; 1994.
75. Hoefer U, Boéttner H, Felske A, Kuéhner G, Steiner K, Sulz G. Thin-film SnO_2 sensor arrays controlled by variation of contact potential a suitable tool for chemometric gas mixture analysis in the TLV range. Sensor Actuat B. 1997;44:429–433.
76. Hoefer U, Boéttner H, Wagner E, Kohl C. Highly sensitive NO_2 sensor device featuring a JFET-like transducer mechanism. Sensor Actuat B. 1998;47:213–217.
77. Hoefer U, Steiner K, Wagner E. Contact and sheet resistances of SnO_2 thin films from transmission-line-measurements. Sensor Actuat B. 1995;26/27:59–63.
78. Hoflund GB, Gardner SD, Schryer DR, Upchurch BT, Kielin EJ. Au/MnO_x catalytic performance-characteristics for low-temperature carbon-monoxide oxidation Appl Catal B. 1995;6:117–126.
79. Hu J, Bando Y, Liu Q, Golberg D. Laser-ablation growth and optical properties of wide and long single-crystal SnO_2 ribbons. Adv Funct Mater 2003;13:493–496.
80. Huck R, Bottger U, Kohl D, Heiland G. Spillover effects in the detection of H_2 and CH_4 by sputtered SnO_2 films with Pd and PdO deposits. Sensor Actuat B. 1989;17:355–359.
81. Iizuka Y, Fujiki H, Yamauchi N, Chijiiwa T, Arai S, Tsubota S, Haruta M. Adsorption of CO on gold supported on TiO_2· Catal Today 1997;36:115–123.
82. Imawan C, Solzbacher F, Steffes H, Obermeier E. Gas-sensing characteristics of modified-MoO_3 thin films using Ti-overlayers for NH_3 gas sensors. Sensor Actuat. 2000;64:193–197.
83. Jin CJ, Yamazaki T, Shirai Y, Yoshizawa T, Kikua T, Nakatani N, Takeda H. Dependence of NO_2 gas sensitivity of WO_3 sputtered films on film density. Thin Solid Films 2005;474:255–260.
84. Klç Ç, Zunger A. Origins of coexistence of conductivity and transparency in SnO_2. Phys Rev Lett 2002;88:095501–095504.
85. Kim CK, Choi SM, Noh IH, Lee JH, Hong C, Chae HB, Jang GE, Park HD. A study on thin film gas sensor based on SnO_2 prepared by pulsed laser deposition method. Sensor Actuat B. 2001;77:463–467.
86. Klumper-Westkamp H, Beling S, Mehner A, Hoffmann F, Mayr P. Semiconductor TiO_2 gas sensor for controlling nitrocarburizing processes. Met Sci Heat Treat 2004;46(7–8):305–309.
87. Kohl D The role of noble-metals in the chemistry of solid-state gas sensors. Sensor Actuat B. 1990;1:158–165.
88. Kolmakov A, Moskovits M. Chemical sensing and catalysis by one-dimensional metal-oxide nanostructures Annu. Rev. Mater. Res. 2004;34:151–180.
89. Kolmakov A, Zhang Y, Cheng G, Moskovits M. Detection of CO and O_2 using tin oxide nanowire sensors. Adv Mater. 2003;15:997.
90. Kronik L, Shapira Y (1999). Surface photovoltage phenomena: theory, experiment, and applications. Surf Sci Rep 37:1–206.
91. Lampe U, Muller J. Thin-film oxygen sensors made of reactively sputtered ZnO. Sensor Actuat B. 1989;18:269–284.

92. Law M, Kind H, Messer B, Kim F, Yang PD. Photochemical sensing of NO_2 with SnO_2 nanoribbon nanosensors at room temperature. Angew Chem Int Ed 2002;41:2405–2408.
93. Li C, Zhang DH, Liu XL, Han S, Tang T. In_2O_3 nanowires as chemical sensors. Appl Phys Lett 2003;82:1613–1615.
94. Li Y, Wlodarski W, Qu W. Carbon monoxide response of molybdenum oxide thin films deposited by different techniques. Sensor Actuat B. 2000;68:168–174.
95. Lieber CM, Wang ZL. Nanowires as building blocks for bottom-up nanotechnology. MRS BULLETIN 2007;32:99–104.
96. Liu Y, Zhu W, Tan OK, Shen Y. Structural and gas sensing properties of ultrafine Fe_2O_3 prepared by plasma enhanced chemical vapor deposition. Mater Sci Eng B. 1997;47:171–176.
97. Lu JG, Chang P, Fan Z. Quasi-one-dimensional metal oxide materials—synthesis, properties and applications. Mater. Sci. Eng. R. 2006;52:49–91.
98. Luca D. Preparation of TiOx thin films by reactive pulsed-laser ablation. J Optoelectron Adv M. 2005;7:625–630.
99. Lüth H. Surfaces and interfaces of solid materials. 3rd ed. Berlin: Springer; 1988.
100. Maccagnani P, Angelucci R, Pozzi P, Dori L, Parisini A, Bianconi M, Benedetto G. Thick porous silicon thermo-insulating membranes. Sensor Mater. 1999;11: 131–147.
101. Mädler L, Kammler HK, Mueller R, Pratsinis SE. Controlled synthesis of nanostructured particles by flame spray pyrolysis. J Aerosol Sci. 2002;33:369–389.
102. Mädler L, Pratsinis SE. Bismuth oxide nanoparticles by flame spray pyrolysis. J Am Ceram Soc 2002;85:1713–1718.
103. Madou MJ, Morrison SR. Chemical sensing with solid state devices. Academic Press Inc.: San Diego; 1989.
104. Martel R, Schmidt T, Shea HR, Hertel T, Avouris P. Single- and multi-wall carbon nanotube field-effect transistors. Appl Phys Lett. 1998;73:2447–2449.
105. Menzel R, Goschinick J. Gradient gas sensor microarrays for on-line process control- a new dynamic classification model for fast and reliable air quality assessment. Sensor Actuat B. 2000;68:115–122.
106. Mönch W. Semiconductor surfaces and interfaces. 3rd ed. Springer: Berlin; 2001.
107. Montmeat P, Pijolat C, Tournier G, Viricelle JP. The influence of a platinum membrane on the sensing properties of a tin dioxide thin film. Sensor Actuat B. 2002;84:148–159.
108. Morrison SR. The chemical physics of surfaces. Plenum Press: New York; 1978.
109. Morrison SR. Selectivity in semiconductor gas sensors. Sensor Actuat. 1987;12:425–440.
110. Müller R, Lange E. Multidimensional sensor for gas-analysis. Sensor Actuat. 1986;9:39–48.
111. Nanto H, Minami T, Takata S. Zinc-oxide thin-film ammonia gas sensors with high-sensitivity and excellent selectivity. J Appl Phys 1986;60:482–484.
112. Nelli P, Faglia G, Sberveglieri G, Cereda E, Gabetta G, Dieguez A, Romano-Rodriguez A, Morante JR. The aging effect on SnO_2-Au thin film sensors: electrical and structural characterization. Thin Solid Films 2000;371:249–253.
113. Nicoletti S, Dori L, Cardinali GC, Parisini A. Gas sensors for air quality monitoring: realisation and characterisation of undoped and noble metal-doped SnO thin sensing films deposited by the pulsed laser ablation. Sensor Actuat B. 1999;60:90–96.
114. Ohtomo A, Tsukazaki A. A pulsed laser deposition of thin films and superlattices based on ZnO semiconductor. Sci Technol. 2005;20:S1–S12.
115. Orton JW, Goldsmith BJ, Chapman JA, Powell MJ. The mechanism of photoconductivity in polycristalline cadmium sulphide layers. J Appl Phys. 1982;53:1602–1614.
116. Orton JW, Powell MJ. The Hall effect in polycristalline and powdered semiconductors. Rep. Prog Phys. 1980;43:1265–1306.
117. Ostrick B, Fleischer M, Meixner H, Kohl D. Investigation of the reaction mechanisms in work function type sensors at room temperature by studies of the cross-sensitivity to oxygen and water: the carbonate-carbon dioxide system. Sensor Actuat B. 2000;68:197–202.

118. Oyabu T. Sensing characteristics of SnO_2 thin-film gas sensor. J Appl Phys. 1982;53:2785–2787.
1199. Pan QY, Xu JQ, Dong XW, Zhang JP. Gas-sensitive properties of nanometer-sized SnO_2. Sensor Actuat B. 2000;66:237–239.
120. Pan ZW, Dai ZR, Wang ZL. Nanobelts of semiconducting oxides. Science 2001;291:1947–1949.
121. Papadopoulos CA, Vlachos DS, Avaritsiotis JN. Comparative study of various metal-oxide-based gas-sensor architectures. Sensor Actuat B. 1996;32:61–69.
122. Park ED, Lee JS. Effects of pretreatment conditions on CO oxidation over supported Au catalysts. J Catal. 1999;186:1–11.
123. Petriz RL. Theory of photoconductivity in semiconductor films. Phys Rev. 1956;104:1508–1516.
124. Pijolat C, Riviere B, Kamionka M, Viricelle JP, Breuil P. Tin dioxide gas sensor as a tool for atmospheric pollution monitoring: problems and possibilities for improvements. J Mater Sci 2003;38:4333–4346.
125. Popp U, Herbig R, Michel G, Muller E, Oestreich C. Properties of nanocrystalline ceramic powders prepared by laser evaporation and recondensation. J Eur Chem Soc 1998;18:1153–1160.
126. Prudenziati M. Thick-film sensors. Elsevier: Amsterdam; 1994.
127. Qu W, Meyer JU. Thick-film humidity sensor based on porous $MnWO_4$ material. Meas Sci Technol 1997;8:593–600.
128. Raiblea I, Burghardb M, Schlecht U, Yasuda A, Vossmeyer T . V2O5 nanofibres: novel gas sensors with extremely high sensitivity and selectivity to amines. Sensor Actuat B. 2005;106:730–735.
129. Ramamoorthy R, Kennedy MK, Nienhaus H, Lorke A, Kruis FE, Fissan H. Surface oxidation of monodisperse SnOx nanoparticles. Sensor Actuat B. 2003;88:281–285.
130. Rao GST, Rao DT. Gas sensitivity of ZnO based thick film sensor to NH_3 at room temperature. Sensor Actuat B. 1999;55:166–169.
131. Ruiz A, Calleja A, Espiell F, Cornet A, Morante JR. Nanosized Nb-TiO_2 gas sensors derived from alkoxides hydrolization. IEEE Sens J. 2003;3:189–194.
132. Ruiz AM, Sakai G, Cornet A, Shimanoe K, Morante JR, Yamazoe N. Cr-doped TiO_2 gas sensor for exhaust NO_2 monitoring. Sensor Actuat B. 2003;93:509–518.
133. Sahm T, Mädler L, Gurlo A, Barsan N, Pratsinis SE, Weimar U. Flame spray synthesis of tin dioxide nanoparticles for gas sensing. Sensor Actuat B. 2004;98:148–153.
134. Sakai G, Matsunaga N, Shimanoe K, Yamazoe N. Theory of gas-diffusion controlled sensitivity for thin film semiconductor gas sensor. Sensor Actuat B. 2001;80:125–131.
135. Samson S, Fonstad CG. Defect structure and electronic donor levels in stannic oxide crystal. J Appl Phys. 44:4618–4621.
136. Sauvan M, Pijolat C. Selectivity improvement of SnO_2 films by superficial metallic films. Sensor Acutat B. 1999;58:295–301.
137. Sberveglieri G. Classical and novel techniques for the preparation of SnO_2 thin-film gas sensors. Sensor Actuat B. 1992;6:239–247.
138. Sberveglieri G, Faglia G, Groppelli S, Nelli P. Methods for the preparation of NO, NO_2 and H_2 sensors based on tin oxide thin-films, grown by means of the rf magnetron sputtering technique. Sensor Actuat B. 1992;8:79–88.
139. Sberveglieri G, Faglia G, Groppelli S, Nelli P, Taroni A. A novel PVD technique for the preparation of SnO_2 thin-films as C_2H_5OH sensors. Sensor Actuat B. 1992;7:721–726.
140. Sberveglieri G, Hellmich W, Muéller G. Silicon hotplates for metal oxide gas sensor elements. Microsyst Technol 1997;3:183–190.
141. Schweizer-Berberich M, Strathman S, Weimar U, Sharma R, Seube A, Peyre-Lavigne A, Gopel W. Strategies to avoid VOC cross-sensitivity of SnO_2-based CO sensors. Sensor Actuat B 1999;58:318–324.

142. Seth M, Kohl CD, Fleischer M, Meixner H. A selective H_2 sensor implemented using Ga_2O_3 thin-films which are covered with a gas-filtering SiO_2 layer. Sensor Actuat B. 1996;36:297–302.
143. Seto JYW. The electrical properties of polycristalline silicon films. J Appl Phys 1975;46:5247–5254.
144. Shapira Y, Cox SM, Lichtman D. Photodesorption from powder zinc oxide. Surf Sci. 1975;50:503–514.
145. Shapira Y, Cox SM, Lichtman D. Cemisorption, photodesorption and conductivity measurements on ZnO surfaces. Surf Sci. 1976;54:43–59.
146. Shapira Y, McQuistan RB, Lichtman D. Relation between photodesorption and surface conductivity in ZnO. Phys Rev B. 1977;15(4):2163–2169.
147. Shukla S, Agrawal R, Cho HJ, Seal S, Ludwig L, Parish C. Effect of ultraviolet radiation exposure on room-temperature hydrogen sensitivity of nanocrystalline doped tin oxide sensor incorporated into microelectromechanical systems device. J Appl Phys. 2005;97:54307.
148. Siegel RW. Synthesis and properties of nanophase materials. Mater Sci Eng A. 1993;168:189–197.
149. Sinkkonen J. DC conductivity of a random barrier network. Phys Status Solid B. 1980;102:621–627.
150. Solis JL, Saukko S, Kish LB, Granqvist CG, Lantto V. Nanocrystalline tungsten oxide thick-films with high sensitivity to H_2S at room temperature. Sensor Actuat B. 2001;77:316–321.
151. Solisa JL, Saukko S, Kisha L, Granqvista CG, Lantto V. Semiconductor gas sensors based on nanostructured tungsten oxide. Thin Solid Films. 2001;391:255–260.
152. Sunu SS, Prabhu E, Jayaraman V, Gnanasekar KI, Gnanasekaran T. Gas sensing properties of PLD made MoO_3 films. Sensor Actuat. B 2003;94:189–196.
153. Sysoev V, Kiselev I, Frietsch M, Goschnick J. Temperature gradient effect on gas discrimination power of a metal-oxide thin-film sensor. Microarray Sens 2004;4:37–46.
154. Sze SM. Physics of semiconductor devices. John Wiley and sons: New York; 1981.
155. Tan OK, Cao W, Hu Y, Zhu W. Nanostructured oxides by high-energy ball milling technique: application as gas sensing materials. Solid State Ionics. 2004;172:309–316.
156. Tournier G, Pijolat C, Lalauze R, Patissier B. Selective detection of CO and CH_4 with gas sensors using SnO_2 doped with palladium. Sensor Actuat B. 1995;26:24–28.
157. Trinchi A, Kaciulis S, Pandolfi L, Ghantasala MK, Li YX, Wlodarski W, Viticoli S, Comini E, Sberveglieri G. Characterization of Ga_2O_3 based MRISiC hydrogen gas sensors. Sensor Actuat B. 2004;103:129–135.
158. Tsuda N, Nasu K, Fujimori A, Siratori K. Electronic conduction in oxides. 2nd ed. Springer: Berlin; 2000.
159. Vlachos DS, Papadopoulos CA, Avaritsiotis JN. Characterisation of the catalyst-semiconductor interaction mechanism in metal-oxide gas sensors. Sensor Acutat B. 1997;44:458–461.
160. Vomiero A, Della Mea G, Ferroni M, Martinelli G, Roncarati G, Guidi V, Comini E, Sberveglieri G. Preparation and microstructural characterization of nanosized Mo:TiO_2 and Mo-W-O thin films by sputtering: tailoring of composition and porosity by thermal treatment. Mater Sci Eng B. 2003;101:216–221.
161. Wan Q, Li QH, Chen YJ, Wang TH, He XL, Li JP, Lin CL. Fabrication and ethanol sensing characteristics of ZnO nanowire gas sensors. Appl Phys Lett. 2004;84: 3654–3656.
162. Wang N, Cai Y, Zhang RQ. Growth of nanowires. Mater. Sci. Eng. R. 2008;60:1–51.
163. Wang X, Yee SS, Carey WP. Transition between neck-controlled and grain-boundary-controlled sensitivity of metal-oxide gas sensors. Sensor Actuat B. 1995;24–25: 454–457.
164. Watson J. A note on the electrical characterization of solid-state gas sensors. Sensor Actuat B. 1992;8:173–177.

165. Weisz PB. Effects of electronic charge transfer between adsorbate and solid on chemisorption and catalysis. J Chem Phys. 1953;21:1531.
166. Williams DE, Pratt KFE. Theory of self-diagnostic sensor array devices using gas-sensitive resistors. J Chem Soc. 1995;91:1961–1966.
167. Williams EW, Lawor CM, Keeling AG, Gould RD. Novel room-temperature carbon-monoxide sensor utilizing rate of change of resistance in thick-films of tin oxide. Int J Electon. 1994;76:815–820.
168. Williams G, Coles GSV. Gas sensing properties of nanocrystalline metal oxide powders produced by a laser evaporation technique. J Mater Chem. 1998;8:1657–1664.
169. Wolkenstein T. Electronic processes on semiconductor surfaces during chemisorption. Plenum press: NY; 1991.
170. Xiangfeng C, Caihong W, Dongli J, Chenmou Z. Ethanol sensor based on indium oxide nanowires prepared by carbothermal reduction reaction. Chem Phys Lett. 2004;399:461–464.
171. Yamazoe N. New approaches for improving semiconductor gas sensors. Sensor Actuat B. 1991;5:7.
172. Zhang D, Li C, Han S, Liu X, Tang T, Jin W, Zhou C. Ultraviolet photodetection properties of indium oxide nanowires. Appl Phys. 2003;77:163–166.
173. Zhang D, Liu Z, Li C, Tang T, Liu X, Han S, Lei B, Zhou C. Detection of NO_2 down to ppb levels using individual and multiple In_2O_3 nanowire devices. Nano Lett. 2004;4:1919–1924.
174. Zhang DH, Liu ZQ, Li C, Tang T, Liu XL, Han S, Lei B, Zhou CW. Detection of NO_2 down to ppb levels using individual and multiple In_2O_3 nanowire devices. Nano Lett. 2004;4:1919–1924.
175. Zhang DJ, Li C, Liu XL, Han S, Tang T, Zhou CW. Doping dependent NH_3 sensing of indium oxide nanowires. Appl Phys Lett. 2003;83:1845–1847.
176. Zhang Y, Kolmakov A, Lilach Y, Moskovits M. Electronic control of chemistry and catalysis at the surface of an individual tin oxide nanowire. J Phys Chem B. 2005;109:1923–1929.
177. Zhu BL, Zeng DW, Wu J, Song WL, Xie CS. Synthesis and gas sensitivity of In-doped ZnO nanoparticles. J Mater Electron;14:521–526.

Chapter 3
Capacitive-Type Relative Humidity Sensor with Hydrophobic Polymer Films

Yoshihiko Sadaoka

3.1 Introduction

Up to now, there has been a great interest in the development of humidity sensors for air-conditioning systems or industrial equipments, etc. Many types of materials, such as electrolytes [1], ceramics [2], and polymers [3], have been proposed for humidity sensors by exploiting the variations in their electrical parameters. Humidity sensors using polymer thin films are divided into two categories, i.e., the resistive and capacitive types. For the former, conductivity is enhanced by the sorption of water and by the dissociation of mobile carriers. Polyelectrolyte is an excellent material for electrical resistive-type humidity sensors, because their electrical conductivity varies with water sorption. However, they are unstable at high humidity, since most polyelectrolyte is soluble in water. Improving their electrical resistance to water is one of the main projects for this type of sensor [3, 4, 5, 6]. For the latter, the impedance is approximately $1/\omega C$ and the capacitance is enhanced by the sorption of water. Less hygroscopic polymers, such as cellulose derivatives [7, 8], polyimide [9, 10, 11, 12], engineering plastics [13], and poly-methylmethacrylate [14], have been used as the materials for this type of sensor.

The capacitive-type humidity sensor utilizing the polymer as a sensing film is used for various applications in the field of the building automation and the factory automation etc., and it is said that capacitive-type sensor will be used in the market of 70% or more [15]. This is because the capacitive-type humidity sensors are relatively cheap and have the features, which can be used under a wide range of temperature and humidity, compared with humidity sensors based on other principles.

Many researchers have also been adopting the measurement of the apparent dielectric constant for humidity sensing [8, 16, 17]. The capacitive-type sensor is superior in linearity of sensor output and in stability at high humidity compared

Y. Sadaoka
Department of Materials Science and Biotechnology, Graduate School of Science and Engineering, Ehime University, Matsuyama 790-8577, Japan
e-mail: sadaoka@eng.ehime-u.ac.jp

E. Comini et al. (eds.), *Solid State Gas Sensing*,
DOI: 10.1007/978-0-387-09665-0_3, © Springer Science+Business Media, LLC 2009

to the electrical resistive-type sensor. However, many problems, i.e., hysteresis, stability at high temperature and in a humid ambience, and durability to some kinds of organic vapors, etc., still remained. In order to achieve a practical capacitive-type humidity sensor, selecting materials suited for this type of sensor become important and interesting. A series of our works for the development of relative humidity sensor based on hydrophobic polymers have been reviewed.

3.2 Fundamental Aspects

3.2.1 Sorption Isotherms of Polymers

In the case of the sorption of water on the hydrophilic polymers, which have strong sorption sites, the sorption isotherm shows a general sigmoidal shape [18]. There is more water sorbed by hydrophilic polymer than other hydrophobic polymer over the entire relative humidity. The sorption isotherm is similar to type II. The shape of the sorption curve gradually changed from type II to type III (Brunauer' classification [19]) with decreasing sorption ability of the polymers. When the attractive forces between the adsorbed gas and adsorbent are smaller than the attractive forces between the adsorbed gases, a type III curve is observed. This isotherm that is linear at low activities and curved upward more rapidly than predicted by Henry' law is sometimes called a Flory–Huggins isotherm [20, 21, 22]. The sorption equilibrium can be discussed in terms of solution theory by treating the sorption process as mixing of the solvent and polymer.

$$\ln a_1 = \ln \rho_1 + (1 - r^{-1})\rho_2 + \chi(\rho_2)^2 \tag{3.1}$$

where a_1 is the solvent activity, ρ_1 and ρ_2 are volume fractions of solvent and polymer, respectively, and χ is the empirical polymer–solvent interaction parameter. The symbol r stands for the ratio of molar volumes of both components and, in the first approximation, may be replaced by the ratio of molecular weights. The term r^{-1} is usually neglected with high-molecular-weight polymers and resulted in

$$\ln a_1 = \ln \rho_1 + (1 - \rho_1) + \chi(1 - \rho_1)^2 \tag{3.2}$$

The Flory–Huggins theory (lattice model) assumes random mixing in polymer–solvent mixtures, and obviously it cannot hold for a multilayer sorption process; it remains, however, of definite interest to point out any concentration dependence of the apparent interaction parameter χ. Such dependence is precisely ascribed in most cases to an inhomogeneous distribution of solvent molecules and polymer segment throughout the solution. Inherent in the Flory–Huggins statistics is the idea that there will be some clustering of the water molecules at higher water contents.

3.2.2 Water Sorption Behavior of Polymers

The quartz crystal microbalance (QCM) technique has been widely used to measure the mass change at the molecular level [23, 24]. The QCM is an instrument for accurately measuring the mass change and is useful for observing the sorption change in situ in a thin film state of the humidity sensor. In particular, when water vapor is the adsorbate, dielectric measurements are believed to be an effective means of determining the macroscopic or microscopic state of the water, because of its strong polarity. However, in attempting to understand the origins of differences in hydrophilic behavior, the relative number of atoms present in each polymer that might be expected to form hydrogen bonds in water requires consideration. Polarity and/or hydrogen-bond acidity of polymers cannot be judged from the chemical structure or other analyses. Spectroscopic techniques are useful for revealing the state of sorbed water due to hydrogen-bond formation. Infrared spectroscopy is a basically important technique for the structural study of materials at the molecular level and has also been applied for understanding the structure of sorbed water molecules in the polymer. Kusanagi and Yukawa studied water structures in polymeric materials from the shifts in the ν_a(OH) and ν_s(OH) bands of sorbed water molecules [25]. For another technique, Paley and coworkers applied the solvatochromic method to the characterization of solid-state polymers [26, 27]. They prepared thin polymer films containing certain media-sensitive dyes in these polymeric media and measured the UV–visible absorption spectra of the films. The obtained empirical parameters constitute a more comprehensive measure of the characteristics, especially for polarity, or the hydrogen-bonding ability of polymers than the dielectric constant or any other single physical characteristic. The solvatochromic method is useful in predicting and understanding the solubility and sorption properties for the interaction of polymer with gaseous water molecule [28].

3.2.3 Effects of the Sorbed Water on the Dielectric Properties

Water has a large value of dielectric constant, ε, compared with other liquids and solids. Consequently, the sorption of water in the polymer affects the apparent dielectric constant of the polymer very much. The electrical capacitance increased with increases in the relative vapor pressure in an ambience. The dielectric constant of the sorbed water could be estimated by applying Kurosaki's equation [16]

$$\varepsilon_s = \varepsilon_p + (\varepsilon_w - \varepsilon_p)/V_w \tag{3.3}$$

where ε_w, ε_p, and ε_s are the dielectric constants of the sorbed water, the dried polymer and the polymer at a certain relative vapor pressure, respectively. V_w is the volume fraction of sorbed water. The dielectric constant of the sorbed water is less than that of liquid water, which is 76.6 at 30°C, and also smaller compared with that of polymers [11, 14]. The dielectric constant of sorbed water on polymers had a larger value when the number of sorbed water molecules per monomer became more than unity due to the formation of clusters of sorbed water molecules. This indicates that the water molecules begin to associate with an increase in the amount of sorbed water. Further the polarization of sorbed water can be expected.

A modified Kirkwood's equation [29] for the dielectric constant ε_0 of a binary system is expressed as

$$(\varepsilon_0 - 1)(2\varepsilon_0 + 1)/9\varepsilon_0 = 4N_1(\alpha_1 + g_1\mu_1^2/3kT)/3 + 4N_2(\alpha_2 + g_2\mu_2^2/3kT)/3 \tag{3.4}$$

where N_1 and N_2 are the number of molecules per unit volume, α_1 and α_2 are the polarizability, μ_1 and μ_2 are the dipole moments for the two constituents, respectively, k is the Boltzmann constant, T is the absolute temperature, and g, the correlation parameter, is a measure of the local ordering of the dipole. It has a value of unity if fixing the position of one dipole does not influence the positions of the rest of the dipoles at all. On the other hand, if fixing one dipole tends to make the neighboring dipoles line up in parallel direction; the correlation parameter will be greater than unity and is close to 3 for bulk water. Similarly, if fixing one dipole tends to line up the neighbors in an anti-parallel direction, the correlation parameter will be less than unity. Potentially, valuable information is available about the state of sorbed waters by studying dielectric properties as a function of temperature, frequency, and concentration. By assuming that the number of polymer molecules per unit volume is hardly dependent on the relative vapor pressure, Eq. (3.4) can be rewritten as

$$(\varepsilon - 1)(2\varepsilon + 1)/9\varepsilon = (\varepsilon_p - 1)(2\varepsilon_p + 1)/9\varepsilon_p + 4N_w(\alpha_w + g_w\mu_w^2/3kT)/3 \tag{3.5}$$

where ε_p is the dielectric constant of polymer in a dry atmosphere and the subscript w indicates the component of sorbed water. For an AC electrical field, Eq. (3.6) is used instead of Eq. (3.5)

$$(\varepsilon - 1)(2\varepsilon + 1)/9\varepsilon = (\varepsilon_p - 1)(2\varepsilon_p + 1)/9\varepsilon_p + 4N_w(\alpha_w + P(\omega))/3 \tag{3.6}$$

where $P(\omega)$ indicates the polarization term caused by the orientation of dipole and depends on the applied frequency. When $\omega = \infty$, Eq. (3.6) is rewritten as

$$(\varepsilon - 1)(2\varepsilon + 1)/9e = (\varepsilon_p - 1)(2\varepsilon_p + 1)/9\varepsilon_p + 4N_w\alpha_w/3 \tag{3.7}$$

In this case, the dipole moment is zero or becomes ineffective. Consequently, the value of $(\varepsilon - 1)(2\varepsilon + 1)/9\varepsilon$ measured in the higher frequencies gives the value of the polarizability term and is scarcely depended on the distortion polarization caused by the intermolecular interaction. The difference between $(\varepsilon-1)(2\varepsilon+1)/9\varepsilon$ values observed at low and high frequencies corresponds to $4N_wP(\omega)/3(=\Delta P)$ and relates to the degree of the orientation of dipoles. As the water molecules are associated with more other water molecules with increasing the water content W_g, which is based on grams of sorbed water per one gram of dry polymer, the slope of $(\varepsilon-1)(2\varepsilon+1)/9\varepsilon$versus W_g becomes larger and the difference of $(\varepsilon-1)(2\varepsilon+1)/9\varepsilon$ between the two frequencies becomes large. It seems that one water molecule can be sorbed on the monomer unit by the dipole–dipole interaction and the excess water molecules are sorbed on the water molecule with the hydrogen bonding (cluster formation) in the amorphous region in the polymer film. In the region where the number of the sorbed water per monomer unit of polymer was less than unity, $(\varepsilon-1)(2\varepsilon+1)/9\varepsilon$ for the sorbed water was linearly proportional to the sorbed water. The formation of the cluster leads to the distinct hysteresis in the sorption isotherms. In the sorption process of water, first the water molecules diffuse in the polymer film and are fixed at a critical point by dipole-dipole interaction between the water molecule and the polymer. Second, the sorbed water associates with the neighboring waters and forms the cluster. The sorption processes may be expressed as follows:

$$\mathrm{H_2O\ (atmosphere)} \rightleftarrows \mathrm{H_2O\ (in\ film)\ (diffusion)} \tag{3.8}$$

$$\mathrm{H_2O\ (in\ film)} + \mathrm{film} \rightleftarrows \mathrm{H_2O_{ad}\ film\ (dipole-dipole)} \tag{3.9}$$

$$\mathrm{H_2O_{ad}\ film} + n\mathrm{H_2O} \rightleftarrows (n+1)\ \mathrm{H_2O_{ad}\ film\ (hydrogen\ bonding)} \tag{3.10}$$

3.3 Characterization of Polymers

3.3.1 Sorption Isotherms

The χ values were derived from experimental data obtained at various water activities a_1 according to Eq. (3.2), where a_1 is nearly equal to % RH (relative humidity)/100. The ρ_1 was obtained by transforming the gravimetric water sorption data obtained for various polymer thin films in Fig. 3.1 to volume fraction using the presumed density of polymers (= 1.2).

The volume fraction of sorbed water dependence of estimated χ–value is shown in Fig. 3.2.

The χ value is positive for all polymers and becomes larger as the polymer becomes less hydrophilic. The χ value was previously reported as 1.18 in the cellulose–water system [30]. Although the chemical structure was not specified in that study, the results obtained in this work seem to be reasonable values. The χ-value is related to the differences between the water binding energies and increases from negative to positive in the order of decreasing binding energies,

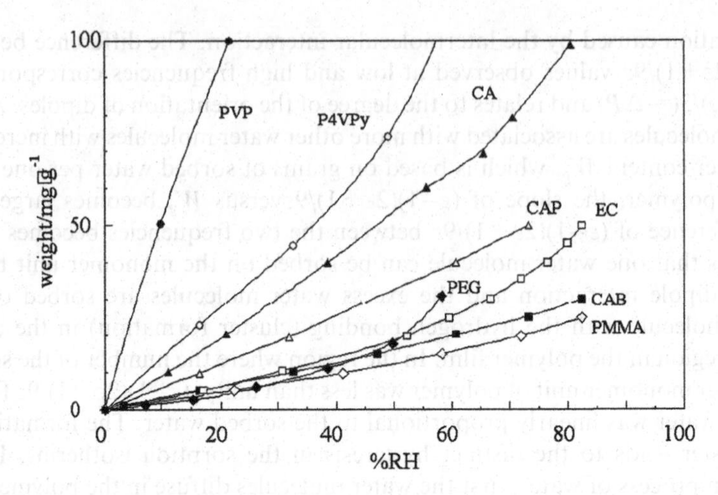

Fig. 3.1 Sorption isotherms measured at 30°C: (●) PVP, (○) P4VPy, (▲) CA, (△) CAP, (■) CAB, (□) EC, (◆) PEG, (◇) PMMA

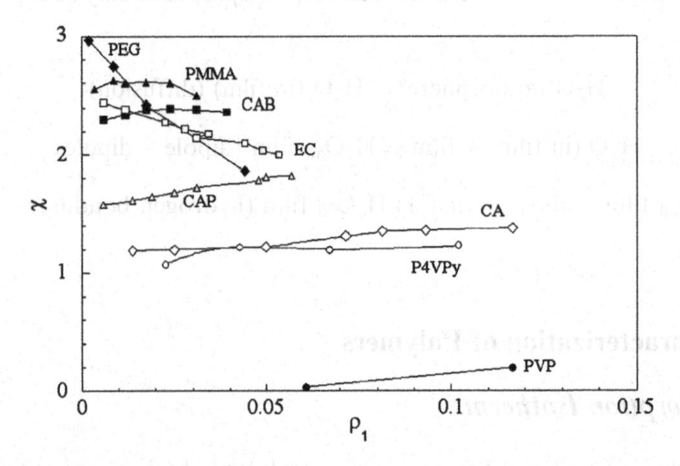

Fig. 3.2 Dependence of χ parameter on the volume fraction of sorbed water measured at 30°C; symbols are same as in Fig. 3.1

i.e. site binding (negative values of χ) > indirect binding > water–water interaction in bulk water. Since less hydrophilic polymers (PEG, EC, CAB, and PMMA) do not have highly specific sorption sites, the interaction between a polymer and sorbed water is very small. This is the reason why the χ value has a large positive value. Molecular structures are illustrated in Fig. 3.3. For hydrophilic polymers, the interaction parameter is a slightly increasing function of the water content and approaches a constant value.

The slight increase in χ for hydrophilic polymers can be explained in terms of predominant site binding at low water concentrations followed by nonspecific

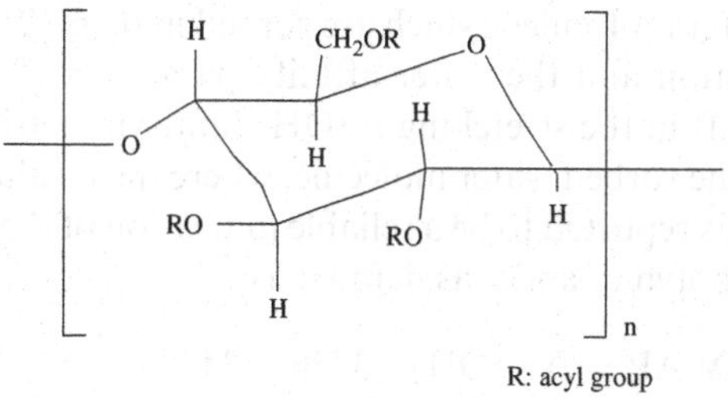

R: acyl group

CAB $[C_6H_7O_2(OH)_{0.3}(OCOCH_3)_{1.0}(OCOC_3H_7)_{1.7}]_n$
CA $[C_6H_7O_2(OH)_{0.8}(OCOCH_3)_{2.2}]_n$
EC $[C_6H_7O_2(OH)_{0.5}(OC_2H_5)_{2.5}]_n$
CAP $[C_6H_7O_2(OH)_{1.0}(OCOCH_3)_{0.2}(OCOC_2H_5)_{1.8}]_n$

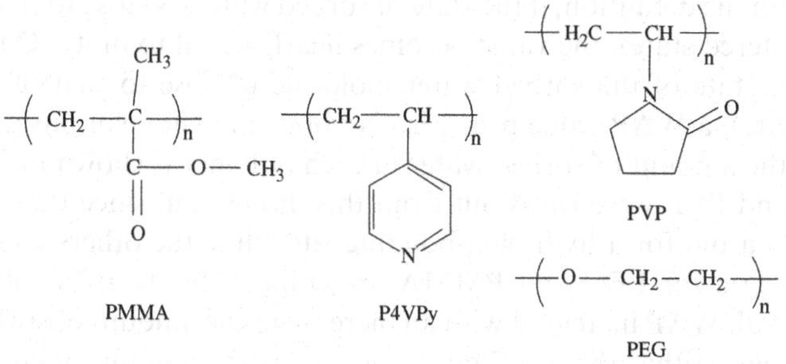

PMMA P4VPy PVP PEG

Fig. 3.3 Chemical structure of polymers

and slightly looser indirect binding for higher ones. In the case of CAB and PMMA, the value of χ is almost constant over the entire water content range or slight decreasing at higher water contents. The strength of interaction seems to be only slightly changed with the variation of water contents in the polymer. It is noteworthy that the values of χ monotonously decreased with an increase in water content for EC and PEG. It was reported that the χ-value was 0.43 when PEG was diluted by water [31]. This means that the sorbed water in EC or PEG starts forming clusters during the initial sorption stage. Some papers report that EC has a high ability to make water clusters [30, 32]. It is suggested from this result that ether groups affect the sorption behavior differently from ester groups especially with respect to the clustering tendency. The trend in the clustering appears to be related to the hydrogen-bonding propensity of the polar groups in the polymer as well as to the concentration of the polar groups.

3.3.2 FT–IR Measurement

Kusanagi and Yukawa reported that infrared spectroscopy is one of the effective techniques to study the intermolecular interactions between various solid polymers and water molecules, which are sorbed in them [25]. The heats of hydrogen-bond formation and the water affinity parameter (WAP) could be estimated using the shift in the stretching ν_a(OH) (antisymmetric) and ν_s(OH) (symmetric) bands of the sorbed water molecule, where the ν values are given in units of cm^{-1}. WAP was reported to be a reliable indication of the site affinity to H_2O molecules in the polymer and was defined as

$$\mathrm{WAP} = \Delta\nu_a(\mathrm{OH})/(3756 - 3430) \tag{3.11}$$

where $\Delta\nu_a(\mathrm{OH}) = 3756 - \nu_a(\mathrm{OH})$, and 3756 and 3430 are the frequencies of the stretching anti-symmetric bands of H_2O (gas) and of H_2O (liquid), respectively. Using this parameter, it is possible to clarify the water structures in solid polymer matrices, i.e., the cluster formation propensity of the sorbed waters. Judging from the definition, if the state of sorbed waters is close to that of liquid waters (clustered state), the value becomes nearly equal to unity. On the other hand, if the state of the sorbed water molecule is close to that of the vapor (isolated state), a WAP value near zero is obtained. The relationship between WAP and the amount of sorbed water in each polymer is shown in Fig. 3.4.

P4VPy and PVP were ruled out from this discussion, since these polymers contain a N atom for a hydrogen-bonding site while the others contain an O atom. The data for PEG and PMMA are comparable to those obtained by Kusanagi et al. WAP increased with an increase in the amount of sorbed water. This may suggest that the water cluster grows as the amount of sorbed water increases. It was observed that PEG and EC have a large WAP value even at low humidity. The sorbed water seems to be clustered irrespective of their lower water sorption ability. On the other hand, the cluster formation of sorbed water did not appreciably occur for PMMA and CAB. These results are consistent with those obtained by the sorption analysis.

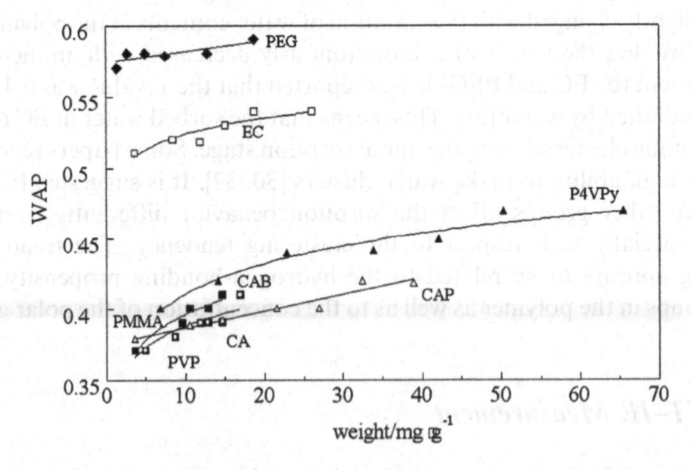

Fig. 3.4 Dependence of WAP value on the amount of sorbed water measured at 30°C; symbols are same as in Fig. 3.1

3.3.3 Solvatochromism

As the third method for characterizing the water sorption behavior of polymers, we measured the UV–visible spectra of composite films doped with four indicators (Fig. 3.5) and examined the effect of relative humidity on the longest wavelength (λ_{max}) of the UV–visible is absorption band (the solvatochromic band). According to Harris and coworkers, solvatochromic parameters were given using the equations listed below [26]. Dipolarity/polarizability, π^*_1, is determined from the longest wavelength UV–visible absorption band maximum (the solvatochromic band) of 4-nitro-*N*,*N*-dimethylaniline (1) in the polymer – Eq. (3.12). In this equation, and all the following ν_{max} values are given in unit of 10^3 cm^{-1}.

$$\pi_1^* = 8.006 - 0.2841\,\nu_{max} \tag{3.12}$$

Hydrogen-bond acidity, α, is determined from the longest wavelength absorption maxima of 4-nitroanisole (**3**) (a poor hydrogen-bond base) and Reichardt's betaine, 2,6-diphenyl- 4-(2,4,6-triphenylpyridinio) phenolate (**4**) (a good hydrogen-bond base) using Eq. (3.13).

$$\alpha = (\nu_{max}(4) + 1.873\ \nu_{max}(3) - 74.58)/6.24 \tag{3.13}$$

Hydrogen-bond basicity, β, is determined from the longest wavelength absorption maxima in the polymer of 4-nitroaniline (**2**) (a good hydrogen-bond acid) and 4-nitro-*N*,*N*-dimethylaniline (**1**) (a poor hydrogen-bond acid) using Eq. (3.14).

$$\beta = (0.984\ \nu_{max}(1) - \nu_{max}(2) - 3.40)/2.67 \tag{3.14}$$

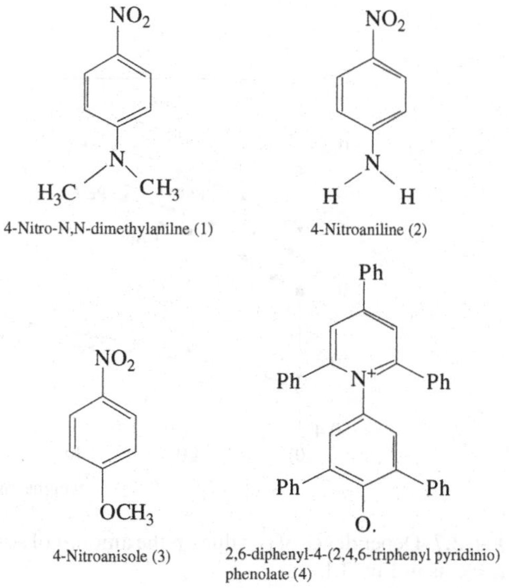

4-Nitro-N,N-dimethylanilne (1)

4-Nitroaniline (2)

4-Nitroanisole (3)

2,6-diphenyl-4-(2,4,6-triphenyl pyridinio) phenolate (4)

Fig. 3.5 Chemical structure of four kinds of indicator dyes

It was proved that these parameters have chemical sense in that they are similar to those found for liquid solvent models [28]. Solvatochromic parameters, π^*_1, α, and β were evaluated as a function of relative humidity using Eqs. (3.12)–(3.14). The results are plotted in Figs. 3.6–3.8.

The parameters depended on the polymers in a dry ambience. This result suggests that the difference in the characteristics of the polymer, such as polarity and/or hydrogen-bond ability, is reflected in these parameters.

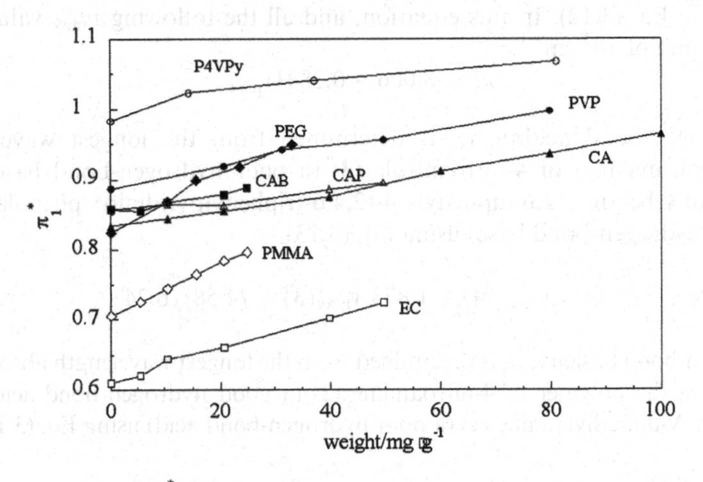

Fig. 3.6 Dependence of π_1^* value on the amount of sorbed water measured at 30°C; symbols are same as in Fig. 3.1

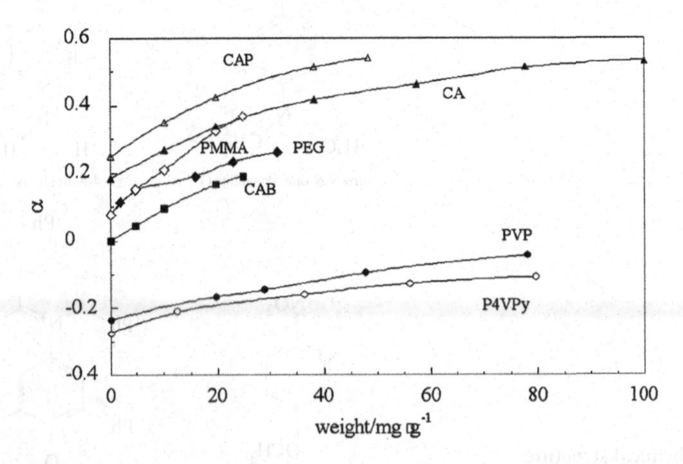

Fig. 3.7 Dependence of α value on the amount of sorbed water measured at 30°C; symbols are same as in Fig. 3.1

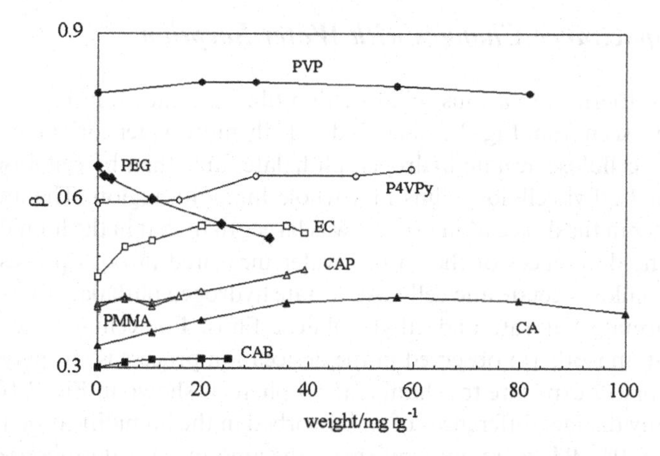

Fig. 3.8 Dependence of β value on the amount of sorbed water measured at 30°C; symbols are same as in Fig. 3.1

Hydrophilic polymers, such as PVP and P4VPy, have large π^*_1 and β values, while α value is small as expected from their chemical structure. It is shown that these polymers have strong basic sites for the hydrogen bonding of water. On the other hand, less hydrophilic polymers such as CAB and PMMA have small values for all parameters. These polymers do not have strong hydrogen-bonding sites and the interaction between polymers and water is expected to be small. PEG and EC are interesting for they have larger β values irrespective of their low water sorption ability. All the parameters were also affected by the relative humidity. The π^*_1 and α values were monotonously increased to the values obtained in bulk water with an increase in the relative humidity. The β value also increased in the lower-humidity region as well. Since a water molecule is well known as an amphoteric hydrogen-bond acceptor–donor, the polarity and/or hydrogen-bonding ability parameters are easily expected to be larger as the polymers sorbs water molecules from the atmosphere. However, for PEG, EC, PVP, and P4VPy, the β value decreased in the higher humidity range and approached the value of bulk water (0.14). This means that the sorbed water in these polymers was changed from a gaseous state to a liquid one with increasing amount of sorbed water. These polymers have hydrogen-bonding basic sites, e.g., ether oxygen for PEG and EC. The hydrogen-bonded water molecules by basic sites become acidic sites, which bind to the basic O site of the water molecule. As a result, the cluster formation is enhanced. The reason why PEG and EC have lower sorption ability might be related to the steric hindrance or higher order structure of these polymers. It was confirmed that the β parameter was effective for the elucidation of the state of the sorbed water, because it first increased and decreased when the cluster formation started.

3.3.4 Capacitance Changes with Water Sorption

Sorption isotherms of various cellulose thin films are shown in Fig. 3.9.

As can be seen from Fig. 3.7, there is decidedly more water sorbed by cellulose acetate and cellulose acetate hydrogen phthalate films than by cellulose acetate butyrate and ethyl cellulose films in a whole humidity region. The amount of sorbed water in the desiccation process was larger than that in the humidification process. The differences of the sorbed water measured in both processes were more for cellulose acetate and cellulose acetate hydrogen phthalate films than for cellulose acetate butyrate and ethyl cellulose films. For cellulose acetate, the water sorption isotherm observed in the desiccation process is strongly affected by the history of exposure to a humid atmosphere as shown in Fig. 3.10.

While any distinct differences of water sorbed in the humidification processes started from 0% RH were not confirmed, the amount of water observed in the desiccation process increase with increase in the highest humidity of exposure. Complex impedance has been employed to analyze the impedance measurements because it seems that the impedance consists of resistive and capacitive components. Complex impedance plots were shown in Fig. 3.11 for cellulose acetate.

In the frequency region from 100 Hz to 1 MHz, the quasi-straight line past the origin was observed in the low-humidity region, while the semicircles past the origin were confirmed in the higher humidity region above ~70% RH. For the semicircles observed in the higher humidity region, the depression angle was estimated to be ~7°. This small value of the depression angle indicates that the equivalent circuit is expressed with the circuit with capacitive and resistive components inserted in parallel to the capacitive component. In addition, in the lower humidity region, the contribution of the resistive component to the impedance may be ignored, i.e., the impedance was approximated to $1/\omega C$. For

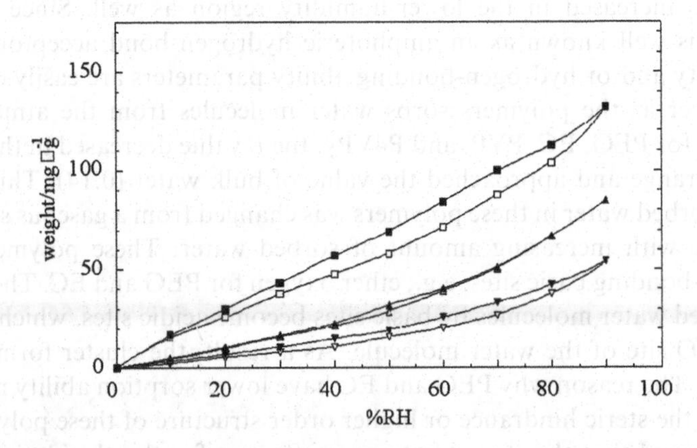

Fig. 3.9 Water sorption isotherms. (□, ■) cellulose acetate hydrogen phthalate. (△, ▲) cellulose acetate butyrate, (▽, ▼) ethyl cellulose. *Open symbols*; humidification. *Closed symbols*; desiccation

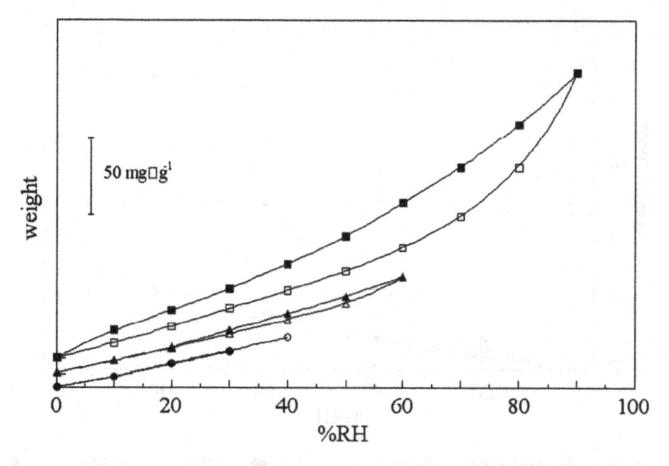

Fig. 3.10 Water sorption isotherms for cellulose acetate. (○, ●) within 40% RH, (△, ▲) within 60% RH, (□, ■) within 90% RH. *Open symbols*; humidification. *Closed symbols*; desiccation

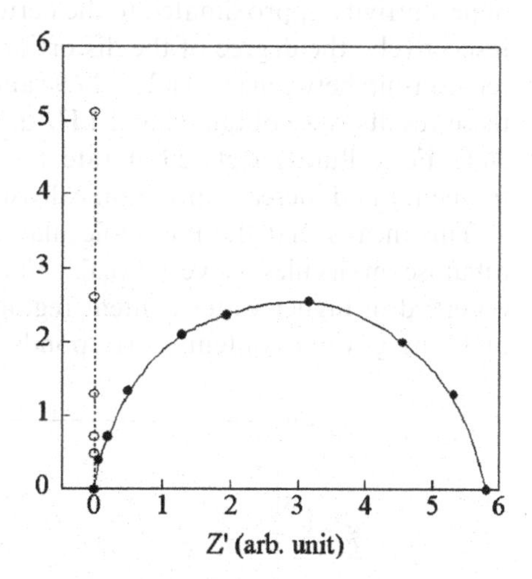

Fig. 3.11 Complex impedance plots for cellulose acetate. (○) 40% RH, (●) 90% RH

cellulose acetate butyrate film, in a whole humidity region, the result was represented with a quasi-straight line in the frequency region from 100 Hz to 1 MHz and the impedance was approximated to $1/\omega C$. The relationship between the permittivity and humidity can be obtained. In Fig. 3.12, the humidity dependence of the permittivity observed at 1 kHz is shown.

The degree of hysteresis for cellulose acetate is larger than that for cellulose acetate butyrate and similar hysteresis was confirmed in the water adsorption isotherms as shown in Fig. 3.10. As mentioned in Section 3.2.3, the value of $(\varepsilon-1)(2\varepsilon+1)/9\varepsilon$ in the higher frequencies gives the value of the polarization term and is scarcely dependent on the distortion polarization caused by the

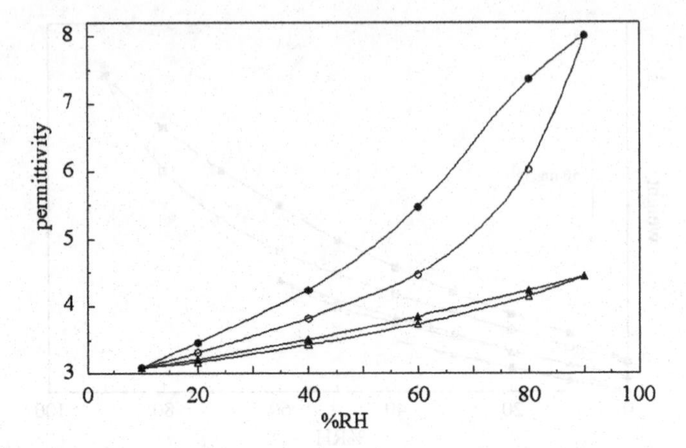

Fig. 3.12 Humidity dependence of permittivity. (○, ●) cellulose acetate, (△, ▲) cellulose acetate butyrate. *Open symbols*; humidification. Closed symbols; desiccation

intermolecular interactions. By assuming that the low- and high-frequency limit of permittivity approximated to the permittivity observed at 1 kHz and 1 MHz, respectively, the degree of the distortion polarization could be estimated. The relationship between $(\varepsilon-1)(2\varepsilon+1)/9\varepsilon$ and ΔW was shown in Fig. 3.13, in which these results were obtained at 1 kHz and 1 MHz in humidification process at 30°C. For cellulose acetate butyrate, $(\varepsilon-1)(2\varepsilon+1)/9\varepsilon$ was not changed with the frequency and increase monotonically with an increase in ΔW.

This means that the intermolecular interaction of water-water and water-cellulose molecules is very weak. For ethyl cellulose, the distinction was observed in higher water content region ($\Delta W \geq 35$ mg/g). The critical value of 83 mg/g water contents corresponds to the fact that the point at which the

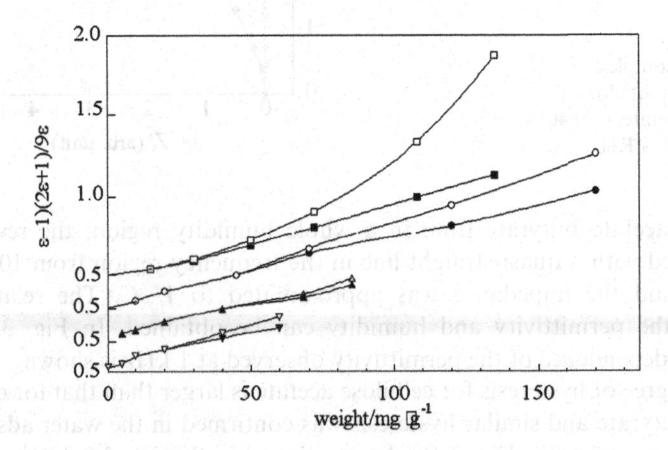

Fig. 3.13 Relationship between $(\varepsilon + 1)(2\varepsilon + 1)/9\varepsilon$ and ΔW. (□, ■) cellulose acetate hydrogen phthalate, (△, ▲) cellulose acetate butyrate. (▽, ▼) ethyl cellulose. *Open symbols*; 1 kHz: *closed symbols*; 1 MHz

mole ratio of water/monomer unit of cellulose is unity, is considerably higher than the water regain at 90% RH. For cellulose acetate and cellulose acetate hydrogen phthalate, a distinction was observed in the region that the mole ratio of water–monomer unit of cellulose was larger than unity, so that it seems that the water molecule is associated with the other water molecules, while the interaction between water and cellulose is weak. In the region where no distinct frequency dependence was observed, the straight line was confirmed in the relation between $(\varepsilon-1)(2\varepsilon+1)/9\varepsilon$ and ΔW. It is expected that the slopes of the straight line and $(\varepsilon-1)(2\varepsilon+1)/9\varepsilon$ are affected by the substituent group of cellulose, but any distinct dependencies were not detected. Therefore, the strength of interaction between water and cellulose may be poorly dependent on the substituent group and the degree of water regain may be influenced by the existence of an amorphous phase and/or the micro-pores. Jefferies [33] reported that the water regain (60% RH) of cellulose was proportional to the amorphous fraction. In this case, the amorphous region was defined as that in which the hydrogen bonding was not regular and which thus gave a broad, featureless, hydroxyl-stretching band. By using the relationship between the amorphous fraction and water regain at 60% RH proposed by Jefferies, the amorphous fraction is estimated and found to be in the order of cellulose acetate (43%) $\geq$ cellulose acetate hydrogen phthalate (43%) > cellulose acetate butyrate (27%) > ethyl cellulose (20%). This order is in qualitative agreement with the order of the water regain at 90% RH. Thus it seems that the degree of water regain is mainly influenced by the amorphous fraction of cellulose rather than the substituent groups in cellulose, especially in higher humidity regions. The difference between $(\varepsilon-1)(2\varepsilon+1)/9\varepsilon$ values observed at 1 kHz and 1 MHz corresponds to $4\ N_wP(\omega)/3$ ($= \Delta P$) and $\Delta P/\Delta W$ indicates the degree of the orientation of dipoles. $\Delta P/\Delta W$at 90% RH is estimated to be 0.0013, 0.0055, 0.0003, and 0.001 and ΔW/(amorphous fraction) is estimated to be 3.95, 3.09, 3.22, and 2.85 for cellulose acetate, cellulose acetate hydrogen phthalate, cellulose acetate butyrate, and ethyl cellulose, respectively. While the value of ΔW/ (amorphous fraction) is slightly dependent on cellulose species, $\Delta P/\Delta W$ at 90% RH is in the order cellulose acetate > cellulose acetate hydrogen phthalate > ethyl cellulose > cellulose acetate butyrate. One of the causes of high values of $\Delta P/\Delta W$ is the existence of micro-pores, i.e., the formation of the associated water molecules. If the bottle-necked micro-pores exist in the film, the amount of sorbed water observed in the humidification process is lower than that observed in the desiccation process, and the differences between the amounts of sorbed water measured in both processes increase with increase in the highest humidity of exposure for cellulose acetate and cellulose acetate hydrogen phthalate in particular. This expectation was confirmed as shown in Fig. 3.10. In addition, it was confirmed that the capacitance value in a higher humidity region decreased gradually with an increase in the annealing temperature up to the glass transition temperature, while the capacitance in a lower humidity region below 60% RH was not affected.

3.3.5 Cross-Linked Polymer

In the case of a polymer that has no strong polar group, the sorption sites of water molecules seem to be the free spaces around the polar nature of the polymer. The contributions of functional groups to the van der Waals volume are quite different and the packing density seems to be dependent on the chemical structure of the polymer [34]. Molecular weight is one of factors that affect the packing density [35, 36]. Further in respect of molecular packing, cross-linking produced a marked increase in water sorption in copolymers of methyl methacrylate and dimethacrylate [23, 36, 37, 38, 39]. The morphology of the materials may vary with the level of the cross-linking and the water sorption may be dependent on the morphology. In identifying the states of molecules sorbed in various substances, some researchers have adopted the measurement of the apparent dielectric constant [8, 11, 17, 28, 40, 41, 42]. The dielectric measurements are believed to be an effective means of identifying the macroscopic or microscopic state of the water, because of its strong polarity.

Four kinds of saturated and unsaturated vinylcarboxylate, i.e., vinyl benzoate (VB), vinyl cinnamate (VCi), vinyl methacrylate (VM), and vinyl crotonate (VCr) were studied [43, 44]. The molecular structures of these polymers are shown in Fig. 3.14.

While VB has one double bond, the others have two double bonds and the possibility to form a cross-linked film. Further, these monomers have a slightly different side group. These differences in the chemical structure of monomers are expected to form different chemical polymer structures. To examine the effects of cross-linked structure of the polymer on the water sorption behavior, it was necessary to discover the relationship between the polymerization condition and the chemical structure of the prepared film. The dielectric relaxation measurement could be adopted for this purpose. It is known that the dielectric relaxation of a polymerization and/or cross-linking reaction proceeds and the film becomes rigid, the segment motion of the main chain seems to be depressed. Consequently, the value of the dielectric loss is lowered. For all samples, dielectric relaxation behavior was affected by the polymerization temperature, t_p, and atmosphere as well as by the BPO content. As the polymerization proceeded, the relaxation peak became lower and shifted to a higher

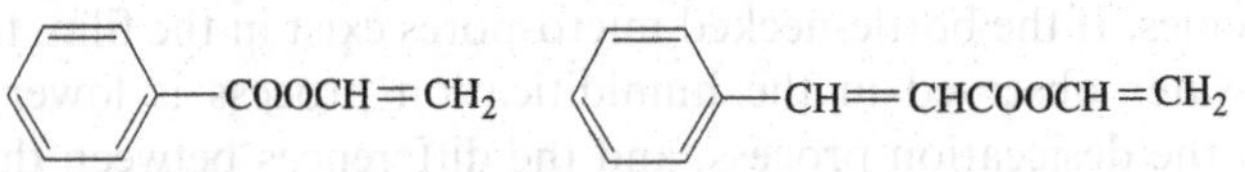

C_6H_5–$COOCH{=}CH_2$ Vinyl Benzoate (VB)

C_6H_5–$CH{=}CHCOOCH{=}CH_2$ Vinyl Cinnamate (VCi)

$CH_2{=}C(CH_3)COOCH{=}CH_2$ Vinyl Methacrylate (VM)

$CH_3CH{=}CHCOOCH{=}CH_2$ Vinyl Crotonate (VCr)

Fig. 3.14 Chemical structure of vinyl carboxylate

temperature. In any case, the dielectric relaxation peak of the film polymerized in a nitrogen atmosphere was observed at a higher temperature compared with that in air. The peak height lowered with polymerizing in nitrogen. Effects of the polymerization temperature and the amount of BPO added to the pre-polymerized monomer on the value of tan δ and the temperature at maximum tan δ, t_{max}, are summarized in Table 3.1.

Addition of 4 mg of BPO per gram of pre-polymerized monomer was effective to make the polymerization proceed. The effects of polymerization temperature were more complicated. Although the optimum polymerization temperature for obtaining film with the lowest value of tan δ depended on the monomers used, the heat treatment in two stages was effective. It seems that the heat treatment at a lower temperature is related to the polymerization and a higher temperature it is related to the acceleration of polymerization and the removal of the residual monomers. The tan δ versus temperature relationship measured for the optimum films at 10 kHz are shown in Fig. 3.15.

In particular, for the films obtained by polymerizing VM or VCr, the peak-height was very small. This behavior may be related to the cross-linking structure of VM and VCr because these compounds have two double bonds. Although VCi also has two double bonds, it is known that the cyclization is likely to occur in preference to the cross-linking reaction. Consequently, the dielectric relaxation peak did not decrease upon polymerization. It is natural that a linear polymer, VB, has a larger dielectric relaxation. Although the degree of cross-linking was not measured directly, the order of the value of tan δ seemed to be related to the order of the degree of cross-linking; i.e., the polymer with higher degree of cross-linking structure showed a lower value of tan δ.

Change in film thickness with heat treatment was examined for VM. Results are shown as shrinkage percentage in Table 3.1. It can be observed that the film shrinks as the value of tan δ becomes lower. This is due to the proceeding of the cross-linking reaction and is consistent with the interpretation of the results of

Table 3.1 Polymerizing condition and some properties of the present films.

Monomer	Sample/No	$t_p/^{\circ}C$	BPO/ $mg{\cdot}g^{-1}$	$\tan\delta\times10^3$	$t_{max}/^{\circ}C$	Schrinkage/%
VB	1	90(1 h)	1	136.0	120	
	2	70(1 h)	1	48.6	90	
	3	70(1 h)	4	55.2	90	4.0
VM	4	120(1 h)	1	27.2	115	
	5	120(1 h)	4	21.3	115	11.3
	6	70(1 h)–120(1 h)	1	23.0	120	
	7	70(1 h)–120(1 h)	4	21.1	135	15.1
	8	90(1 h)		1	78.1	120
VCr	9	70(1 h)–120(1 h)	1	47.7	135	
	10	70(1 h)–120(1 h)	4	17.5	155	
VCi	11	90(0.5 h)–160(1 h)	1	165.5	210	
	12	90(0.5 h)–120(1 h)	4	146.5	215	

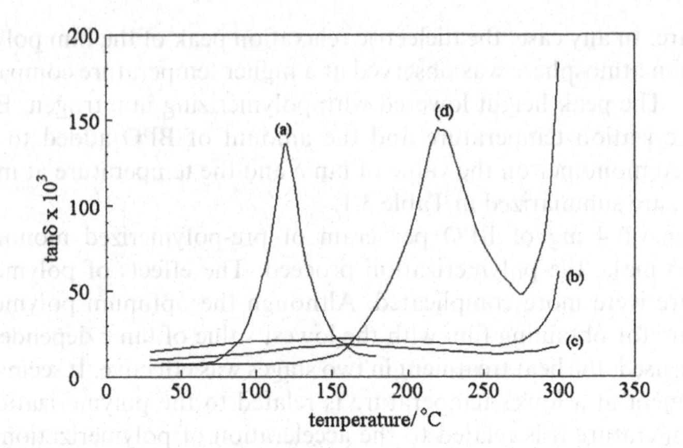

Fig. 3.15 Dependence of tan δ on temperature measured at 10 kHz. (**a**) poly-VB (Sample No. 1), (**b**) poly-VM (No. 7), (**c**) poly-VCr (No. 10), and (**d**) poly-VCi (No. 12)

dielectric relaxation behavior. The measurement of the dielectric relaxation may be considered as a reliable indication of the degree of cross-linking.

The amount of sorbed water was measured for these polymer thin films as a function of relative vapor pressure, p/p_0. The effects of polymerization conditions on the amount of sorbed water were observed to be significant. The water content of the films W_g, which was based on grams of sorbed water per one gram of dry polymer, measured at $p/p_0 = 0.81$ and 30°C is summarized in Table 3.2.

It should be noted that the sorption ability increased with proceeding of the cross-linking reaction for each polymers. Sorption isotherms measured at 30°C for the films are shown in Fig. 3.16. All the curves are almost linear. These polymers were hydrophobic and W_g depended on the polymer species. The difference of sorption ability seems to be derived from the difference of the structure of the films as well as the difference of the side groups of the polymer. It has been reported that the saturation value of water gain of copolymers of MMA with dimethacrylate monomers increases with increasing degrees of cross-linking [23, 37, 38, 39]. The behavior can be interpreted in terms of an

Table 3.2 W_g and W_m measured at $p/p_0 = 0.81$ and 30°C

Monomer	No.[a]	W_g/mg·g^{-1}	W_m
VB	1	4.9	0.02
	2	6.1	0.04
VM	4	28.1	0.08
	6	30.1	0.09
	7	37.4	0.10
VCr	9	12.8	0.04
	10	20.9	0.06
VCi	11	13.3	0.07
	12	14.4	0.07

[a]Sample No. is indicated in Table 3.1.

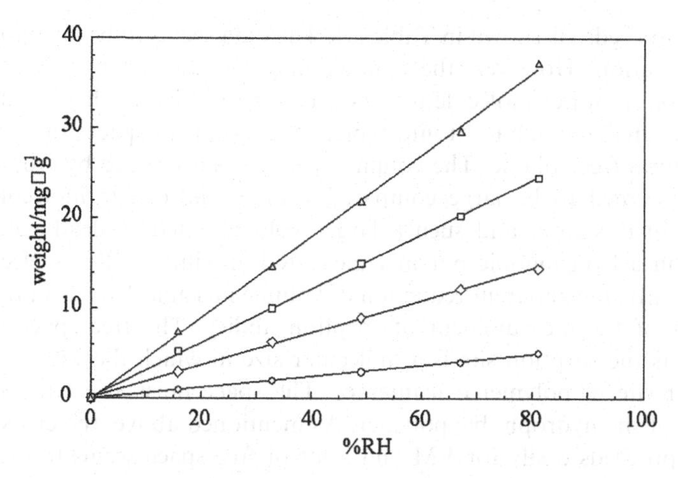

Fig. 3.16 Sorption isotherms measured at 30°C. (○) poly-VB (Sample No. 1), (△) poly-VM (No. 7), (□) poly-VCr (No. 10), and (◇) poly -VCi (No. 12)

increase in the concentration of the hydrophilic groups by introducing the hydrophilic cross-linking agent. In the case of this system, the effects of hydrophilicity of the cross-linking agent need not be considered because this monomer can cross-link by itself. Kalachandra et al., [40] measured the water sorption ability and compared the water uptake with the contact angle for water in glycol dimethacrylate and methylmethacrylate polymers. The weight percent oxygen (WPO) content, which is indicative of the hydrophilic character of these materials, was calculated and correlated with weight percent water uptake. The linear relationship between the WPO and the percentage water uptake was observed in a series of four glycol-dimethacrylate and eight linear methylmethacrylate monomers. The increased water uptake of polymer compared with monomers was attributed to the microvoids present in the polymer matrix. However, the effects of the cross-linking degree were not considered. We estimated another measure of water content W_m instead of W_g on the basis of the mole ratio of sorbed water molecule to oxygen atom in each monomer molecule. Results are shown in Table 3.2. Poly-VM had the largest W_m (0.10 mole water/mole oxygen atom) and decreased in the order of poly-VCr (0.07), poly-VCi (0.07), and poly-VB (0.04) at $p/p_0 = 0.8$. The value of W_m of the cross-linked polymers was greater than that of linear polymers. There were also some differences among the cross-linked polymers. The orders of W_m may be related to the cross-linking degree. The packing density change of the films with the formation of cross-linking structure should be taken into consideration. In the case of hydrophobic polymer without strong sorption sites, the free space around the polar site is likely to be a sorption site. Cross-linked films prepared by polymerizing monomers with short chains seem to have highly strained chain conformations that prevent close chain packing. That leads to an increase in free spaces. Now, it must be noted that the film shrinks as the cross-linking reaction proceeds, as shown in Table 3.1. The data seems to be in conflict with our explanation. However, the conflict may be resolved by the following explanation. A water molecule may be encompassed by a sphere of diameter about 3Å, small enough to fit into typical microvoids or spaces that comprise the structural free volume. The volume, which was decreased by cross-linking reaction, seemed to be large compared with the adsorbate molecule (H_2O molecule in this case) and such a larger volume acted as dead volume for sorption in a hydrophobic polymer measured gravimetrically or electrically. Consequently, the apparent reduction of volume in Table 3.1 was independent of results of the measurement of sorption ability. The free space we now consider as the sorption site is a molecular size in which the water molecule and polar site of polymer can interact. The space must affect the sorption behavior of the hydrophobic polymer. As mentioned above, the cross-linking reaction proceeds easily for VM and a lot of free space seems to exist. This explanation agreed with the order of water content.

The electrical capacitance of present polymers was measured as a function of the relative vapor pressure at 100 kHz. The electrical capacitance increased almost linearly with increases in the relative vapor pressure. The values of C_{90}/C_0 (since the capacitance is affected by the film thickness, we adopted the ratio of capacitance at $p/p_0=0.9$ to that capacitance at $p/p_0=0$ for comparison) are shown in Table 3.3. Dependence of the C_{90}/C_0 on W_g was examined and is shown in Fig. 3.17.

The C_{90}/C_0 increases linearly with an increase in W_g in the low-sorption region, followed by curvature upward. Irrespective of the difference of polymer species and of the degree of cross-linking, the C_{90}/C_0 depended only on W_g. The dielectric constant of the sorbed water was estimated by applying Kurosaki's equation (Eq. (3.3)) [16]. The results are summarized in Table 3.3. The dielectric constant is less than that of liquid water, which is 76.6 at 30°C, and also smaller compared with that of cellulose derivatives [8, 14]. The dielectric constant of sorbed water on cellulose derivatives had a larger value when the number of sorbed water molecules per monomer became more than unity. We have

Table 3.3 C_{90}/C_0 measured at 100 kHz and 30°C and ε_w estimated at $p/p_0=0.8$. 1 kHz and 30°C

Monomer	No.	C_{90}/C_0	ε_w
VB	1	1.04	21.3
	2	1.05	20.4
VM	4	1.27	24.1
	6	1.30	24.9
	7	1.37	25.0
VCr	9	1.12	23.3
	10	1.33	34.5
VCi	11	1.13	24.3
	12	1.12	20.2

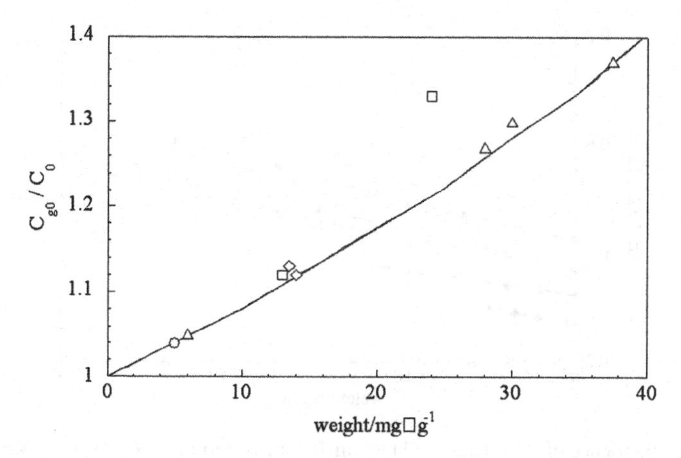

Fig. 3.17 Dependence of the C_{90}/C_0 on the amount of sorbed water measured for film with various cross-linking degree at 100 kHz and 30°C. (○) poly-VB, (△) poly-VM, (□) poly-VCr, and (◇) poly-VCi

explained this behavior as the formation of clusters of sorbed water molecules. The value of the dielectric constant on these polymers is equivalent to the reported poly-methylmethacrylate and cross-linked polyimide, which is too hydrophobic to form clusters [11, 14]. The dielectric constant became slightly larger with increase in the relative vapor pressure. This indicates that the interaction between water molecules begins to associate with increases in the amount of sorbed water. Further the polarization of sorbed water was estimated. The relationship between $(\varepsilon-1)(2\varepsilon+1)/9\varepsilon$ and W_g is shown in Fig. 3.18, in which these results were obtained at 1 kHz and 1 MHz in the humidification process at 30°C.

The value of $(\varepsilon-1)(2\varepsilon+1)/9\varepsilon$ was affected by the applied frequency and increased monotonically with an increase in W_g. The result is slightly different from that of cellulose derivatives [8]. In the case of cellulose derivatives, the value was scarcely affected by the frequency until the molar ratio of water/monomer unit of cellulose became unity. In the region where the number of the sorbed water per monomer unit of the cellulose was less than unity, $(\varepsilon-1)(2\varepsilon+1)/9\varepsilon$for the sorbed water increased linearly with the amount of sorbed water then steeply, especially for the measurement at 1 kHz. The difference between $(\varepsilon-1)(2\varepsilon+1)/9\varepsilon$ values observed at 1 kHz and 1 MHz corresponds to $4\ N_wP(\omega)/3(=\Delta P)$ and indicates the degree of the orientation of dipoles. As the water molecule is associated with more other water molecules with increasing W_g, the slope of $(\varepsilon-1)(2\varepsilon+1)/9\varepsilon$ versus W_g becomes larger and the difference of $(\varepsilon-1)(2\varepsilon+1)/9\varepsilon$ between the two frequencies becomes large. The value of ΔP at certain W_g is shown in Table 3.4.

In the case of these polymers, the difference was very small compared with other polymers [14]; this is due to the small interaction between the polymers and the sorbed water molecules. By considering these results together with the result

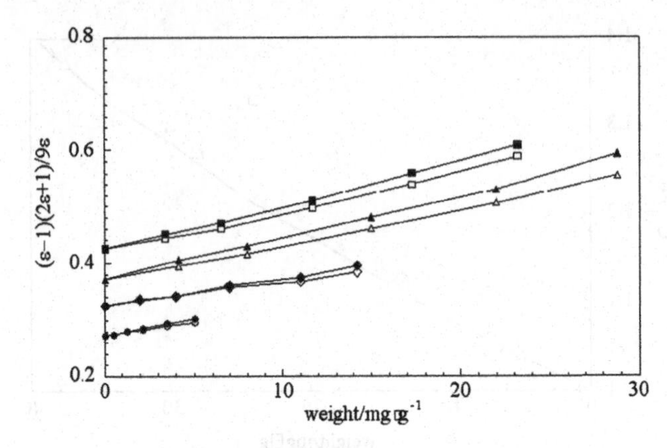

Fig. 3.18 Dependence of $(\varepsilon - 1)(2\varepsilon + 1)/9\varepsilon$ on W_g measured at 30°C. ○) poly-VB (Sample No. 1), (△) poly-VM (No. 7), (□) poly-VCr (No. 10), and (◇) poly-VCi (No. 12). *Open symbols*; 100 kHz and *closed symbols*; 1 kHz

Table 3.4 ΔP estimated at 30°C

W_g	5 mg·g^{-1}	10 mg·g^{-1}	20 mg·g^{-1}
Poly-VB	0.007		
Poly-VM	0.010	0.016	0.027
Poly-VCr	0.007	0.012	0.021
Poly-VCi	0.004	0.007	

of estimated dielectric constant of sorbed water, the sorbed water scarcely formed clusters due to the hydrophobic nature and the space limitations of polymers.

Among examined monomers, VM and VCr formed a cross-linked structure by polymerization while VB and VCr did not. The difference of the structure of polymer, that is, whether the polymer is cross-linking type or linear type and the difference of the degree of cross-linking, affect the sorption ability. The polymer with highly cross-linked structure had a larger amount of water content. The state of sorbed water estimated with dielectric measurement was almost the same for both cross-linked and linear polymer, irrespective of the difference of degree of cross-linking. The sorbed water molecules scarcely formed clusters, because of the hydrophobic properties and space limitation of the polymer.

Consequently, the enhancement of water sorption ability by forming cross-linked structure is attributable to the increase in free space in molecular size around a polar site derived from the highly strained chain conformations.

3.4 Humidity-Sensors-Based Hydrophobic Polymer Thin Films

The important requirements for fabricating capacitive-type humidity sensors are low hygroscopic and the rigid structure of the sensing polymer thin film. The formation of clusters of sorbed water molecules causes the hysteresis, also, the formation of ink-bottled type pores in polymer films. The swelling and/or geometrical deformation of the polymer film causes the large temperature coefficient and lack of long-term stability and reproducibility of the sensing behavior. It seems that the cross-linked hydrophobic polymer films are adequate for a practical capacitive-type humidity sensor.

3.4.1 Poly-Methylmethacrylate-Based Humidity Sensor

3.4.1.1 Initial Performances

Figure 3.19 shows the relationship between the sensitivity measured at 100 kHz and relative humidity for PMMA together with various cellulose derivatives previously reported [8].

Though the sensitivity change of electrical capacitance of PMMA to relative humidity is smaller than that of the cellulose derivatives because of its lower hygroscopicity, it is enough to detect a humidity change. Furthermore this sensor has good linearity over the humidity range of 10–90% RH. The response was very fast (the 90% response time was within 30 s by changing the relative humidity from 20 to 80% RH and vice versa) and the hysteresis was less than ± 1% RH.

3.4.1.2 Temperature Dependence

The temperature dependence of PMMA capacitance is shown in Fig. 3.20.

It was observed that this sensor had a positive temperature coefficient. This temperature dependence is shown to arise from the temperature dependence of PMMA permittivity itself. It is well known that PMMA showed a dielectric

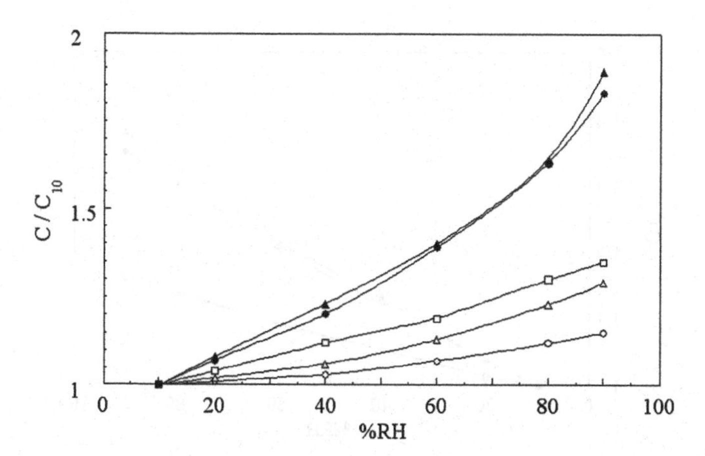

Fig. 3.19 Relation between the sensitivity and relative humidity at 100 kHz, 30°C; (○) PMMA, (△) EC, (□) CAB, (●) CA, and (▲) CAHP

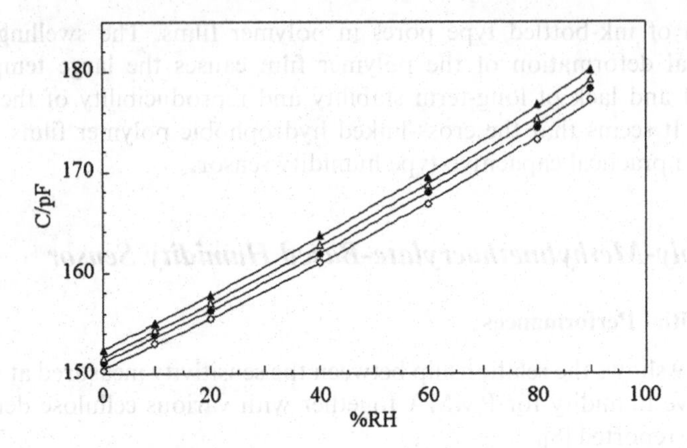

Fig. 3.20 Temperature dependance of PMMA capacitance at 100 kHz; (○) 30, (●) 35, (△) 40, and (▲) 45°C

relaxation peak at ca, 67°C at 1 kHz due to the side-chain motion. In order to eliminate the effect of PMMA dielectric relaxation, the correlations between the sensitivity and relative humidity at each temperature are plotted and shown in Fig. 3.21.

Compared with the results of CAB, in the case of PMMA the temperature dependency of sensitivity is small.

3.4.1.3 Long-Term Stability

The durability of this PMMA sensor towards acetone vapor was tested since acetone is a good organic solvent for PMMA. The results are shown in Fig. 3.22.

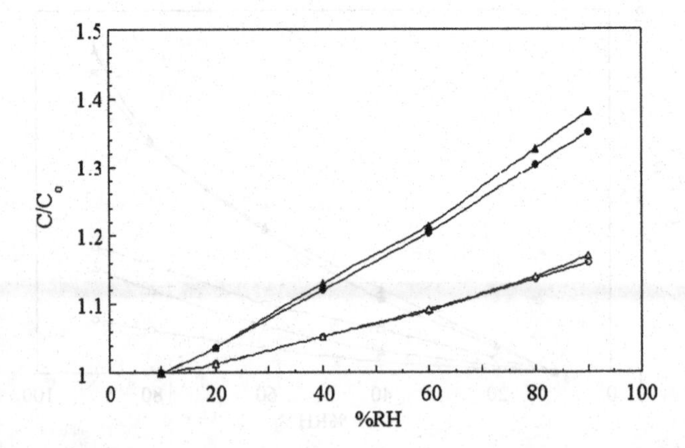

Fig. 3.21 Temperature dependance of PMMA and CAB sensitivity at 100 kHz, MMA: (○) 30 and (△) 45°; CAB: (●) 30 and (▲) 45°

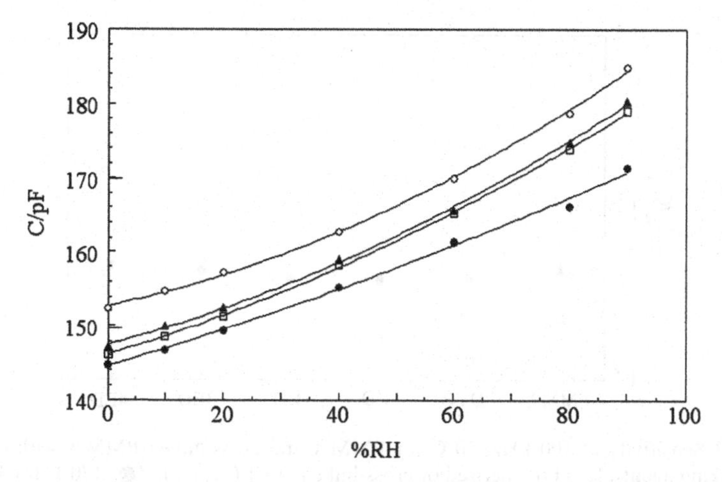

Fig. 3.22 Durability of PMMA for acetone measured at 100 kHz, 30°C. Only the increasing direction of humidity is shown; (●) initial value before exposure to (○) first measurement

This test was performed as followed. After exposure to saturated acetone vapor for 20 min in a closed container, the sensor was quickly transferred to a vessel in which the humidity and temperature were controlled. The sensitivity was measured repeatedly for three cycles. The term "cycle" refers to the process where the humidity was changed from ~0% RH to 90% RH followed by reduction from 90% RH back to ~0% RH. The capacitance value immediately after exposure to acetone vapor was large, and this is due to the sorption of acetone. The increments of 5% of capacitance at ~0% RH and 8% at 90% RH were observed upon exposure to acetone. After repeating the measurements three times, the increments of capacitance decreased to 0.7% at 0% RH and 5% at 90% RH as the acetone was desorbed. However, the value did not return to its original amount.

3.4.2 Characteristics of Cross-Linked PMMA-Based Sensor

3.4.2.1 Initial Performances

There have been reports that the saturation value for water uptake of PMMA increases with increasing the degree of cross-linking [38, 39]. For the cross-linking agent, divinyl benzene (DVB), ethylene glycol dimethacrylate (ED), triethylene glycol dimethacrylate (3ED), tetraethylene glycol dimethacrylate (4ED), and 1.3-butylene dimethacrylate (BD) were used. This behavior can be exhibited in terms of an increase in the concentration of the hydrophilic groups. The sensitivity over the humidity range of 10–90% RH is shown in Fig. 3.23.

When the PMMA was cross-linked at 90°C for 1 h, the sensitivity became larger in the order of 4 ED > 3 ED > PMMA > ED > BD > DVB. However,

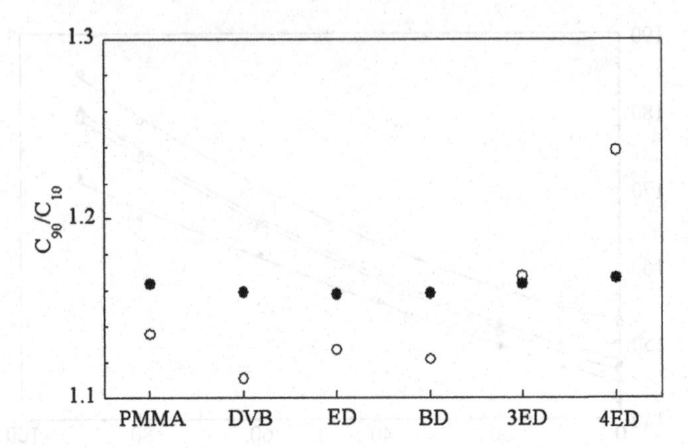

Fig. 3.23 Sensitivity at 100 kHz, 30°C for PMMA and cross-linked PMMA with various cross-linking agents; (○) polymerized or cross-linked at 90°C for 1 h, (●) 170°C for 1 h

with increasing cross-linking temperature up to 170°C, the sensitivity is enhanced except for the 4 ED system and uninfluenced by the cross-linking agent. This result is explained as follows. When the cross-linking temperature is low, cross-linking is insufficient, because the ether bond is long. Consequently, the un-reacted cross-linking agent, which acts as a plasticizer, sorbs more water. When the cross-linking temperature is increasing, the cross-linking was promoted, resulting in an increase in density of the polymer and stiffness of the film. Consequently, the cause of increasing sensitivity by reacting at higher temperature seems to arise from its morphology, i.e., an increase in film stiffness leads to an increase in the free volume.

3.4.2.2 Temperature Dependence

The temperature dependence of sensitivity was also examined. The C_{45}/C_{30} (capacitance at 45 and 30°C, respectively) ratios at ~0% RH for the film cross-linked at 90°C was not improved except for DVB and the temperature coefficient of sensitivity (C_{45}/C_{30} at 90% RH and C_{45}/C_{30} at ~0% RH) also was not improved. This is due to the segment motion of PMMA having insufficient cross-linking. On the other hand, when the PMMA was cross-linked at higher temperature up to 170°C, temperature dependence both at ~0 and 90% RH became small and the temperature coefficient became smaller than PMMA itself. This is due to the depression of segment motion and/or swelling of the polymer by sufficient cross-linking.

3.4.2.3 Durability against Acetone Vapor

Furthermore, we examined the durability of these sensors against acetone vapor. The cross-linked sensor with ED that reacted at 90°C was also unstable

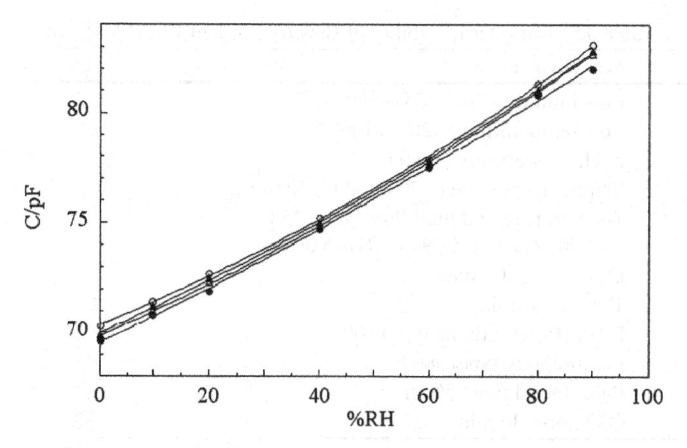

Fig. 3.24 Durability of cross-linked PMMA with ED for acetone measured at 100 kHz, 30°C; symbols represent the some values as shown in Fig. 3.22

like PMMA itself, and the value of the capacitance at a constant humidity was increased by the exposure to acetone and then decreased with increasing measuring cycle. On the other hand, the sensor cross-linked at 170°C was quite stable, showing little effect by acetone vapor and almost recovered to the original value after three cycles of measurements as shown in Fig. 3.24.

The increments of capacitance immediately after exposure to acetone were 1% at ~0% RH and 1.6% at 90% RH and those after three cycles measurements were 0.3% at ~0% RH and 1% at 90% RH. This means that the cross-linked PMMA film simply sorbed acetone without permanent deformation. The result can be explained as follows. When the PMMA was cross-linked at 90°C, the cross-linking was insufficient as previously mentioned, and the acetone was easy to diffuse into the bulk and difficult to desorbed. Further, partial dissolution of the PMMA film is also possible. Consequently, the value of capacitance drifted. As the cross-linking proceeded and the film stiffened, the sorption of acetone was limited to the surface and/or film/electrode interfaces. To confirm this interpretation, we measured the amount of sorbed acetone by the gravimetric method. The amount of sorbed acetone for the sample cross-linked at 170°C (15 mg/g) was reduced to one-half that of the sample cross-linked at 90°C.

3.4.2.4 Long-Term Stability

We have also examined long-term stability under various conditions, e.g., high-temperature and high-humidity, various paint vapors, and tobacco smoke, etc. These results are summarized in Table 3.5. Excellent stability was exhibited in the environments, except that with tobacco smoke, which showed less than 2% RH drift after the tests, though the sensors showed 3% RH drift just after removal from the smoke. Also, the drift returned to its initial value after exposure to the office environment.

Table 3.5 Long-term stability of the cross-linked PMMA sensor

No.	Test environments	Sensor drift(% RH)
1	Low humidity (10% RH) 1000 h	±2
2	Low temperature (−20°C) 1000 h	±2
3	High temperature (+60°C)	
4	Temperature cycle (−20 to 60°C) 50 cycles	±2
5	Temperature and humidity cycle (25°C, 50% RH ~40°C, 90% RH) 5 cycles	±2
6	Dew cycle, 50 cycles	±2
7	Tobacco smoke	±3
8	Paint (Polyurethane type) 500 h	±2
9	Paint (epoxy type) 500 h	±2
10	Paint (acryl type) 500 h	±2
11	Oil vapor, 30 min	±2

3.4.3 Polysulfone-based Sensor

3.4.3.1 Initial Performances

The chemical structure of polysulfone is illustrated in Fig. 3.25.

The typical humidity dependence of the sensor capacitance from 10 to 90% RH at 25°C is shown in Fig. 3.26. Also typical results over a relative humidity

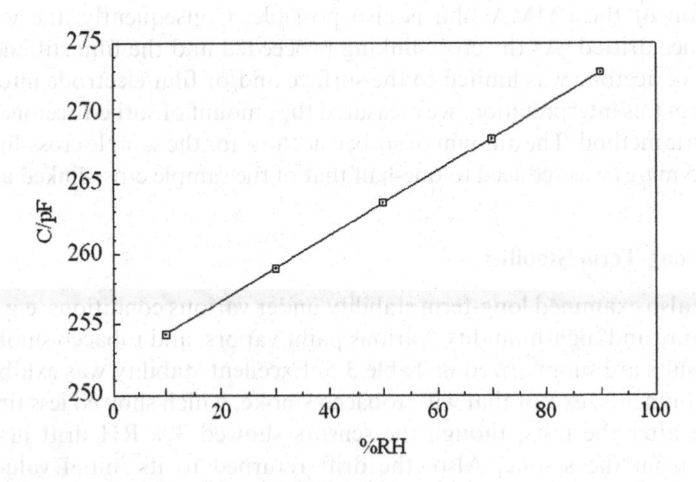

Fig. 3.25 Chemical structure of polysulfone (PSF)

Fig. 3.26 Humidity dependence of sensor capacitance 25°C

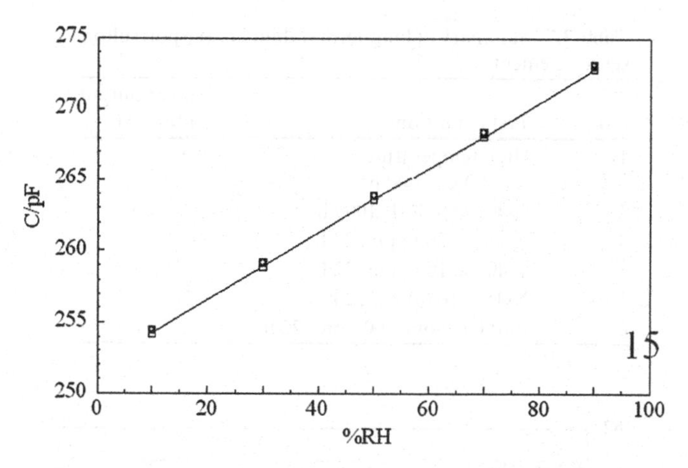

Fig. 3.27 Humidity dependence of sensor capacitance at 10.25 and 40°C

range of 10–90% RH at 10, 25, and 40°C are shown in Fig. 3.27. The curves show the hysteresis and temperature dependence of the sensor capacitance. The temperature dependence of the capacitance is 0.0±0.1% RH/°C and is reproducible. The typical time response at 25°C is less than 30 s in both adsorption and desorption test conditions. A summary of the initial performance of the sensor elements is shown in Table 3.6. Typical hysteresis and terminal linearity are less than 0.5 and +0.8% RH, respectively, at 10–90% RH and 10–40°C.

3.4.3.2 Long-Term Stability

The long-term stability of the sensor elements was evaluated in various harsh environments. The sensor elements were exposed to the testing conditions and tested over a humidity cycle of 10–90% RH at 25°C. Some of the test results are summarized in Table 3.7.

The test results shown in this Table are the maximum value, mostly observed at 90% RH. Excellent stability was demonstrated in the various environments. In particular, the sensor drift was as small as +3% RH after 1000 h of exposure to 40°C, 90% RH conditions, as shown in Fig. 3.28.

Table 3.6 Summary of initial performance of the polysulfone sensor element

Hysteresis at 25°C, 10–90% RH (% RH)	0.5
Terminal linearity at 25°C, 10–90% RH (% RH)	+0.8
90%U response time 10 → 90% RH at 25°C (s)	30
90%U response time 90 → 10% RH at 25°C (s)	30
Dissipation factor at 50% RH, 25°C at 100 kHz	0.008
Temperature coefficient of capacitance 10–40°C, 10–90% RH (%RH °C^{-1})	0±0.1

Table 3.7 Summary of long-term stability of the polysulfone sensor element

No.	Test conditions	Sensor output shift (%U)
1	High temperature +80°C, 1000 h	±2
2	40°C, 90% RH, 1000 h	±3
3	Acetone 200 ppm, 22 h	±2
4	Toluene 100 ppm, 22 h	±2
5	Xylene 100 ppm, 22 h	±2
6	Butyl acetone 200 ppm, 22 h	±2

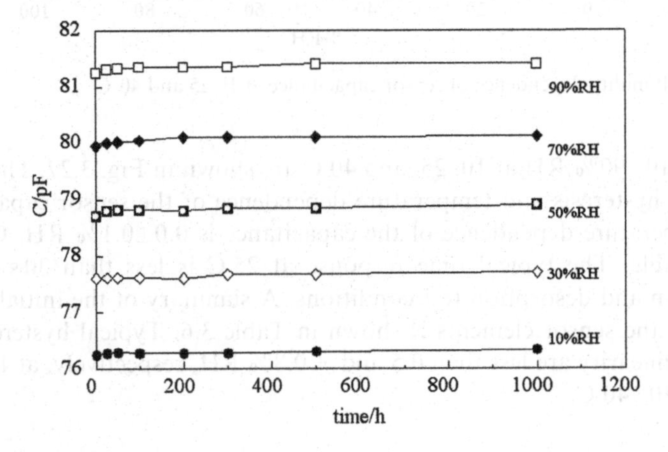

Fig. 3.28 Sensor drift under 40°C, 90% RH

This sensor performance under hot and humid conditions is superior to that of most commercially available capacitance-type sensors.

In addition, the sensor showed stable performance within ±2% RH reproducibility under room conditions, for about one and a half years, as shown in Fig. 3.29.

Other environmental test results stayed within 2% RH reproducibility, demonstrating an excellent overall performance.

3.4.4 Acetylene-Terminated Polyimide-based Sensor

3.4.4.1 Determination of Curing Condition

Molecular structure of acetylene-terminated polyimide is illustrated in Fig. 3.30.

TG measurements ware carried out on the present oligomers from room temperature to 45°C at a heating rate of 5°C/min in flowing air and little

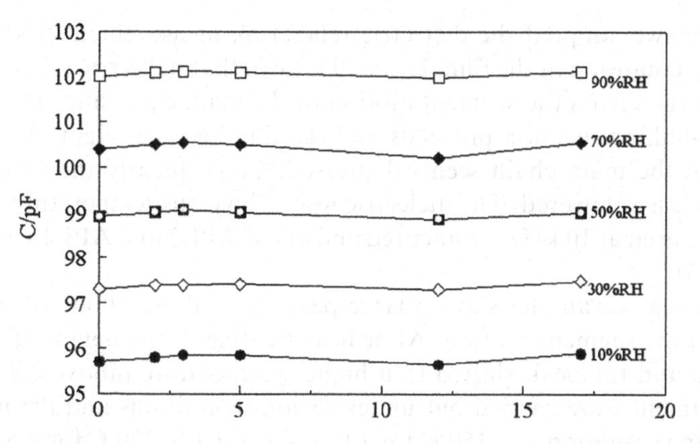

Fig. 3.29 Sensor drift under room conditions

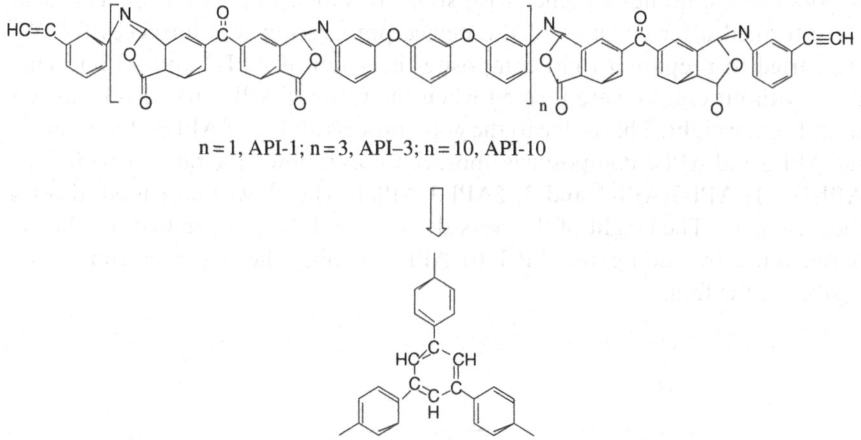

n = 1, API-1; n = 3, API-3; n = 10, API-10

Fig. 3.30 Molecular structure of the acetylen-terminated poly-isoimide oligomer and cross-linked polyimide

weight loss was observed. This result proves that the curing proceeded without by-products. Further, good thermal stability was confirmed. Thermal analysis was carried out on a DSC at 5°C/min in flowing nitrogen. Two large exothermic peaks ware observed at ca. 220 and ca. 260°C. Bott et al. reported that cross-linking readily occurs at temperatures above 200°C [45] and the present result was consistent with their result. Identification of the respective peak was not done. For the sample after heating at 220°C for 2 h and followed by 260°C for 2 h in nitrogen flow, DSC measurements also were performed. In this case, no peaks were observed and the result suggests that this heat treatment is appropriate for reacting the thermal acetylene.

Further, we adopted the dielectric relaxation measurement to determine the curing condition in the film. It is well known that dielectric relaxation of a polymer arises from the segment motion of the main chain and side chain. If the cross-linking reaction proceeds and the film becomes rigid, the segment motion of the main chain seems depressed. Consequently the value of the dielectric loss is lowered. The dielectric loss (ε") versus temperature relationships measured at 10 kHz for uncured and cured API-3 and API-10 are shown in Fig. 3.31.

Both uncured samples show a large peak in ε" above 200°C which corresponds to the segment motion. After heat treatment, the height of the peak decreased and the peak shifted to a higher temperature (above 260°C). The heat treatment ware carried out under various conditions and the most suitable curing condition was 150°C for 1 h, 180°C for 1 h, 220°C for 2.5 h, 260°C for 2 h, and 400°C for 2 h, in flowing nitrogen. Further, the shorter the oligomer chain length, the lower the value of ε". This means that a rigid film is obtained using the oligomer with shorter chains. For API-1, many cracks ware formed after curing and the device preparation was unsuccessful. We then tried to prepare a rigid composite then film of API-3 and API-1. Thin films without cracks were formed when the ratio of API-1 to API-3 was less than 1/2 in weight. This is due to the good process ability of API-3. The name of the API-3 and API-1 composite is abbreviated as follows: the ratio of API-3 and API-1 is 1; API-3-API-1 and 2; 2API-3-API-1. The ε" was measured also for these samples. The height of the peak decrease and the peak shifted to a higher temperature by adding the AP-1 to API-3; adding the oligomer enhances a rigidity of the film.

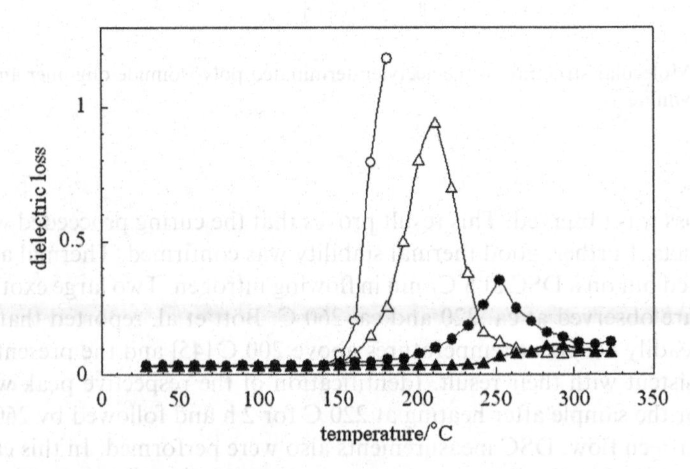

Fig. 3.31 Temperature dependence of ε at 10 kHz for uncured and cured API samples: ○, uncured API-10; ●, cures API-10; △, uncured API-3;and ▲, cured API-3

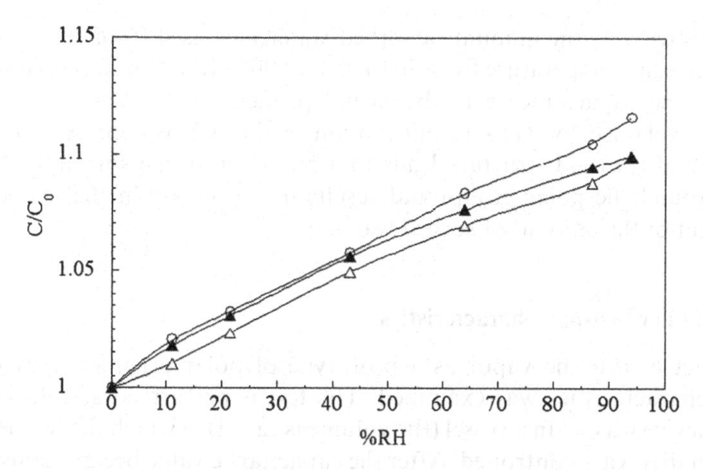

Fig. 3.32 Humidity dependence of the sensitivity at 100 kHz and 30°C for API-10: ○, ●, cured; △, ▲; uncured; *open symbols*, humidification; and *closed symbols*, desiccation

3.4.4.2 Initial Performances

In the uncured sample (Fig. 3.32), hysteresis was observed between the capacitance when measured with increasing humidity and the capacitance when measured with decreasing humidity. Hysteresis was not detected for the cured films because the amount of sorbed water is small and the swelling of the polymer and the formation of clusters of sorbed water are prevented when using the cured sample.

We do not have to consider the capillary condensation as the cause of hysteresis because the hysteresis becomes smaller after the curing. The space we consider is not the geometrical pores but the molecular-sized micro-voids. Further the electrical capacitance changes linearly with relative humidity. These are the preferable characteristics of a humidity sensor. Among the prepared sensors, the sensitivity (capacitance at 90% RH/capacitance at ~0% RH ratio) of the API-3-API-1 sensor was the largest (1.184) and decreased in the order of 2API-3-API-1 (1.176), API-3 (1.171), and API-10 (1.116).

3.4.4.3 Temperature Dependence

The temperature coefficient of the sensor was examined. For the uncured sample, a large temperature coefficient was observed. At ~0% RH, the value of electrical capacitance slightly decreased with increasing temperature. This decrease is due to the segment motion of the API. However, above 10% RH, the electrical capacitance increased with increasing temperature and also with increasing relative humidity (the temperature coefficient is +0.3% RH/°C at 40% RH and +0.5% RH at 80% RH in the temperature range from 30 to 45°C). These result is derived from the temperature coefficient of the amount of sorbed water, *i.e.*, the amount of sorbed water increased from 29 to 49 mg/g with increasing temperature from 30 to 45°C at 90% RH. For the cured sample, the electrical capacitance was almost independent oh the temperature. This result is explained by; (1) segment motion of the API is depressed by cross-linking, and (2) rigid structure leads to a reduction in the sorption sites (the space around the polar group) and results in a decrease in the temperature coefficient of the amount of sorbed water.

3.4.4.4 Other Sensing Characteristics

The effect of acetone vapor as a prototype of polar organic vapor on the sensor characteristics was examined. The test is performed as follows. The sensor device was set in a vessel (the volume is ca. 5 l) in which the temperature and humidity ware controlled. After the capacitance value became constant at each humidity 500 μl of acetone was injected into the vessel. The time when the acetone was injected was defined as zero, and the capacitance change with time was recorded at regular time intervals. The results are converted to relative humidity changes and are shown in Fig. 3.33 for the cured API-10 samples.

The capacitance of both the uncured and cured sample is not affected by the acetone vapor in a dry atmosphere, while the capacitance is affected by the acetone vapor in humid atmospheres, this means that the cured polyimide does not sorbed the acetone vapor. In the humid atmosphere, the sorbed waters

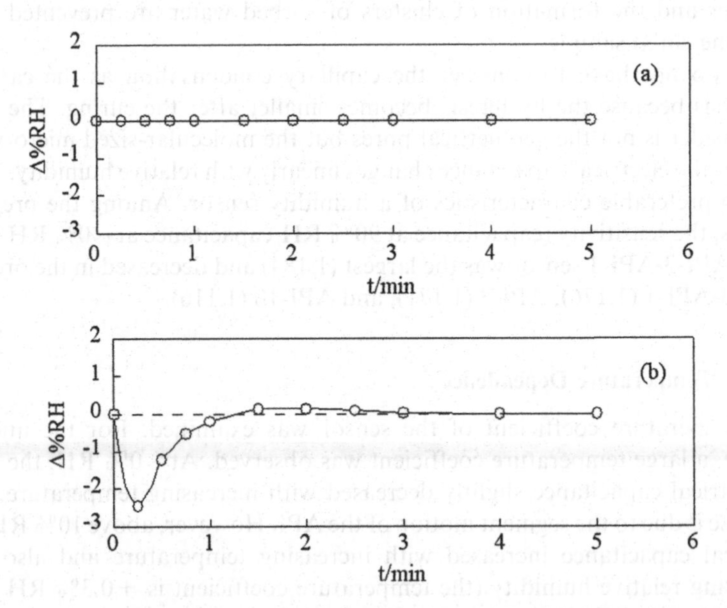

Fig. 3.33 Effect of acetone vapor on the sensor characteristics of thecured API-10 measured at 30°C and 100 kHz: (**a**) 0% RH; and (**b**) 90% RH

accelerate the diffusion of acetone vapor. The capacitance drift recovered to its initial value. For the sensors stored at 40°C and 90% RH, the sensitivity at 30°C was examined as a function of the storage periods at 40°C and 90% RH. The sensitivity drifted within the initial 100 h and approached a constant value after 1000 h; the drift of the API-3 sensor output was +6% RH. The value was very small compared with other polyimide sensor and the stability in high temperatures and in humid atmospheres was improved.

The present acetylene-terminated polyimide resins is polymerized before coating and cured on the substrate. This good workability reflects the reproducibility of the sensor. The sensor characteristics as shown in the upper curve of Fig. 3.32 are reproducible and calibration of each sensor is not needed. Further, we have tested the long-term stability in ambient atmosphere. The drift of sensor output was within 2% RH even after the sensor was allowed to stand for at least 17 months. Consequently, preparation of the reliable capacitive-type humidity sensor was achieved by using the acetylene-terminated polyimide resins.

3.4.5 Cross-Lined Fluorinated Polyimide-Based Sensor

3.4.5.1 Sensor Fabrication

The fabrication process of the sensor chip is as follows. First, the platinum lower electrode is evaporated in vacuum on a glass substrate. The fluorinated polyimide oligomer is dissolved in a solvent and then spin-coated onto the glass substrate. The wafer is then dried at 150°C, followed by heat treatment in a programmable furnace with a final temperature of 400°C. This heat treatment makes the oligomer react, resulting in a cross-linked polymer structure. After this heat treatment the polymer is etched and patterned to open the contact pads for soldering, followed by gold upper-electrode evaporation. The finished glass wafer is then cut into sensor chips. Leads with hermetically sealed metal base are soldered to the platinum contact pads. Finally a metal can is welded to the metal base and a pair of filters is attached on the open windows of the can.

3.4.5.2 Initial Performances

Typical results over the humidity from 10 to 90% RH at 10, 25, and 40°C are plotted in Fig. 3.34.

The curves show hysteresis and temperature dependence of the sensor capacitance is 0±0.1% RH/°C and is reproducible, as shown in Fig. 3.35.

The typical time response at 25°C is shown in Fig. 3.36, with adsorption and desorption conditions. The fast response in both testing conditions was observed.

Sensor hysteresis, terminal linearity, 90% response time, dissipation factor and temperature coefficient of the capacitance are summarized in Table 3.8. Typical hysteresis and terminal linearity were both less than 1% RH at 10–90%

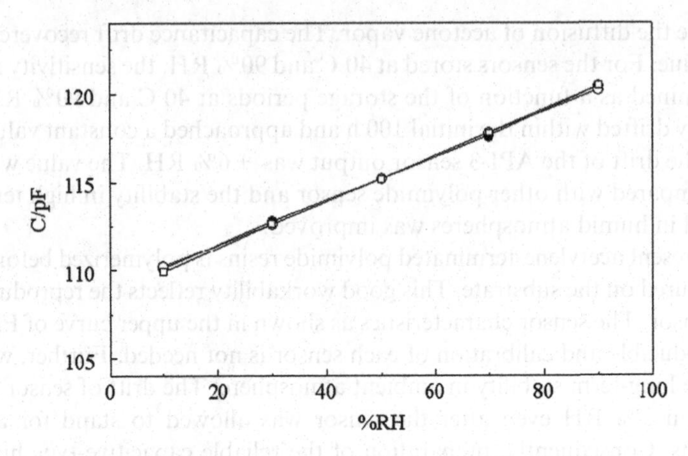

Fig. 3.34 Humidity dependence of sensor capacitance at 10, 25 and 40°C

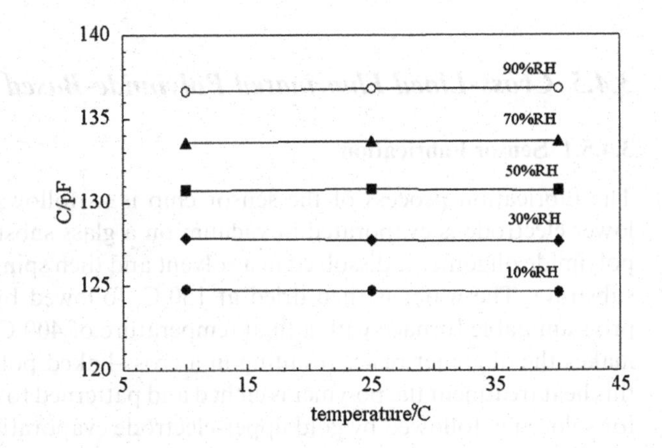

Fig. 3.35 Temperature dependence of the sensor capacitance

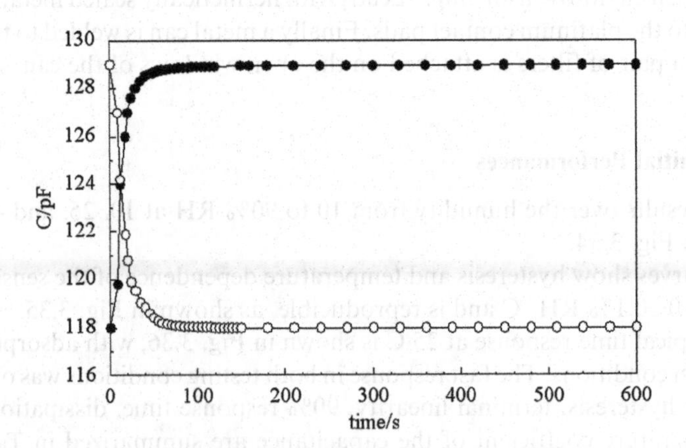

Fig. 3.36 Time response of sensor at 25°C

Table 3.8 Summary of the initial performance

Hysteresis at 25°C, 10–90% RH (% RH)	0.5
Terminal linearity at 25°C, 10–90% RH (% RH)	0.6
90% Response time 10 → 90% RH at 25°C (s)	30
90% Response time 90 → 10% RH at 25°C (s)	30
Dissipation factor at 50% RH, 25°C at 100 kHz	0.005
Temperature coefficient of capacitance 10–40°C, 10–90% RH (% RH/°C)	0±0.1

RH and 10–40°C. These test results demonstrate the excellent initial performance of the sensor.

3.4.5.3 Long-Term Stability

Long-term stability of the sensor elements was evaluated under various hostile environments. The sensor elements were exposed to the testing conditions and tested over a humidity cycle of 10–90% RH at 25°C. Some of the test results are summarized in Table 3.9. The test results shown in this Table are the maximum value, mostly observed at 90% RH. An excellent stability was exhibited in the various environments. In particular, the sensor showed as small as +3% RH drifts after 1000 h exposure to 40°C, 90% RH conditions. This stability under hot and humid conditions is essential for opening up a wide variety of applications. Other environmental test results proved to be within ±2% RH reproducibility showing an overall superior performance compared with conventional polymer-based humidity sensors.

3.4.6 Improvements Using MEMS Technology

When the capacitive-type humidity sensor is used for a long term under hot and humid atmosphere, it is known that the output of the sensor gradually

Table 3.9 Long-term stability of the fluorinated polyimide sensor element

Test environments	Sensor output shift (% RH)
Low temperature −40°C, 1000 h	±2
High temperature +80°C, 1000 h	±2
Temperature cycle (−40 to +80°C) 150 cycles	±2
Forst cycle 5 cycles	±2
40°C, 90% r.h., 1000 h	±3
Tobacco smoke (two cigarettes)	±2
Oil vapor 30 min	±2
Salt mist cycle (salt mist 2 h →40°C 90% RH. 22 h, 3 cycles)	±2

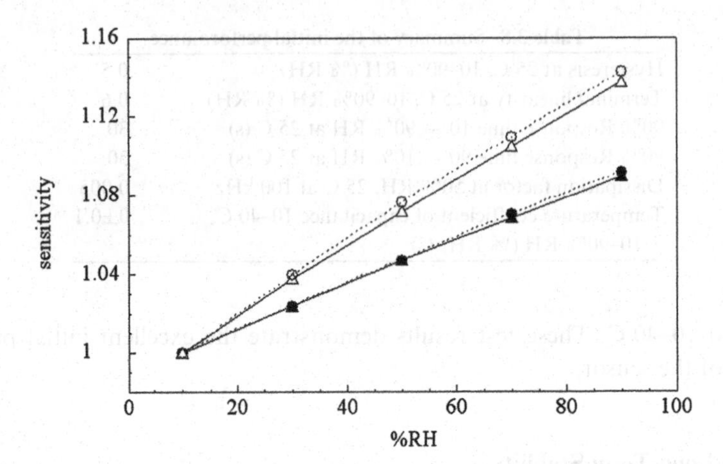

Fig. 3.37 Sensitivity before and after the exposure to 40°C and 90% RH atmosphere for two types commercially available sensors. *solid*; before the exposure, *broken;* after the exposure

increases. As shown in Fig. 3.37, when a sensor is exposed to 40°C and 90% RH atmosphere for a long time, sensitivity of the humidity sensor increases.

The research on this deterioration phenomenon has been done so far, and it was found that the performance shift is partially caused by the swelling of the sensing polymer film resulting in the change of the state of the sorbed water [12, 46, 47, 48, 49, 50]. To depress the deterioration and suppress the expansion of a sensing polymer film, the use of the structure having vertically tall interdigital microelectrodes using MEMS technology was considered.

3.4.6.1 Sensor Preparation

A new capacitive-type humidity sensor having interdigital microelectrodes is shown in Figs. 3.38–3.40.

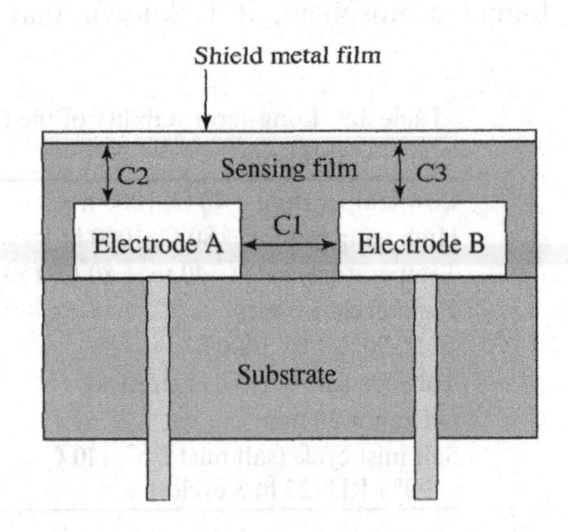

Fig. 3.38 Schematic illustration of the sensor structure

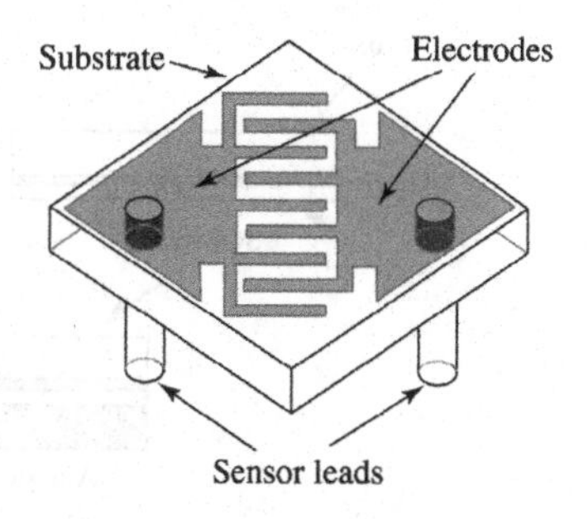

Fig. 3.39 Cross section of the sensor and capacitance components

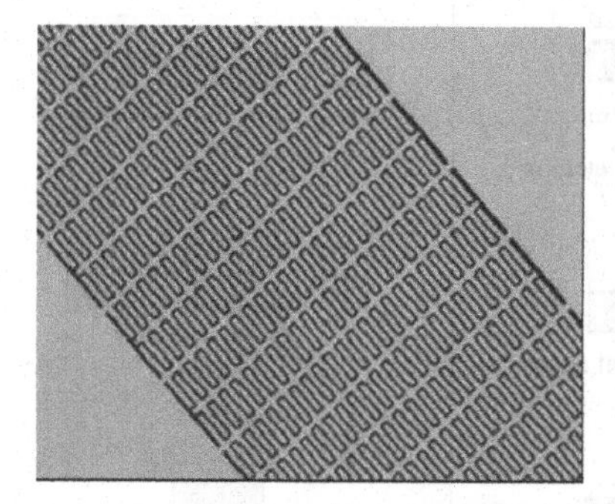

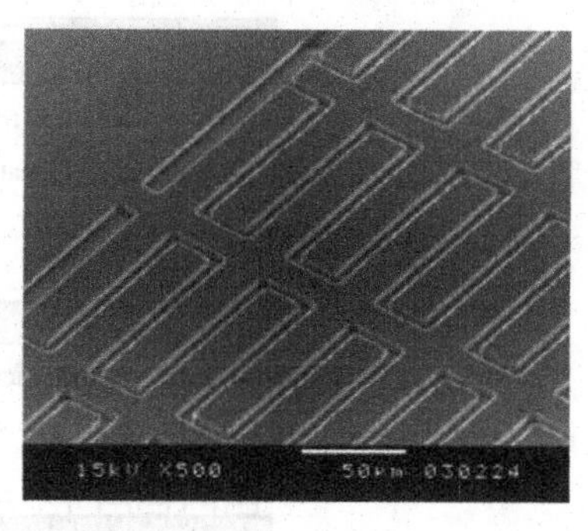

Fig. 3.40 SEM photograph of microelectrode patterns

The sensor was prepared for a prototype using MEMS technology such as the anodic bonding, photolithography and the ICP (inductively coupled plasma) etching, etc. [51]. To remove drift caused by water surface adsorption, shield, metal film was formed on the sensing film. The cross-linked polyimide was used for the sensor film as previously reported [47]. Figure 3.41 shows the manufacturing process flow.

First, SOI wafer and the glass substrate with holes are attached by the anodic bonding, and silicon is processed to a thin film by polishing and chemical etching. After SiO_2 is removed by chemical etching, the silicon electrode is formed with the ICP (Inductively Coupled Plasma etching) and the photolithography. Then polymers is coated and heat-treated onto the electrode, and sensor film is formed. Next, metal shield film is formed over the sensor film by vacuum evaporator. At the end, sensor leads are taken out after the dicing and dividing the wafer to the chips.

3.4.6.2 Sensing Characteristics

Figure 3.42 shows performance of the prototype sensor.

Sensitivity increased slightly before and after the exposure to 60°C and 90% RH atmosphere. Figure 3.43 shows the comparison of the drift performances between the microelectrode sensor and the conventional sensor.

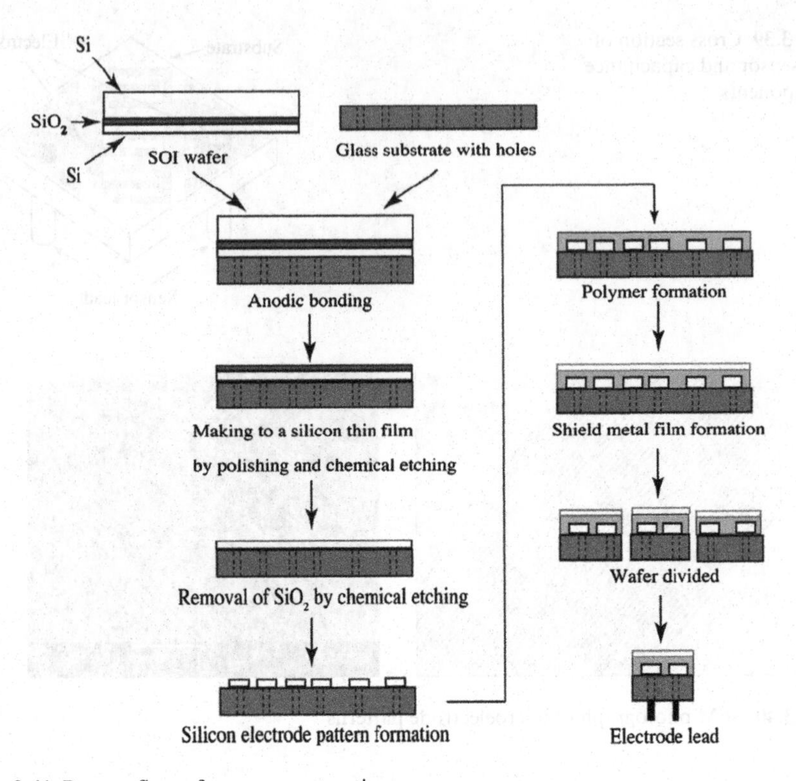

Fig. 3.41 Process flow of sensor construction

Drift of the microelectrode sensor has decreased about a half compared with one of the conventional sensor. This phenomenon is thought that the microelectrodes suppressed swelling of a sensing film. The sensor capacitance is formed with capacitance C_1, C_2, and C_3 (C_1: capacitance between electrode A and electrode B, C_2: capacitance between electrode A and shield metal film, C_3: capacitance between electrode B and shield metal film). The films of C_2 and C_3 swell in humid atmosphere, because they are not suppressed by microelectrodes. As a result, it is estimated that capacitance C_2 and C_3 cause the drift. Capacitance C_1 can be measured by grounding the metal shield electrode. The total capacitance of the sensor was almost 20 pF and capacitance, C_1, was almost 9 pF. Therefore, the ratio of capacitance C_1 and capacitance C_{23} in design is about 1:1 (C_{23} combined capacitance C_2 and capacitance C_3). The capacitive-type humidity sensor having a new structure was successfully made as prototype utilizing MEMS technology. Drift of new sensor has decreased about a half compared with of conventional type sensor after the exposure to hot and humid atmosphere.

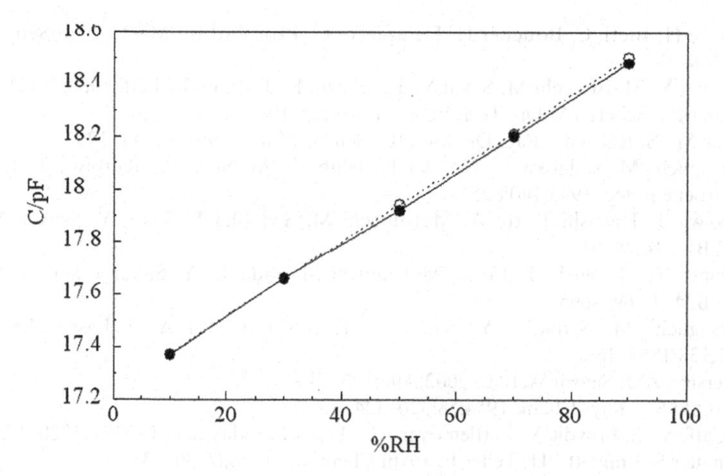

Fig. 3.42 Performance of prototype sensor before (solid) and after (broken) the exposure to 60°C and 90% RH atmosphere

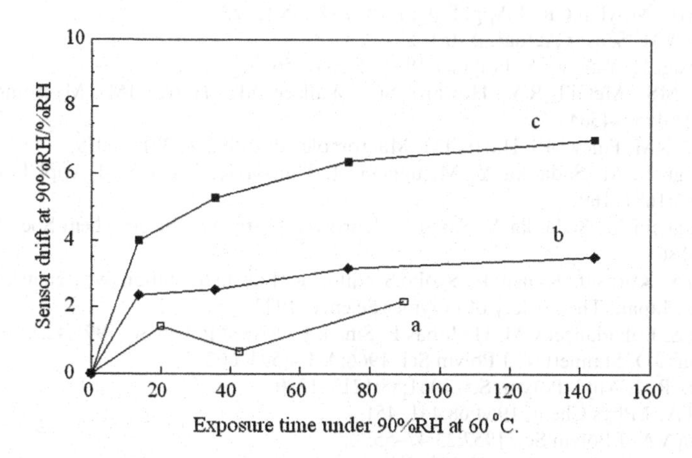

Fig. 3.43. Comparison of drift performance of sensors. a this work used MEMS, **b** sensor with the same polymer in this work and a simple sandwich-type structure, **c** commercial sensor with sandwich-type structure

References

1. Dunmore FW. J Res Natl Bur Stand. 1938;20:723.
2. Traversa E. Sensor Actuat. 1995;B23:135–156.
3. Sakai Y. Sensor Actuat. 1993;B13–14:82–85.
4. Hijikigawa M, Miyoshi S, Sugihara T, Jinda A. Sensor Actuat. 1983;4:307–315.
5. Sakai Y, Sadaoka Y, Matsuguchi M. J Electrochem Soc. 1989;136:171–174.
6. Sakai Y, Sadaoka Y, Matsuguchi M, Rao VL, Kamigaki M. Polymer 1989;30:1068–1071.

7. Grange H, Bieth C, Boucher H, Delapierre G. Proc 2nd Int Meet Chem Sens, 1986. pp. 368–371.
8. Sadaoka Y, Matsuguchi M, Sakai Y, Takahashi K. J Mater Sci Lett. 1988; 7:121–124.
9. Glenn MC, Schetz JA. Dig Tech Pap Transducer. 1985;85, 217–220.
10. Jaafar MAS, Ralsten ARK, Denton DD. Sensor Mater. 1991;32:111–125.
11. Matsuguchi M, Sadaoka Y, Nosaka K, Ishibashi M, Sakai Y, Kuroiwa T, Ito A. J Electrochem Soc. 1993;140:825–829.
12. Kuroiwa T, Hayashi T, Ito A, Matsuguchi M, Sadaoka Y, Sakai Y. Sensor Actuat. 1993;B13–14:89–91.
13. Kuroiwa T, Miyagishi T, Ito A, Matsuguchi M, Sadaoka Y, Sakai Y Sensor Actuat. 1995;B24–25:692–695.
14. Matsuguchi M, Sadaoka Y, Sakai Y, Kuroiwa T, Ito A. J Electrochem Soc. 1991;138:1862–1865.
15. Rittersma ZM. Sensor Actuat. 2002;A96:196–210.
16. Kurosaki S. J Phys Chem. 1954;58:320–324.
17. McCafferty E, Pravdic V, Zettlemoyer AC. Trans Faraday Soc. 1970;66:1720–1731.
18. Brunauer S, Emmett PH, Teller E. J Am ChemSoc. 1938;60:309–319.
19. Brunauer S, Deming LS, Deming WE, Teller E. J Am Chem Soc. 1940;62:1723–1732.
20. Brown GL. Clustering of water in polymers. In: Rowland SP, editor. Water inPolymers. Washington, DC: American Chemical Society; 1940. pp. 441–450.
21. Kawasaki K, Sekita Y. J Appl Phys Jpn 1957;26:678–680.
22. Bueche F. J Polym Sci. 1954;14:414–416.
23. Best ME, Moylan CR. J Appl Polym Sci. 1992;45:17–23.
24. Mecea VM. Sensor Actuat. A40:1–27.
25. Kusanagi H, Yukawa S. Polymer 1994;35:5637–5640.
26. Paley MS, McGill RA, Howard SC, Wallace SE, Harris JM. Macromolecules. 1990;23:4557–4564.
27. McGill RM, Paley MS, Harris JM. Macromolecules 1992;25:3015–3019.
28. Matsuguchi M, Sadaoka Y, Mizuguchi H, Umeda K, Sakai Y. J Appl Polym Sci. 1997;63:1681–1691.
29. Matsuguchi M, Sadaoka Y, Sakai Y, Kuroiwa T, Ito A. J Electrochem Soc 1991;138: 1862–1865.
30. Sekita Y. Kurasuta Kansu. In: Syobo S. editor, Kobunshi to Suibun (written inJapanese). Tokyo, Japan: The Society of Polymer Science; 1982., pp. 27–29.
31. Petrik S, Bohadanecky M, Hadobas F, Simek L. J Appl Polym Sci 1991;42:1759–1765.
32. Wellons JD, Stannett V. J Polym Sci. 1966;A-1 4:593–602.
33. Jeffries R. J. Appl. Polym. Sci. 1961;88:1213–1220.
34. Bondi A. J Phys Chem. 1964;68:441–451.
35. Kargin VA. J Polym Sci. 1957;23:47–55.
36. Turner DT. Polymer. 1987;28:293–296.
37. Atsuta M, Hirasawa T, Masahara E. J Jpn Soc Dent Appar Master. 1969;10:52.
38. Barton JM. Polymer. 1979;20:1018–1024.
39. Turner DT, Abell AK. Polymer. 1987;28:297–302.
40. Kalachandra S, Kusy RP. Polymer. 1991;32:2428–2434.
41. Baldwin MG, Morrow JC. J Phys Chem. 1962;36:1591–1593.
42. McCafferty E, Zettlemoyer AC. Discuss. Faraday Soc. 1970;52:239–254.
43. Matsuguchi M, Sadaoka Y, Shinmoto M, Sakai Y. Bull Chem Soc Jpn. 1994;67:46–51.
44. Matsuguchi M, Sadaoka Y, Nuwa Y, Shinmoto M, Sakai Y. J Electrochem Soc. 1994;141:614–618.
45. Bott RH, Taylor LT, Ward TC. ACS poly. Div Prep. 1986;27:72–73.
46. Ralston ARK, Klein CF, Thoma PE, Denton DD. Sensor Actuat. 1996;B34:343–348.
47. Matsuguchi M, Kuroiwa T, Miyagishi T, Suzuki S, Ogura T, Sakai Y. Sensor Actuat. 1998;B52:53–57.

48. Matsuguchi M, Hirota E, Kuroiwa T, Obara S, Ogurra T, Sakai Y. J Electrochem Soc. 2000;147:2796–2799.
49. Matsuguchi M, Takahashi Y, Kuroiwa T, Ogura T, Obara S, Sakai Y. J Electrochem Soc. 2003;150:H192–H195.
50. Matsuguchi M, Yoshida M, Kuroiwa T, Ogura T. Sensor Actuat. 2004;B102:97–101.
51. Ogura T, Sadaoka Y, Kuroiwa T. IEEJ Trans Sensor Micromach 2004;124:81–86.

Chapter 4
FET Gas-Sensing Mechanism, Experimental and Theoretical Studies

Anita Lloyd Spetz, Magnus Skoglundh, and Lars Ojamäe

The chemical gas sensor area has gained large improvements from the nanoscience technology, e.g., more reproducible processing including control on the nanoscale of annealing procedures, which implies improved long-term stability. Analytical tools have developed towards detection and investigation of ever smaller size phenomena. This has been of importance for the key problem of chemical gas sensors, the detailed understanding on the nanoscale level of the gas-sensing mechanism. In this chapter, which deals with FET (field effect transistor) gas sensor devices, we will review analytical tools that provide information about the detection mechanism with special emphasise on the FET sensor area. The DRIFT, diffuse reflection infrared Fourier transform, spectroscopy as a rather new and very important tool is reviewed. Theoretical modelling will speed up the process to provide further details in the mechanistic studies. Examples will be given in this chapter. A number of other important analytical tools will be briefly described.

4.1 Introduction

The field effect transistor device is one of the most common electrical components in electronics and therefore very attractive as a transducer for chemical sensors, providing standard mass production processing for example. Since the invention of first the ISFET, ion-selective field effect transistor, by Bergvelt in 1970 and the Pd gate Si-FET hydrogen gas sensor in 1975 by Lundström et al., the FET sensor area has generated a large amount of knowledge recently reviewed in Bergveld and Lundström et al. [16, 47], respectively. Other semiconductor materials like GaN or diamond have been introduced for ISFET devices and catalytic activity of enzymes and cell action potentials were recorded by AlGaN/GaN FETs [12, 75]. The FET gas sensor area was

A.Lloyd Spetz
Department of Physics, Chemistry and Biology, Linköping University, SE-581 83
Linköping, Sweden
e-mail: spetz@ifm.liu.se

E. Comini et al. (eds.), *Solid State Gas Sensing*,
DOI: 10.1007/978-0-387-09665-0_4,

expanded to the use of different catalytic metals, like porous metal layers of Pt and Ir, and other semiconductors, especially SiC, reviewed in [7, 47, 86]. Commercialisation are realised through spin-off companies like Adixen Sensistor AB, AppliedSensor AB and recently SenSiC for commercialisation of Si- or SiC-based FET gas sensors [3, 9, 72]. In this book chapter mainly the FET gas sensor devices are reviewed, However, the development of the fundamental understanding of the sensing mechanism of FET gas sensors have moved them closer to the ISFET devices, which is commented further in Section 4.3.

The studies of the detection mechanism of FET devices have improved considerably from the new improvements in spectroscopic methods by the introduction of nanotechnology in the solid-state gas sensor area. UHV, ultra high vacuum, studies may give basic information like activation energies, temperature for dissociation of molecules, basic differences between different catalytic metal surfaces. However, in order to get more details in the mechanistic puzzle, it is necessary to use as realistic conditions as possible like gas flows at atmospheric pressure and realistic temperatures. Of great importance is the possibility to analyse sensor surfaces during more sensing like conditions, and *in situ* studies during sensor operation, operando studies, is now introduced by some sensor groups.

Spectroscopic studies using scanning electron microscopy (SEM), transmission electron microscopy (TEM), and atomic force microscopy (AFM), have developed over time. Where sensing films deposited on special substrates had to be studied by TEM, it is now possible to use SEM studies on as mounted sensor devices. The influence of operation conditions can easily be studied. Mass spectrometry equipped with capillary has enabled studies at atmospheric conditions, which are of special importance to develop understanding on the nanoscale level of the sensing mechanism. Recently, studies using DRIFT spectroscopy have been very successful to reveal species present on the sensor surface during gas detection. Since we think this method has special potential it will be one of the focus points in this chapter. Furthermore, theoretical modelling, a powerful method to speed up the process of revealing mechanistic details, will gain large attention in this chapter.

4.2 Brief Summary of the Detection Mechanism of FET Devices

Different FET devices, Schottky diodes, capacitors and transistors, have been employed as gas sensors (see Fig. 4.1). The Schottky diode is processed by the least number of process steps of the three and is operated by similar simple electronics as for the transistor device. Both Schottky diodes and transistors are operated in the constant current mode, for which the DC gate voltage change is reflected in the response of the output signal (see Fig. 4.2). The response of the Schottky diode may not always be compared to the response of a transistor device. In the Schottky diode the gas species that react with the sensing layer

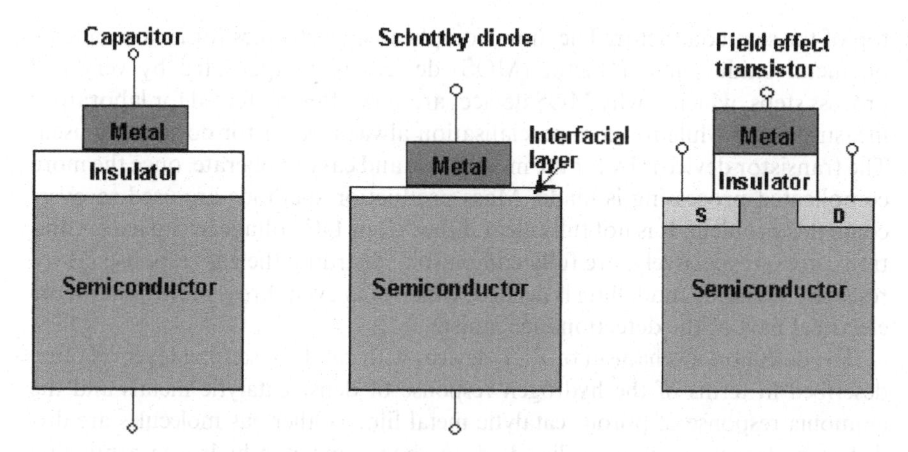

Fig. 4.1 Schematic figure of FET devices

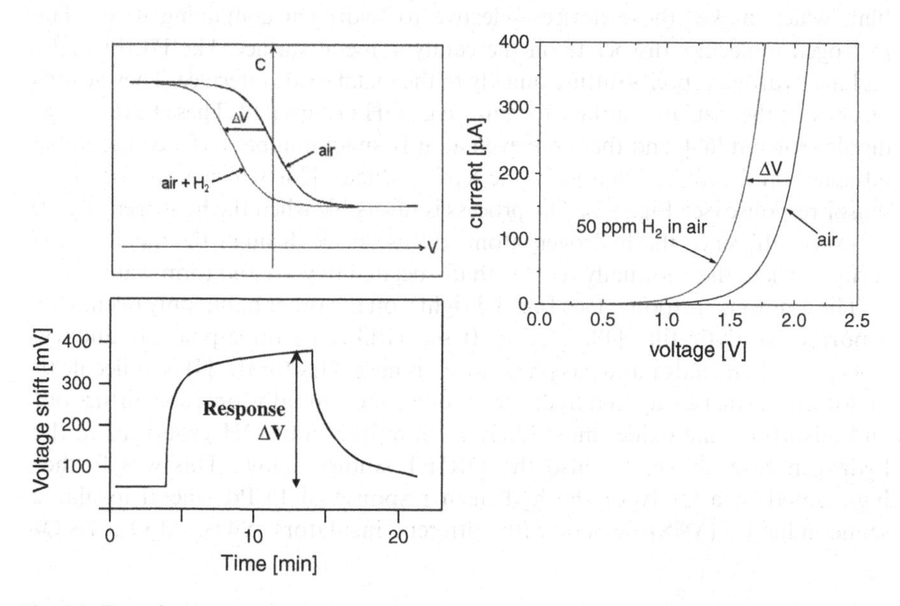

Fig. 4.2 Capacitance voltage, CV, curve, current voltage, IV, curve and the sensor signal, the voltage, at a constant capacitance or constant current

may induce both a resistivity change and a field effect. If the gate contact of the Schottky diode has large resistivity, this part of the response may be larger than the induced field effect [53]. The capacitor on the other hand (see Fig. 4.1), only give a response to gas species that give rise to field effect changes of the gate metal, which makes the response, i.e., the drop in the capacitance voltage, C-V, curve directly comparable to that of the transistor device. The drawback of the capacitor is the AC operation, which puts high requirements on both the conductivity of the gate metal and the insulating properties of the insulator on top of the semiconductor. The more complicated electronics for measurements of, metal oxide semiconductor (MOS) devices is compensated by very few process steps, which is why MOS devices are very often preferred for laboratory measurements while for commercialisation always transistor devices are used. The transistor device is by far the most stable and easy to operate, once the more complicated processing is made. Mass production methods are used to overcome this problem. It is not fully clear if the AC or DC voltage in capacitors and transistors, respectively, are fully compatible regarding the gas response. More research including modelling is needed, which may even throw more light on the electrical part of the detection mechanism.

The detection mechanism of FET devices with catalytic sensing layers is often described in terms of the hydrogen response of dense catalytic metals and the ammonia response of porous catalytic metal films. Other gas molecules are divided in hydrogen containing, like hydrocarbons, and non-hydrogen-containing gases like CO or NO/NO_2. Only hydrogen atoms diffuse through a dense catalytic film, which makes those devices selective to hydrogen containing gases. The hydrogen molecules dissociate on the catalytic metal surface, like Pd, Pt or Ir, and the hydrogen species diffuse quickly to the metal oxide interface. They adsorb on sites on the insulator surface forming, e.g., OH groups [80]. These have a large dipole moment [68], and therefore even a fairly small number has a considerable influence on the mobile charges in the semiconductor [28], which gives rise to the sensor response (see Fig. 4.2). The process is reversible when the hydrogen supply is turned off, since the hydrogen atoms diffuse back through the metal to the surface, where they normally react with dissociated oxygen and form water.

The ammonia response (see Fig. 4.3 right), on the other hand, only occurs for a porous catalytic film [49, 73, 74]. It seems like the three-phase boundary, where metal, insulator and gas phase meet, is necessary for the NH_3 molecule to dissociate in such a way that hydrogen atoms, or eventually ions, can diffuse out and adsorb on the oxide, most likely forming the same OH groups as in the hydrogen case above, see also the DRIFT studies below. This was further highlighted in a study of the hydrogen response of Pt/Pd metal insulator semiconductor (MIS) devices with different insulators, SiO_2, Al_2O_3, Ta_2O_5

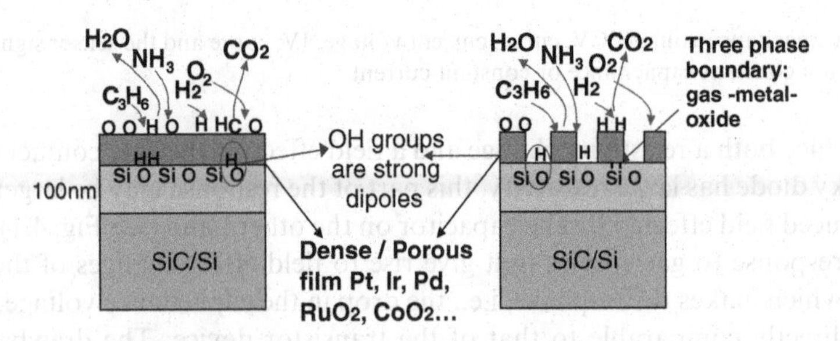

Fig. 4.3 The gas response of a FET with a dense metal (to the *left*) and a porous metal (to the *right*)

and Si_3N_4, which scaled towards the oxygen concentration of the insulator surface [26]. It was also shown that even for the Schottky diode an insulator is necessary for the gas sensing capabilities of the device. The quality of the interfacial layer of 1 nm native oxide of SiC Schottky diodes was studied by ellipsometry [89]. Also, for Schottky diodes based on Pd-GaN, a thin oxide considerably improved the hydrogen response [84]. It has also been shown by the SLPT technology, described below, that the hydrogen atoms or ions, decomposed from NH_3, diffuse both on the oxide surface of the exposed oxide outside the metal but also to some extent diffuse in under the metal from the three-phase boundary [48]. This is easier to understand in view of other recent results, which show that for a metal deposited onto an oxide the nanostructure of the metal surface facing the oxide influences the size of the response to hydrogen [93]. It was also shown that a catalytic metal deposited onto the oxide, e.g., at a high temperature, which makes the film stick very well to the surface, exhibits a very low response to hydrogen [93]. Also, the detection of ammonia is reversible via the metal surface, since the hydrogen complexes, OH groups, decompose and hydrogen forms water with oxygen most likely via the metal surface. The presence of OH groups on the insulator surface in the FET devices makes the similarity to the detection mechanism of ISFET sensors apparent. The catalytic metal in the FET gas sensor may even be regarded as an integrated reference electrode in an ISFET device: It has also been suggested that oxygen plays a role in the detection mechanism [54, 70].

Many studies have been performed to reveal the details in the detection mechanism and several of these are reviewed below. However, still numerous details are to be investigated which is important for the possibility to process reproducible sensor devices with tailor made properties.

4.3 UHV Studies of FET Surface Reactions

Studies in ultra high vacuum using techniques like ultraviolet photoelectron spectroscopy (UPS), electron energy loss spectrometry (EELs), and x-ray photoelectron spectroscopy (XPS), have generated basic information about solid-state gas sensors like adsorption and dissociation probabilities of molecules and the dependence of surface structure and temperature, adsorption, desorption and reaction energies, and the energies of active surface sites. However, there has always been a problem to relate the results to real operation conditions for sensors. Therefore, the development of methods like the DRIFT spectroscopy and mass spectrometry for atmospheric pressure, as described later on in this chapter, have considerably improved the knowledge about detection mechanism of solid-state gas sensors. Some UHV investigations of FET sensor surfaces are reviewed here.

Sulfur contamination of a Pt-gate SiC field effect gas sensor was studied in UHV at 527°C [32]. Hydrogen sulfide showed a serious sulfur poisoning effect of the Pt surface reducing the response due to a change between hydrogen and oxygen atmosphere by 70%. Oxygen exposure fully restored the sensor surface and the gas-sensing capability, except for at very high sulfur coverages.

The hydrogen responses of Pd-SiO_2-Si and Pt-SiO_2-Si MIS devices were compared in UHV studies [68]. Qualitatively the results agree but there are quantitative differences. The hydrogen sensitivity of the Pd device showed to be two third higher as compared to the Pt device. It was shown that the number of active sites at the sensor interface – see the detection mechanism as outlined above – was about three times larger for Pd, 2×10^{18} m^{-2} as compared to $7\times10^{17} m^{-2}$ for Pt. The dipole moment of the polarised hydrogen species at the interface was estimated to 0.6–0.7 D.

Another interesting very sensitive surface analysis tool is XPS by which the chemical nature of adsorbates may be revealed. Here, one example of a metal oxide sensor component is given. Tin dioxide and copper oxide composite films showed high sensitivity regarding resistance change to hydrogen sulphide (> 10^4 for 50 ppm H_2S) [40]. The films were thermally evaporated in a three layer stack of Sn-Cu-Sn, which was oxidised at 250°C for 30 min followed by 800°C for 60 min in an oxygen atmosphere. The obtained SnO_2-CuO composite sensors were very selective to H_2S, a large number of different gases were tested but there was no response except for the response to H_2S. The sensors were prone to drift, degraded into lower sensitivity and slower response at long-term operation in industrial environment. XPS showed that due to interaction with moisture $Cu(OH)_2$ was formed, which is responsible for the drift. Either exposure to a high concentration of H_2S (>100 ppm) or annealing at > 400°C could restore the sensor.

Palladium-nickel and palladium-silver alloys were studied in UHV since they are interesting as sensing layers for high hydrogen concentrations in hybrid suspended gate field effect (HSGFET) devices [71]. Films were produced by co-evaporation in a UHV chamber. The Pd-Ni alloy showed to be stable, without any blister formation, for hydrogen concentrations up to 2% even at room temperature, with a corresponding sensor response in dry air of 500 mV. Humid carrier gas reduced the sensor response but speeded up desorption time. On the other hand, the Pd-Ag alloy was not stable as a sensing material for high hydrogen concentrations.

It was shown in UHV studies that CO has no direct influence on the CV characteristics of a Pd-MOS device, but that CO, on the other hand, influences the hydrogen response of the device [25]. The measurements are performed at 187°C and RT using EELS and mass spectrometry. Two different phenomena are tentatively assigned to the influence on the hydrogen response: (a) CO efficiently consumes oxygen, which in turn gives an oxygen free surface very sensitive to extremely small concentrations of hydrogen like the background hydrogen pressure; (b) at coverages of about 0.3 ML (monolayers) CO starts to block the hydrogen dissociation as well as association. Therefore, the hydrogen response is inhibited and hydrogen may even be locked into the structure by an adsorbed layer of CO. The CO acted like a lid in the sensor device for a low operating temperature, here room temperature (RT), locking in already adsorbed hydrogen at the interface. The time needed to reach equilibrium increases for the CO blocked surface. It seems like the activation energies are increased for both dissociation and association of hydrogen. These results are further investigated in the Section 4.7.

Studies in UHV using UPS, EELS and work function measurements, of the Pd-MOS device with silicon as the semiconductor, i.e., a capacitor device, was performed during exposure to oxygen, hydrogen, nitric oxide and carbon monoxide over the temperature range 323–523 K [20]. Oxygen and NO and of course H_2 dissociates on the sensor surface in this temperature range while CO does not. The C-V signal shows the similarity between NO and O2 exposure, see Fig. 4.4, while only a very small influence is seen from the CO exposure. It is concluded that it is the interplay with background hydrogen in the UHV chamber that give rise to the sensitivity to oxygen of the C-V signal. Since the Pd-MOS structure is extremely sensitive to hydrogen over a large pressure range (10^{-11} – 20 Torr), i.e., a Tempkin adsorption isotherm [see ref. in 47], the background hydrogen will always fill some adsorption sites at the metal oxide interface.

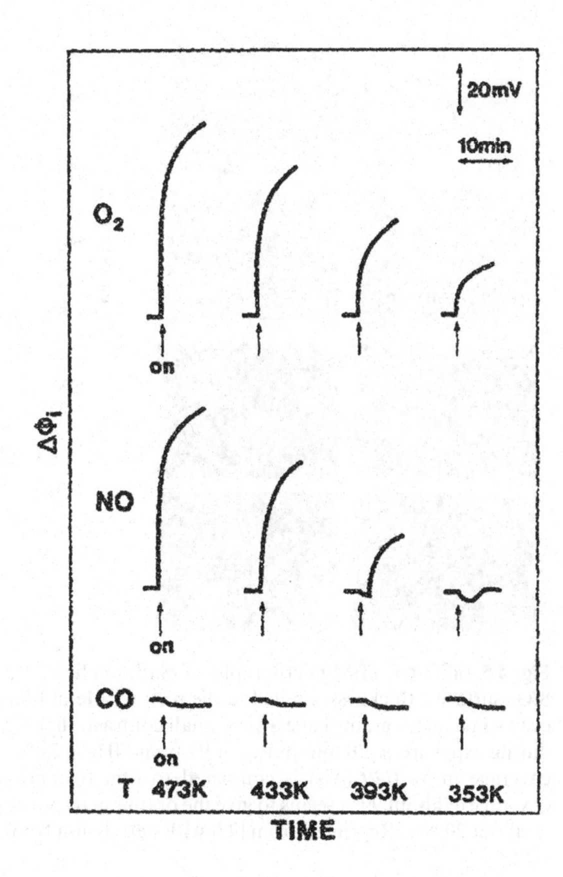

Fig. 4.4 Measurements of the CV response due to CO, NO, and O_2 exposure of the Pd-MOS structure at different temperatures in UHV. Reprinted with permission from [20]. Copyright [1989], American Institute of Physics

4.4 TEM and SEM Studies of the Nanostructure of FET Sensing Layers

Scanning and transmission electron microscopy is of course of large importance for mechanistic studies of solid-state gas sensors. The performance of these techniques has improved a lot over the years. Almost 20 years ago, when the ammonia sensitive silicon based Pt and Ir MOS devices were investigated to find the origin of the sensitivity to ammonia, special TEM substrates had to be developed for this purpose. TEM membranes of SiO_2 or Si_3N_4 were produced by etching a silicon wafer from the rear side. The ammonia sensitive Pt or Ir films were simultaneously deposited onto these TEM substrates and on the gas sensor surfaces. It was revealed that even Pt or Ir films that looked as dense metals for the eye had a large porosity coupled to the ammonia sensitivity [74] (see Fig. 4.5a). The size of the gas sensor response to NH_3 was correlated to the porosity through the size of the area of the pores in the films (see Fig. 4.5b). It

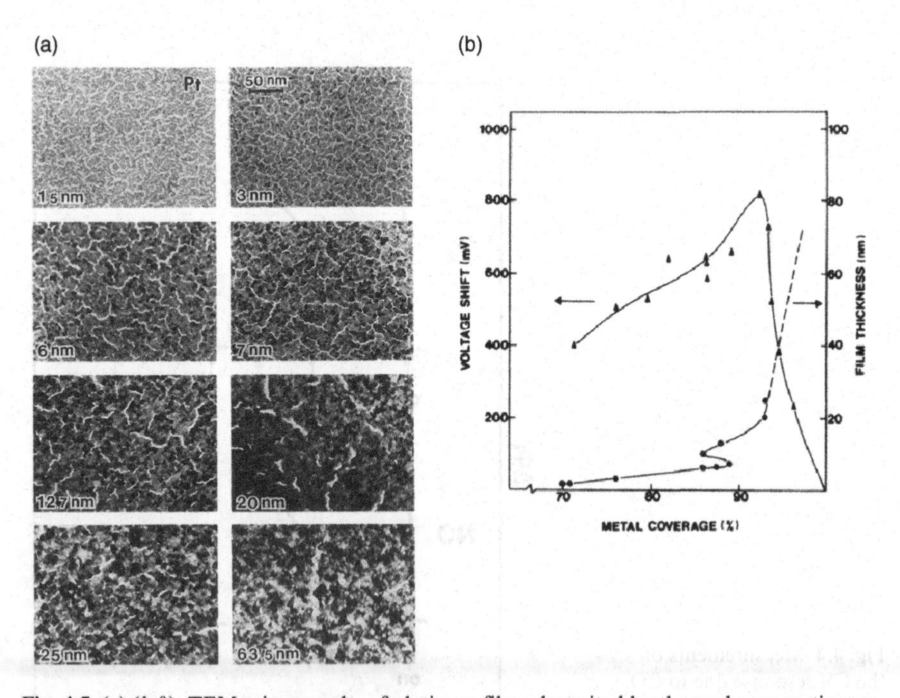

Fig. 4.5 **(a)** (*left*). TEM micrographs of platinum films deposited by thermal evaporation on SiO_2 surfaces, thicknesses noted in the picture. Metal film, dark colour. In the 63.5 nm film even some metal grains have a very small contrast. **(b)** (*right*). Pt-MOS sensor measurements during exposure to 10 min pulses of 97 ppm NH_3 in 20% O_2/Ar. The response versus metal coverage in % (*left* axis) is compared to film thickness versus metal coverage. A metal coverage of about 90% seems to give the optimum response and that relates to a film thickness of about 20 nm. Reprinted from [75] with permission from Elsevier

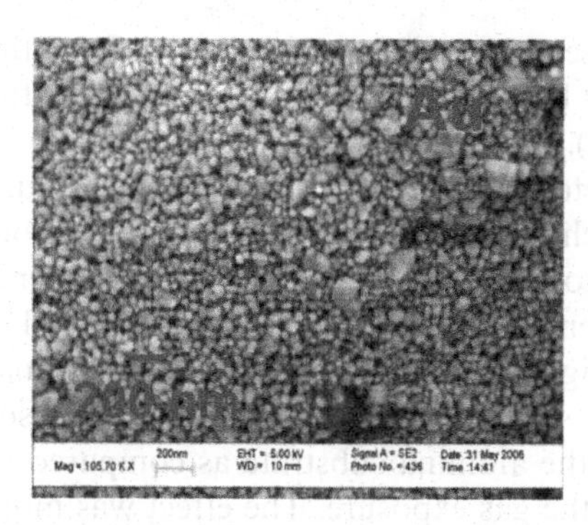

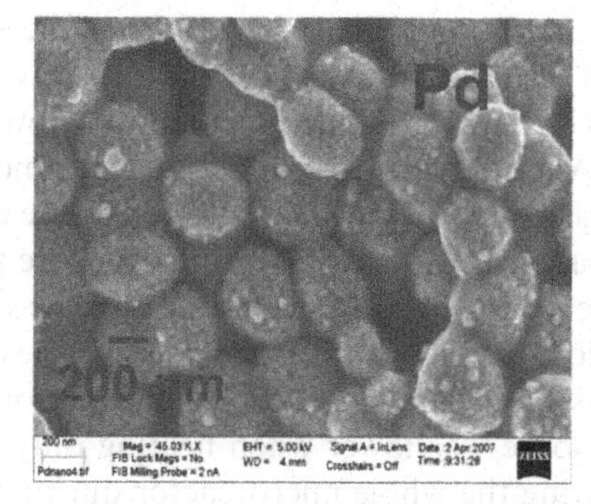

Fig. 4.6 SEM image of Au nanoparticles annealed at 200°C (to the *left*) and Pd nanoparticles annealed at 300°C (to the *right*)

was also possible to study the influence on film structure of different gas exposure and annealing procedures.

Later the SEM technique improved in resolution and it was possible to study the nanostructure of catalytic metals on FET devices even on already mounted samples before and after annealing and operation of the devices. For example, SEM images of Pd and Au nanoparticles used in Si capacitor devices showed the agglomeration of the particles due to annealing at 300 and 200°C, respectively (see Fig. 4.6)[17]. For the Pd nanoparticles an interesting fractal structure is revealed, the 2–300 nm grains are built up of nanoparticles a few nanometres in size.

4.5 Mass Spectrometry for Atmospheric Pressure Studies

The use of capillaries for gas sampling has made it possible to perform mass spectrometry analysis at atmospheric pressures. The first studies on FET gas sensors revealed very important phenomena regarding mass transport limitation during exposure of FET devices to test gases [2]. It was shown that a large catalytic surface may consume more gas molecules in its vicinity than the amount which diffuse into the sensor surface from the surrounding gas atmosphere [1]. Later it was shown that a square catalytic surface larger than 0.25 mm^2 and a reactive sticking coefficient higher than 10^{-4} reduced the sensor response significantly [39]. The studies have been expanded to three dimensions [38].

The capillary equipped mass spectrometer has also been used to study the production of gas species from chemical reactions on chemical gas sensor surfaces during operation. For example, the detection mechanism of CO on silicon carbide based FET devices was studied using simultaneous gas sensor measurements and mass spectrometry, see Fig. 4.9 [6]. The influence of temperature (100–400°C), oxygen and hydrogen on the CO response was studied in detail. The studies confirm the influence of back ground hydrogen when CO efficiently consumes oxygen from the sensor surface, as suggested from the UHV studies above. These studies have been combined with results from the DRIFT spectroscopy (see section below).

Another version of a mass spectrometer for atmospheric pressure was built by a group in Hungary [14]. Resistance changes of $SrTiO_3$ thin film gas sensors on alumina substrates during exposure to ethanol and acetone in the temperature range 25–900°C were studied. The sensor measurements were performed in a flow through quartz microreactor designed also for mass spectrometry analysis by a capillary close to the sensor surface. They found that the sensor response was lower when heating only the alumina substrate as compared to heating the whole microreactor during the gas exposure. The effect was more pronounced for acetone than for ethanol.

4.6 The Scanning Light Pulse Technology

It is, of course, very attractive to find analytical methods to study the species assumed to spill out onto the oxide surface in the vicinity of the metal oxide interface in FET devices, the phenomena described above. One method that has been used for this is the scanning light pulse technology, SLPT, [33]. This method makes use of a light pulse, which, when focused onto the surface of a MIS device, generates electron–hole pairs in the depletion layer of the semiconductor, which in turn give rise to a current in an external circuit. The current generated by the pulsed light beam is proportional to the width of the depletion region in the semiconductor. Since the charges produced by gas molecules that reacts with the catalytic metal in a FET gas sensor device modulates the size of the depletion region, the SLPT technology can be used to study the gas response of FET devices. It is possible to study gas interaction with a thin metal film, which is enough transparent for light. The SLPT has also been used to study gas-induced phenomena directly on the insulator in the vicinity of the metal. The measurements indicated that positive charges are spread on the oxide outside a metal contact for a positive bias on the contact. Two different species were observed, one "fast" and one "slow" mover on the oxide. It was the slow process that dominated during the spill-over investigations. It was concluded from calculations that even if only a small fraction of the sites on the oxide were occupied, this has a large influence on the charges in the semiconductor. This may be due to the spill-over hydrogen forming OH groups as identified by the DRIFT spectroscopy below, and these OH groups exhibit strong dipole moments. Later SLPT measurements indicated that hydrogen diffuses both out onto the exposed oxide surface but also in under the metal, probably into microvoids or even nanovoids under the metal (see above and [48]).

The SLPT method has several other interesting advantages like the possibility to measure a large continuous surface with, e.g., a gradient in metal thickness or temperature [49]. In this way sensing layers of varying properties in two dimensions can quickly be evaluated [41].

4.7 DRIFT Spectroscopy for In Situ Studies of Adsorbates

Time resolved DRIFT spectroscopy, offers an excellent tool to identify molecular species adsorbed on gas sensor surfaces in operando/in situ studies or operation like conditions [44]. Molecules in gas phase can be separated from molecules on the gas sensor surface and different adsorption sites on the metal or on the insulator can be identified. Below we review a major part of the work performed in this area.

The first publication found using DRIFT spectroscopy for studies of the detection mechanism of gas sensors is from 1996 [15]. Simultaneous conductance and DRIFT spectroscopy measurements of a CdGeON oxygen sensor were performed in synthetic air or nitrogen atmosphere. The CdGeON powder was screen printed on an Al_2O_3 support with Au contacts for conductivity measurements and a Pt coil was screen printed on the backside for heating and temperature control. The DRIFT spectroscopy revealed incorporation of oxygen ions in the CdGeON framework as the reason for the change in conductance. There were two suggestions for the conductance mechanism: (i) moving of the un-ionic vacancies through the solid and blocking of the vacancies by oxygen (ii) the extra oxygen ions in the network modifies the electronic band structure and the band gap.

Barium carbonate has been suggested as gas-sensing material for work function based CO_2 gas sensors [58]. Kelvin probe measurements were used to measure the influence of CO_2 of screen printed films on alumina substrates with a Pt heater on the backside. A large range of carbonates was investigated and $BaCO_3$ showed to be the superior material. For detection of CO_2 (0.003–5%) in wet synthetic air, optimum sensitivity was found at 50°C. However, already the response at room temperature was high enough (50–150 meV). The authors concluded dimeric HCO_3^- species ($Ba(HCO_3)_2$), identified by DRIFT spectroscopy, as the cause of the work function change. Adsorbed water activates the surface reaction and the response to CO_2 disappears in a dry atmosphere.

Kelvin probe was also used to investigate TiN in sputtered films on the same alumina substrates as above or with Si wafers as the substrate [59]. The TiN functions as a work function changing sensing material for ammonia in humid atmosphere. Exposure to ammonia (400 ppb–100 ppm) leads to a reversible work function change at room temperature with a sensitivity of –30 meV per decade of ammonia concentration. The sensitivity decreases by increasing temperature as investigated up to 100°C and the cross-sensitivity to humidity is low. DRIFT studies showed that adsorption of ammonia correlates with a decrease of OH species on the sensor surface. It is suggested that the TiN may be used in FET devices for ammonia sensing at room temperature.

For SnO_2, probably the most studied sensor material in the gas sensor community, DRIFT spectroscopy has been used both to understand the sensitivity to NO and NO_2, and to CO. For the former case TPD, temperature programmed desorption, was combined with the DRIFT studies [46]. At room temperature monodentate nitrate species were identified on the SnO_2 surface during exposure to NO_2. Above 100°C some of the nitrate groups decompose to NO_2 whereas the remaining groups are reorganised to more stable bidentate nitrate species. The most stable nitrate groups remain on the surface up to 500°C, which is beyond the operation temperature for these devices as NO_x sensors. Therefore, the presence of nitrato species on the SnO_2 surface is assigned to the change of the resistivity, which constitutes the sensor response. It was also shown that in the presence of surface hydroxyl groups, hydrogen nitrates are formed.

The method has also been employed for metal oxide gas sensors by Emiroglu et al. [24]. The adsorption of CO was studied on model sensor surfaces of SnO_2 powder, by DRIFT spectroscopy at room temperature. In situ measurements showed that CO reacts with adsorbed water and surface OH groups, and that surface carbonate ions dissociate due to acidic intermediate products on the surface. Covalently bound chelating and bridging bidentate carbonates and carboxylates were identified on the surface [31].

In order to correlate the DRIFT results with the electrical measurements of SnO_2 devices as CO sensors, simultaneous DRIFT and electrical measurements, operando studies, at increased temperature were performed [44]. The operando DRIFT and gas-sensing studies were performed on screen printed SnO_2 sensors on alumina substrates with Pt electrodes and heaters annealed at two different temperatures, 500 or 700°C. It was shown that an increased annealing temperature increases the concentration of oxygen vacancies at the sensor surface, which in turn most likely increases the amount of adsorbed anionic oxygen species, O_2^- and O^-. Oxygen vacancies are considered to be the sites for these oxygen surface species.

At a certain operation temperature and oxygen pressure the ratio O_2^-/O^- is constant. Furthermore it is concluded that O^- is more active than O_2^-. From the combined experimental data of DRIFT studies and sensor measurements in this study it is suggested that CO annealed at 500°C interacts predominantly with O_2^- and CO annealed at 700°C interacts predominantly with O^- surface species. This is thus suggested as the origin of the electrically measured resistance changes during CO exposure of these sensors.

A specific study of the water–oxygen interplay on the tin dioxide surface was also performed at 330 and 400°C [43]. The reactions on the sensor surface were monitored as the change in surface species by DRIFT spectroscopy and simultaneous conductance change sensor measurements. The DRIFT studies gave experimental evidence for the suggested reaction between pre-adsorbed O^- ions and water forming Sn-OH groups and release of an electron to the conduction band. This can also explain that the sensitivity to CO is generally higher in dry conditions for the SnO_2 sensor device.

The CO sensing mechanism of Pt-, Pd-, and Au-doped SnO_2 sensors in H_2-rich atmospheres was investigated by DRIFT spectroscopy [88]. These atmospheres relate to reforming processes in H_2 production for fuel cell applications. Pt-doped SnO_2 showed highest sensitivity and at 220°C also the cross-sensitivity to H_2O, H_2, CO_2 and MeOH was very low. The DRIFT spectra were recorded at 110°C and water, OH-groups, methoxy groups and carbonates were identified on the sensor surface. Due to the result of the DRIFT spectra and support by numerical simulations based on simple Langmuir-type competitive adsorption a detection mechanism is suggested. The CO exposure results in a rapid increase of the formate coverage, which then is mainly responsible for the increase in resistance during CO exposure, since it overrules the reducing effect of hydrogen.

Chromium-doped WO_3 nanochrystalline powders were investigated as ammonia sensing material [37]. Consumption of oxygen gives the change in resistivity due to ammonia. DRIFT and TPD (temperature programmed desorption) studies were applied to chromium-doped or pure WO_3 powders. In the presence of chromium it was shown that ammonia is oxidised to nitrogen and nitrous oxide on chromium centers, and that the transition occurs to a higher extent as compared to on the WO centers. That means that the chromium sites are more catalytically active, which is favourable for the magnitude of the sensor response to ammonia.

The CO_2 sensing mechanism in LaOCl nanopowders was investigated by simultaneous DRIFT studies and sensor measurements [51]. These studies were combined with PLS (partial least square) data processing and CO_2 TPD experiments. This showed that probably both hydroxycarbonates and bicarbonates are present on the sensor surface during the gas detection and that adsorbed OH^- groups play a role in the sensing mechanism.

Model sensor surfaces consisting of metal oxide powder, SiO_2, which is the top insulator layer in the FET gas sensor devices, were used for DRIFT studies of FET gas sensors. The silica is impregnated by catalytic metals, Pt or Ir [80–82]. Studies have been performed during exposure to H_2 and NH_3 over a temperature interval of 50–300°C. The adsorption at 50°C of NH_3 on Pt-SiO_2 powder showed adsorption on both platinum sites and silanol groups on the SiO_2 surface. When the temperature was increased to 150°C ammonia remained adsorbed on platinum but desorbed from the silanol groups. Thus the adsorption on the metal is stronger than on the silanol groups. At 150°C an increased amount of NH_2 and gas phase N_2O appeared showing that NH_3 starts to oxidise in oxygen at this temperature and that in turn improves the NH_3 dissociation. The results also strongly indicate that spillover of hydrogen to the insulator surface occurs during the gas exposure for both hydrogen and ammonia. During the H_2 or NH_3 exposure of the model sensor surfaces, Pt/SiO_2 or Ir/SiO_2, new OH groups as well as perturbed OH groups appeared on the insulator surface. No such changes of the OH group concentration were observed without the Pt or Ir metal on the model sensor surface. It was also shown for H_2 exposure that the process of OH group formation is reversible. The results were confirmed by exposure to D_2 [81]. Density functional theory (DFT) calculations were used to support the adsorption band assignment in the DRIFT spectra. Furthermore, these calculations were used to model the adsorption of NH_3 and NH_2 on a Pt_4 cluster (see Fig. 4.7). This now supports

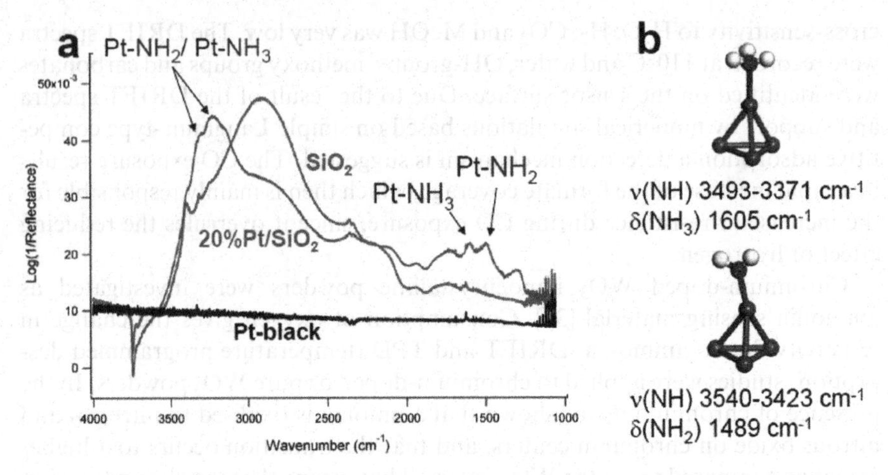

Fig. 4.7 (**a**) DRIFT spectra of model sensors, SiO_2, 20% Pt/SiO_2 and Pt in situ recorded during 1000 ppm NH_3 exposure in air at 50°C. (**b**) Geometry optimized structures from DFT calculations of the adsorption of NH_3 (*upper*) and NH_2 (*lower*) on a Pt_4 cluster

the model for the detection mechanism of FET sensors to H_2 and NH_3 locating the detected species to the insulator surface as described in Section 1.2. It was also shown for the Pt-SiO_2 surface that the density of formed OH groups followed the coverage of Pt of the SiO_2 surface. This is interpreted such that the OH groups are actually formed in the vicinity of the metal oxide border. This shows that more details are left to reveal in the jigsaw puzzle of the detection mechanism of the FET devices to H_2 and NH_3. It even suggests that electrostatic forces may be into play, and this should also be further studied.

The detection mechanism of FET devices towards non-hydrogen-containing gases like CO and NO/NO_2 is of crucial interest both due to the practical interest to monitor these gases, which are of great environmental concern, and due to fundamental reasons. To reveal details in the detection mechanism of gases that do not contain hydrogen may also give more general insight in the gas detection mechanism of FET gas sensor devices. DRIFT studies were performed for CO exposure of Pt-SiO_2 model sensors with and without hydrogen exposure [13]. This was compared to similar, but not yet simultaneous, sensor measurements [6]. The sensor sensing studies were combined with mass spectrometry [13]. The DRIFT spectroscopy when exposing the model sensor to CO at 100°C showed that the Pt crystallites were quickly covered by CO even at the lowest concentration of CO and very small or no CO_2 production occurred, at this low temperature. Nor did hydrogen exposure show any effect at this temperature.

The DRIFT spectra recorded at 150°C shows that CO adsorbs on Pt both linearly and bridge bonded, see Fig. 4.8. Now exposure to hydrogen reduces these peaks and after about 5 min they have disappeared completely. These studies were compared to similar sensor measurements of porous Pt-gate

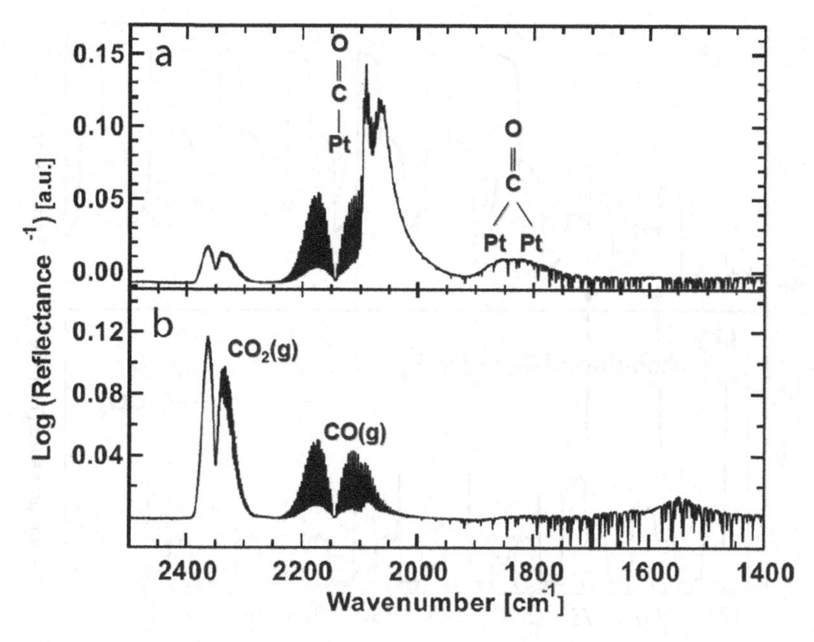

Fig. 4.8 In situ DRIFT spectra of $PtSiO_2$ model sensor at 150°C exposed to (**a**) 250 ppm CO in 5% O_2 for 5 min and (**b**) 5 min after the addition of hydrogen to the gas flow in (**a**)

capacitors simultaneous with mass spectrometer measurements of the gas stream that had passed the sensors. Comparison was made to sensor measurements performed at 175°C, since, due to experimental conditions, most likely the temperature on the sensor surface is lower by about 10–20°C. Exposure to 125 ppm of CO gives a low sensor signal and high CO_2 production and the DRIFT spectra show low CO coverage on the sensor surface. Addition of hydrogen to the exposure of the sensor surface decreases the CO_2 production, as measured by mass spectrometry, and increases the sensor signal. It seems like the CO_2 oxidation is inhibited by H_2. Increasing the CO exposure to 250 ppm increases the sensor response and decreases the CO_2 production and now addition of hydrogen has the effect of decreased sensor response and increased CO_2 production, see Fig. 4.9. This indicates strongly that hydrogen breaks the self-poisoning of the CO oxidation.

At 200°C the oxidation rate, as studied by the mass spectrometer increases, as could be expected and the CO coverage on the Pt surface decreases even further. More details from these studies are found in the reference [13].

Regarding the sensing mechanism, the DRIFT spectroscopy studies show an increase in CO coverage of Pt for an increase in sensor response. However, the combined DRIFT spectra, sensor response and mass spectrometry data from the simultaneous hydrogen and CO exposure point towards an indirect response mechanism, as discussed in Section 1.3. Oxygen removal seems less likely, while background hydrogen, which can diffuse to the active sites on the

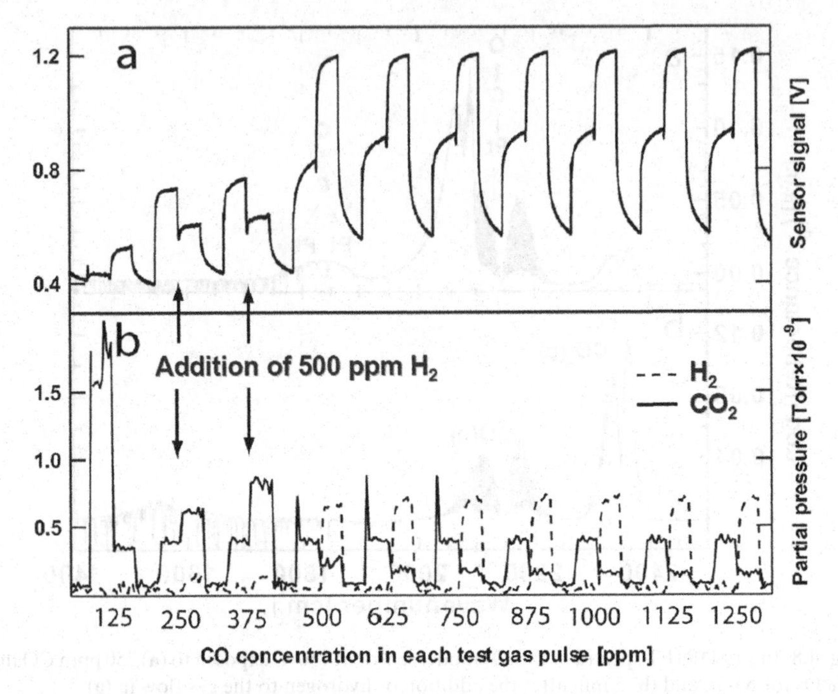

Fig. 4.9 Field effect capacitor device with a porous Pt contact as the sensing layer. (**a**) The sensor signal during exposure to CO in air with addition of H_2 (**b**) The simultaneously recorded mass spectrometer signals for H_2 and CO_2

oxide surface when the water production is inhibited by CO coverage of the Pt surface, appears as the most likely detection model.

4.8 Atomistic Modelling of Chemical Reactions on FET Sensor Surfaces

A versatile technique for analyzing and elucidating the gas-sensing processes at the FET sensor surface at an atomistic, or molecular, level is theoretical computational-chemistry modelling. In the perspective of FET surfaces, most often different electronic structure methods have been used, which give a solution to the electronic Schrödinger equation for the electrons and nuclei of the system [36]. These calculations, in principle, can provide all properties of the system, although nuclei dynamics (or temperature) effects are usually neglected. Very detailed information on, e.g., adsorption structures, adsorption energy magnitudes, electron redistributions, spin densities and even reaction paths can be obtained in this way, thus enabling a more thorough understanding of the microscopic behaviour.

Other computational-chemistry methods include force-field methods, which utilise pre-parameterised analytical potential energy expressions that represent the interactions between atomic centres, and molecular dynamics (MD) simulations, that do incorporate nuclei dynamics through the classical Newton's equations of motion. Force-field methods can provide equilibrium geometries and interaction energies at low computational cost, but appear to be more seldom used in computations on FET-related systems, and this apply also to force-field-based MD simulations. Increasingly, the MD method is instead used in conjunction with electronic-structure methods in simulations termed ab initio MD (AIMD) [77].

The specific brands of electronic structure method used vary. For high-level computations, density functional theory is the most commonly adopted, although computations based on the ab initio Hartree-Fock method (or hybrids between ab initio and DFT methods) are also frequent, especially in supermolecular calculations. In the supermolecular approach, the gas molecule surface is modelled by cutting out a segment of atoms from a crystalline surface, which is then subjected to computations. To ease the influence from the unphysical boundaries of the extracted cluster, it may be surrounded by point charges to mimic the reminder of a crystalline surface, by atoms such as H and F that saturate broken bonds, or by atomic entities treated by force-field methods. A more direct route which avoids abrupt boundary effects is to use periodic models, where a slab of the surface, periodic in two dimensions, interacts with the gas molecules, which also by necessity will be periodically repeated [65]. Care must then be taken to choose a sufficiently large surface unit cell in order to avoid overlap between the periodically distributed gas molecule images or surface defective sites.

If we first focus on computations of relevance to FET gas sensors of the metal oxide type, then there is an abundance of studies that have been published. For example, the interaction between NO and the $RuO_2(110)$ surface have been studied [34], where the adsorption characteristics of NO and the formation of N_2 and N_2O was analysed, by plane-wave DFT calculations. Similarly, for this surface CO oxidation was studied [60], oxygen adsorption and the energetics dependence on the degree of surface coverage [83], hydrogen adsorption and hydroxyl formation [76], hydrogen transfer reactions [42], and hydrogen interaction with oxygen defects (as an hypothesis for explaining the large FET H_2 sensitivity) and ammonia dissociation, see Fig. 4.10, [69].

MgO is maybe the archetype of an apparently simple metal oxide structure, but its surfaces possess an evident chemical activity, largely due to the presence of various surface defects. Ojamäe and Pisani used a perturbed-cluster method to investigate surface defects consisting of di-vacancies with trapped electrons (see Fig. 4.11) that act as reactive sites which can participate in various surface reactions, and discussed these in relation to other surface defect models [57]. D'Ercole et al. similarly studied the interaction between hydrogen and an oxygen vacancy and the formation of a colour centre at the MgO surface [21]. Ealet et al. found by plane-wave DFT calculations that water dissociates

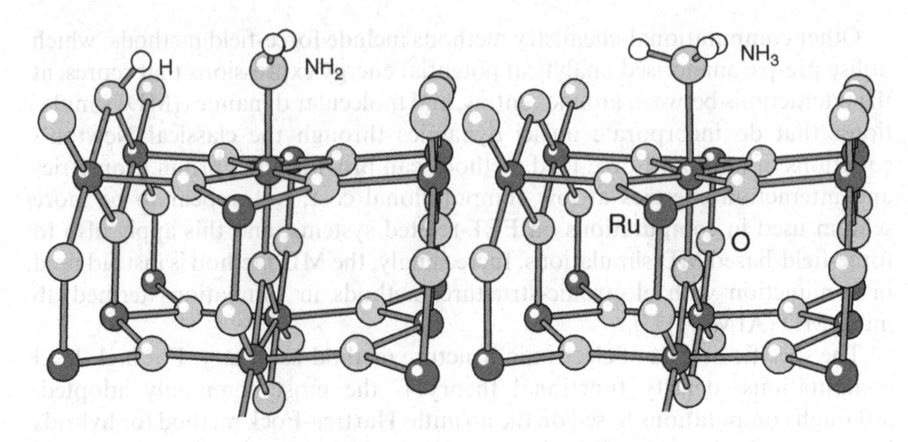

Fig. 4.10 An ammonia molecule chemisorbed in a dissociative (*left*) and non-dissociative (*right*) fashion at the $RuO_2(110)$ surface according to plane-wave DFT

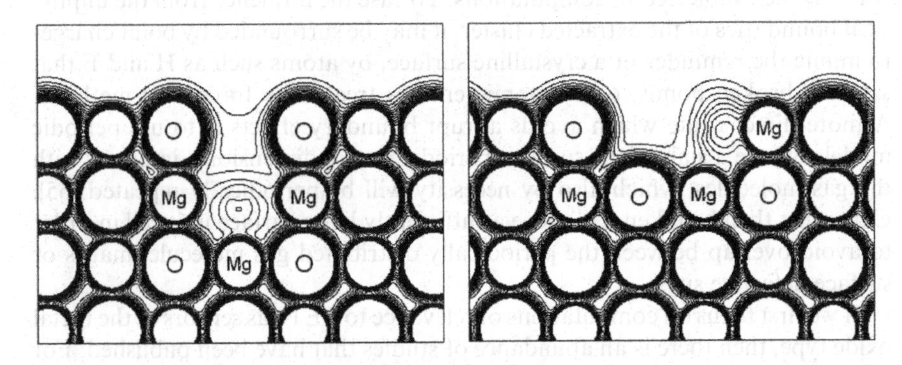

Fig. 4.11 Electron densities of two MgO surfaces with electron-capturing ion di-vacancies. The density is plotted on a plane that is perpendicular to the surface and which intersects the midpoints of the di-vacancy and of the neighbouring Mg and O ions. The trapped electron is clearly visible and can be said to constitute a "pseudo-atom", i.e. an atom lacking a nucleus

spontaneously at di-vacancy defects [23]. Water adsorption on MgO was the subject of the study by Alfè and Gillan using AIMD simulations [5].

Titanium dioxide is a material with a wide range of applications, besides as gas sensors for example in solar cell devices and as biocompatible materials, and there is a widespread interest in modelling chemical reactions at TiO_2 (both the anatase and rutile phases) surfaces [22]. Reactions of oxygen molecules and water with oxygen vacancies on the $TiO_2(110)$ rutile surface was studied using plane-wave DFT [85], likewise for oxygen molecules but by a cluster approach [91]. Adsorption of chlorine and oxygen atoms was modelled [52] as well as NO adsorption [50], and reaction pathways was mapped out for CO oxidation on this surface [64]. The adsorption of HCOOH was studied by several authors

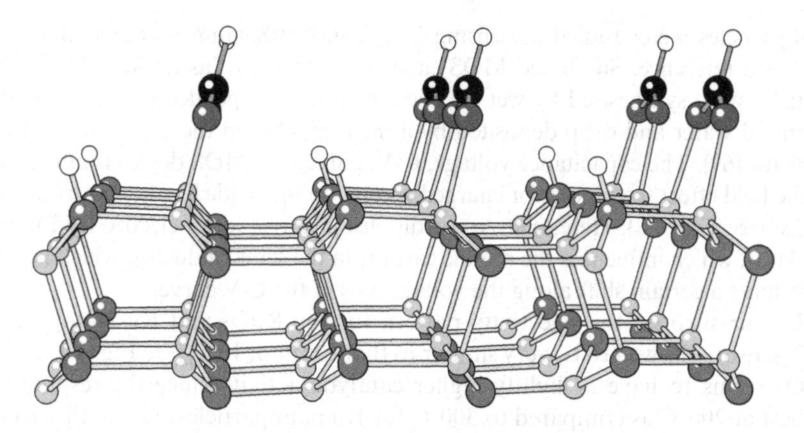

Fig. 4.12 Dissociatively chemisorbed HCOOH with a ¼ coverage at the ZnO(10-10) surface. Periodic Hartree-Fock computations

[4, 10, 45, 56], using AIMD simulations for an anatase surface [30], as well as the catalysed decomposition reaction of the molecule [78].

The molecular interactions at the surfaces of the slightly more ionic ZnO crystal has been investigated (e.g. [11, 19]) using a combined quantum-chemical and force-field cluster method for the H_2, CO, CO_2, H_2O and NH_3 molecules. Formic acid chemisorption and decomposition on ZnO has been studied [55, 88] using cluster models and using periodic ab initio computations, see Fig. 4.12 [61, 62].

Among other metal oxide computations are studies of molecular reactions at the SnO_2(110) surface [18] and the drift-diffusion model (although not an atomistic model) of SnO_2 nanobelt gas sensors [8]. Gavartin et al. [29] studied defects in the HfO_2 crystal consisting of various nitrogen compounds, which can enhance the electric properties and stability of high-dielectric materials. Other semiconductor studies include analysis of ammonia adsorption on Pt/SiO2 cluster model systems [81] and modelling of AlN nanowires and their ammonia sensing capability [91]. Diamond is a promising sensor material, and Petrini and Larsson investigated the interface between a diamond surface and a water adlayer in order to explain the peculiar high surface conductivity of diamond [63].

4.9 Nanoparticles as Sensing Layers in FET Devices

Especially the findings that three phase boundaries are important in the detection mechanism of FET gas sensor devices made nanoparticles as sensing layers interesting. It was shown that SnO_2 nanoparticles of controlled grains size, 6–16 nm, showed a higher sensitivity to H_2S for larger grains and thinner films, which was explained by a gas diffusion-reaction theory [79]. For the FET devices conducting metal oxide nanoparticles were attractive new kinds of sensing material as compared to catalytic metals like Pd, Pt and Ir studied earlier. Ruthenium oxide nanoparticles were studied as sensing layers and ruthenium nanoparticles were used as a reference. SiC-based MOS capacitors were used as devices. The nanoparticles were synthesised by wet chemical methods, suspended in methanol and deionised water and drop deposited by a micropipette on the top contact of the capacitor [69]. The capacitance voltage, C-V, curve of a MOS device is influenced by the field effect from polar or charged species on the oxide like for the transistor device (see Fig. 4.2). Here, charges or dipoles give rise to an electric field in the insulator, which influences the mobile carriers in the semiconductor, which in turn introduces a certain shift along the voltage axis in the C-V curve.

To our surprise the selectivity pattern for the RuO_2 and Ru nanoparticle FET sensors showed to be very similar to that for Pt or Ir [7] (see Fig. 4.13). The RuO_2 seems to have a slightly higher catalytic activity since the response is highest at 200°C as compared to 300°C for Ru nanoparticles. For both sensing layers high sensitivity to hydrogen, hydrocarbons and ammonia and low sensitivity to carbon monoxide and nitric oxide is very similar to the response pattern from Pt or Ir sensing layers. The test gases are chosen as the typical components in diesel exhausts.

Now we turned to gold nanoparticles in order to investigate a different, non-catalytic (in the case of Au films), material. We know that an evaporated polycrystalline Au film on a Si MOS device is sensitive to NO [27]. For the Au nanoparticles an electrochemical synthesising method, the Reetz sacrificial anode electrolysis, was used with gold as the sacrificial anode and Pt as the cathode [66, 67]. With a proper current density and surfactant in the electrolyte the size of the nanoparticles were accurately controlled around 2–5 nm with a standard deviation of the core diameter $\leq 20\%$. Quarternary ammonium salts in the electrolyte acts as surfactants to stabilise the nanoparticles in the solution, which also contained tetrahydrofuran and acetonitrile, described in more detail in [35]. Capacitor devices, based on Si MOS devices, were processed and the gold nanoparticles drop deposited from a suspension as described above. The nanoparticles were annealed at a temperature higher than the operation

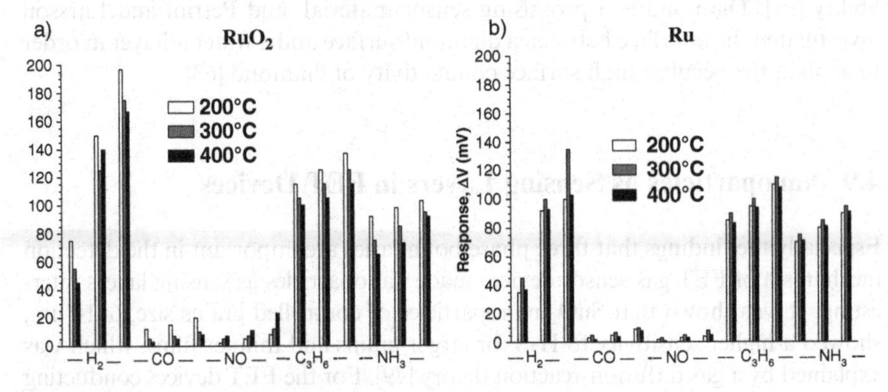

Fig. 4.13 Sensor response of a capacitor with RuO2 nanoparticles (to the *left*) and Runanoparticles (to the *right*)

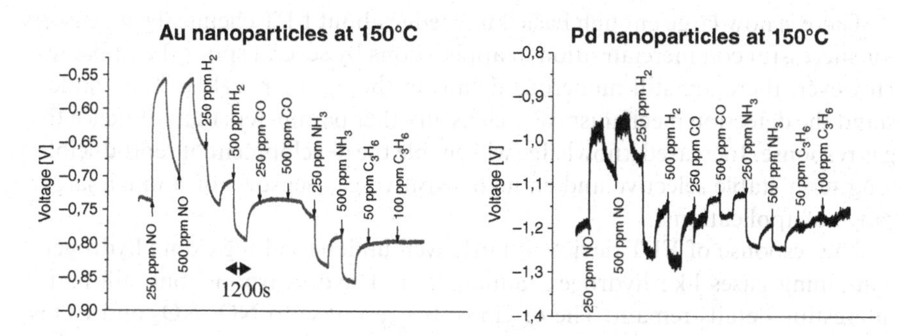

Fig. 4.14 Sensor response of a capacitor with Au nanoparticles (*left*) and Pd nanoparticles (*right*)

temperature before the measurements in order to get rid of the solvent and also to avoid restructuring during the measurements. This induced agglomeration of the nanoparticles, which, however, retained the nanostructured nature (see Fig. 4.6). Surface analysis by XPS showed that the Au nanoparticles after this annealing procedure were of totally metallic character. Now the selectivity pattern dramatically changed as compared to Pt or Ir. The gold nanoparticles show an increased response to nitric oxide and nitric dioxide, which were both tested (see Fig. 4.14, left). A smaller, and in the opposite direction, response was also shown to ammonia and hydrogen but very little to hydrocarbons and carbon monoxide.

We also tested palladium nanoparticles produced by the same electrochemical method but with Pd as the anode (see Fig. 4.14, right). Here the annealing performed at 300°C induced agglomeration with a fractal character, larger grains compost by smaller grains (see Fig. 4.6). The XPS analysis here indicated the presence of some PdO together with Pd for the annealed material. To our surprise also these nanoparticles showed a high response to nitric oxide and nitrogen dioxide but also as expected a high response to hydrogen and ammonia and for hydrocarbons and carbon monoxide a smaller but clear response (see Fig. 4.14, right). This will indeed be very interesting to study by the DRIFT spectroscopy.

4.10 Summary and Outlook

New and improved analytical tools develop in order to acquire nanoscale knowledge about the parameters that influence the gas response of solid-state chemical sensors. For the FET devices methods like DRIFT spectroscopy, atmospheric pressure mass spectroscopy and SLPT has contributed valuable information. This relates to both structural information about sensing layers, as well as to molecular information about chemical reactions and adsorbates on sensing surfaces. Atomistic modelling of chemical reactions is a powerful tool to reveal details in the detection mechanism.

There is now large enough basic knowledge about FET chemical gas sensors for successful commercialisation in applications by several spin-off companies. However, there are still numerous details in the jigsaw puzzle to fully understand the detection mechanism as well as all other parameters that influence the gas response. Increased knowledge will enable the development of reproducible, long term stable selective and enough sensitive gas sensors for a much larger range of applications.

The response of FET devices if fairly well understood regarding hydrogen-containing gases like hydrogen, ammonia and hydrocarbons, but still many interesting details remain. The origin of the response to NO, NO_2 and CO is especially interesting since these molecules contain no hydrogen. Also, interesting is the difference between Au and Pd on the one hand and Pt, Ir, RuO_2 and Ru on the other hand, why do we get a response to NO and NO_2 for the former ones but not for the latter ones? Which are the adsorbing species and chemical reactions on the surfaces? Operando DRIFT studies, as has been performed for metal oxide solid-state gas sensors, are under development also for FET devices. We are facing problems due to the small surface area of FET devices; however, recent development has considerably improved the possibility to perform operando/in situ studies on FET sensor devices. Development in atomistic modelling has made this to a most valuable tool to explain and support experiments and analytical investigations.

Of considerable importance is also the electrostatic influence on the gas response, since the FET devices operate with an applied voltage. It seems like the OH groups that are identified as responsible for the gas response to both hydrogen and ammonia appears mainly close to the metal insulator border. This makes it likely that electrostatic interaction plays a role. This should preferably be investigated both experimentally and by electrostatic modelling.

Acknowledgments Grants are acknowledged from the Swedish Research Council.

References

1. Ackelid U, Fogelberg J, Petersson L-G. Local gas sampling and surface hydrogen detection in catalysis on planar surfaces. Vacuum. 1991;42(14):889–895.
2. Ackelid U, Petersson L-G. How a limited mass transfer in the gas phase may affect the steady-state response of a Pd-MOS hydrogen sensor. Sens Actuators B. 1991;3:139–146.
3. Adixen Sensistor AB, http://www.sensistor.se (2008-08-17)
4. Ahdjoudj J, Minot C. A theoretical study of HCO_2H adsorption on $TiO_2(110)$. Catalysis Letters. 1991;46:83–91.
5. Alfè D, Gillan MJ. Ab initio statistical mechanics of surface adsorption and desorption. I. H_2O on MgO(001) at low coverage. J. Chem. Phys. 2007;127:114709. doi: 10.1063/1.2772258.
6. Andersson M, Wingbrant H, Lloyd Spetz A. Study of the CO Response of SiC based Field Effect Gas Sensors, Proc. IEEE Sensors 2005, Irivine, USA, October 31–November 2, 2005, pp. 105–108.

7. Andersson M, Wingbrant H, Petersson H, Unéus L, Svenningstorp H, Löfdahl M, Holmberg M, Lloyd Spetz A. Gas sensor arrays for combustion control. In: CA, Grimes EC Dickey, editors. Encyclopedia of Sensors. Stevenson Ranch, Ca, USA: American Scientific Publishers; 2006. vol. 4, pp. 139–154.
8. Andrei P, Fields LL, Zheng JP, Cheng Y, Xiong P. Modeling and simulation of single nanobelt SnO_2 gas sensors with FET structure. Sensors and Actuators B. 2007;128:226–234. doi: 10.1016/j snb 2007.06.009.
9. Applied Sensor AB, http://www.appliedsensor.com (2008-08-17).
10. Bates S P, Kresse G, Gillan MJ. The adsorption and dissociation of ROH molecules on TiO_2 (110). Surf Sci. 1998;409:336–349.
11. Batista Lopes Martins J, Longo E, Rodríguez Salmon OD, Espinoza V AA, Taft CA. The interaction of H_2, CO, CO_2, H_2O and NH_3 on ZnO surfaces: an Oniom study. Chem Phys Lett. 2004;400:481–486. doi: 10.1016/j cplett 2004.10.150.
12. Baur B, Howgate J, von Ribbeck H-G, Gawlina Y, Bandalo V, Steinhoff G, Stutzmann M, Eickhoff M. Catalytic activity of enzymes immobilized on AlGaN/GaN solution gate field-effect transistors. Appl Phys Lett. 2006;89(18):183901-1–183901-3.
13. Becker E, Skoglundh M, Andersson M, Lloyd Spetz A. In situ DRIFT study of hydrogen and CO adsorption on Pt/SiO2 model sensors. In Proc IEEE Sensors 2007, Atlanta, USA, Oct 28–31, 2007. pp. 1028–1031.
14. Bene R, Kiss G, Perczel I V, Meyer FA, Réti F. Application of quadrupole mass spectrometer for the analysis of near-surface gas composition during DC sensor-tests. Vacuum. 1998;50(3–4):331–337.
15. Benítez JJ, Centeno MA, Lois dit Picard C, Merdrignac O, Laurent Y, Odriozola JA. In situ diffuse reflectance infrared spectroscopy (DRIFTS) study of the reversibility of CdGeON sensors towards oxygen. Sens Actuators B. 1996;31:197–202.
16. Bergveld P. Thirty years of ISFETOLOGY what happened in the past 30 years and what may happen in the next 30 years. Sens Actuators B. 2003;88:1–20.
17. Buchholt K, Ieva E, Torsi L, Cioffi N, Colianni L, Söderlind F, Käll P-O, Lloyd Spetz A. A comparison between the use of Pd- and Au-nanoparticles as sensing layers in a field effect NO_x-sensitive sensor. In: Proc. 2nd Int Conf on Sensing Techn November 26–28, Palmerston, North New Zealand, 2007. pp 87–92.
18. Calatayud M, Andrés J, Beltrán A. A theoretical analysis of adsorption and dissociation of CH_3OH on the stochiometric SnO_2(110) surface. Surf Sci. 1999;430:213–222.
19. Casarin M, Maccato C, Vittadini A. Theoretical investigation of the chemisorption of H_2 and CO on the ZnO(10–10) surface. Inorg Chem. 1998;37:5482–5490. doi: 10.1021/ic980443s.
20. Dannetun H, Petersson L-G. NO dissociation on polycrystalline palladium studied with a Pd-metal-oxide-semiconductor structure. J Appl Phys. 1989;66(3):1397–1402.
21. D'Ercole A, Giamello E, Pisani C, Ojamäe L. Embedded-cluster study of hydrogen interaction with an oxygen vacancy at the magnesium oxide surface. J Phys Chem B. 1999;103:3872–3976. doi: 10.1021/jp990117d.
22. Diebold U. The surface science of titanium dioxide. Surface Sci Rep. 2002;48:53–229.
23. Ealet B, Goniakowski J, Finocchi F. Water dissociation on a defective MgO(100) surface: Role of divacancies. Phys Rev B. 2004;69:195413. doi: 10.1103/PhysRevB.69.195413.
24. Emiroglu S, Bârasan N, Weimar U, Hoffman V. In situ diffuse reflectance infrared spectroscopy study of CO adsorption on SnO_2. Thin Solid Films. 2001;391:176–185.
25. Eriksson M, Ekedahl L-G. The influence of CO on the response of hydrogen sensitive Pd-MOS devices. Sens Actuators B. 1997;42:217–223.
26. Eriksson M, Salomonsson A, Lundström I. The influence of the insulator surface properties on the hydrogen response of field-effect gas sensors. J Appl Phys. 2005;98:034903-1–034903-6.
27. Filippini D, Weiss T, Aragón R, Weimar U. New NO_2 sensor based on Au gate filed effect devices. Sens Actuators B. 2001;78:195–201.

28. Fogelberg J, Eriksson M, Dannetun H, Petersson L-G. Kinetic modeling of hydrogen adsorption/absorption in thin films on hydrogen-sensitive field-effect devices: Observation of large hydrogen-induced dipoles at the Pd-SiO_2 interface. J Appl Phys. 1995;78(2):988–996.
29. Gavartin J L, Schluger A L, Foster A S, Bersuker G I. The role of nitrogen-related defects in high-k dielectric oxides: density-functional studies. J Appl Phys. 2005;97:053704. doi: 10.1063/1.1854210.
30. Gong X-Q, Selloni A, Vittadini A. Density functional theory study of formic acid adsorption on anatase TiO_2(001): Geometries, energetics, and effects of coverage, hydration, and reconstruction. J Phys Chem B. 2006;110:2804–2811. doi: 10.1021/jp056572t.
31. Hahn S H, Bârsan N, Weimar U, Ejakov SG, Visser JH, Soltis RE. CO sensing with SnO2 thick film sensors: role of oxygen and water vapour. Thin Solid Films. 2003;436:17–24.
32. Ho Kahng Y, Tobin RG, Loloee R, Ghosh RN. J Appl Phys. 2007;102:064505-1–064505-9.
33. Holmberg M, Lundström I. A new method for the detection of hydrogen spillover. Appl Surf Sci. 1996;93:67–76.
34. Hong S, Rahman TS, Jacobi K, Ertl E. Interaction of NO with RuO_2(1110) surface: a first principles study. J Phys Chem. 2007;111:12361–12368. doi: 10.1021/jp072063a.
35. Ieva E, Buchholt K, Colianni L, Cioffi N, Sabbatini L, Capitani GC, Lloyd Spetz A, Käll P-O, Torsi L. Au nanoparticles as gate material for NO_x Field Effect Gas Sensors, Sensor Letters. 2008;6(4):577–584.
36. Jensen F. Introduction to Computational Chemistry. John Wiley & Sons, reprinted 2001 ISBN 0-471-98425-6. 1999.
37. Jiménez I, Centeno MA, Scotti R, Morazzoni F, Arbiol J, Cornet A, Morante JR. NH_3 interaction with chromium-doped WO_3 nanocrystalline powders for gas-sensing applications. J Mat Chem. 2004;14:2412–2420.
38. Johansson M, Ekedahl L-G. The water formation rate on platinum and palladium as a function of the surface hydrogen pressure from three-dimensional hydrogen pressure distributions. Appl Surf Sci. 2001;180:27–35.
39. Johansson M, Loyd D, Lundström I. Influence of mass transfer on the steady-state response of catalytic –metal-gate sensors. Sens Actuators B. 1997;40:125–133.
40. Katti VR, Debnath AK, Muthe KP, Kaur M, Dua AK, Gadkari SC, Gupta SK, Sahni VC. Mechanism of drifts in H2S sensing properties of SnO2:CuO composite thin film sensors prepared by thermal evaporation. Sens Actuators B. 2003;96:245–252.
41. Klingvall R, Lundström I, Löfdahl M, Eriksson M. A combinatorial approach for field-effect gas sensor research and development. IEEE Sens J. 2005;5:995–1003.
42. Knapp M, Crihan D, Seitsonen AP, Over H. Hydrogen Transfer Reaction on the Surface of an Oxide Catalyst. J Am Chem Soc. 2005;127:3236–3237. doi: 10.1021/ja043355h.
43. Koziej D, Bârsan N, Weimar U, Szuber J, Shimanoe K, Yamazoe N. Water-oxygen interplay on tin dioxide surface: Implication on gas sensing. Chem Phys Lett. 2005; 410:321–323.
44. Koziej D, Thomas K, Bârsan N, Thibault-Starzyk F, Weimar U. Influence of annealing temperature on the CO sensing mechanism for tin dioxide based sensors – Operando studies. Cat today. 2007;126:211–218.
45. Käckell P, Terakura K. First-principle analysis of the dissociative adsorption of formic acid on rutile TiO_2 (110). Applied Surf Sci. 2000;166:370–375.
46. Leblanc E, Perier-Camby L, Thomas G, Gibert R, Primet M, Gelin P. NO_x adsorption onto dehydroxylated or hydroxylated tin dioxide surface. Application to SnO_2-based sensors. Sens Actuators B. 2000;62:67–72.
47. Lundström I, Sundgren H, Winquist F, Eriksson M, Krantz-Rülcker C, Lloyd-Spetz A. Twenty-five years of field effect gas sensor research in Linköping. Sens Actuators B. 2007;121:247–262.
48. Löfdahl M, Thesis, Linköping studies in Science and Technology, dissertation No. 696, Linköping, Sweden 2001, pp. 93–111.

49. Löfdahl M, Utaiwasin C, Carlsson A, Lundström I, Eriksson M. Gas response dependence on gate metal morphology of field-effect devices. Sens Actuators B. 2001;80:183–192.
50. Markovits A, Mguig B, Calatayud M, Minot C. Spin localization for NO adsorption on surface O atoms of metal oxides, Catalysis Today. 2006;113:201–207. doi: 10.1016/j.cattod.2005.11.081.
51. Marsal A, Centeno MA, Odriozola JA, Cornet A, Morante JR. DRIFTS analysis of the CO_2 detection mechanisms using LaOCl sensing material. Sens Actuators B. 2005;108: 484–489.
52. Ménétrey M, Markovits A, Minot C. Adsorption of chlorine and oxygen atoms on clean and defective rutile-TiO_2(110) and MgO(100) surfaces. J Mol Struct: THEOCHEM. 2007;808:71–79. doi: 10.1016/j.theochem.2006.12.044.
53. Nakagomi S, Lloyd Spetz A. Gas sensor device based on Catalytic metal-Metal oxide-SiC structure. In: Grimes CA, E. Dickey C, editors. Encyclopedia of Sensors. Stevenson Ranch, Ca, USA: American Scientific Publishers 2006. vol. 4, pp. 155–170.
54. Nakagomi S, Tobias P, Baranzahi A, Lundström I, Mårtensson P, Lloyd Spetz A. Influence of carbon monoxide, water, and oxygen on high temperature catlaytic metal – oxide – silicon carbide structures. Sens Actuators B. 1997;45/3:183–191.
55. Nakatsuji H, Yoshimoto M, Umemura Y, Takagi S, Hada M. Theoretical study of the chemisorption and surface reaction of HCOOH on a ZnO(10-10) surface. J Phys Chem. 1996;100:694–700.
56. Ojamäe L, Aulin C, Pedersen H, Käll P-O. IR and quantum-chemical studies of carboxylic acid and glycine adsorption on rutile TiO_2 nanoparticles. J Colloid and Interface Sci 2006;296:71–78. doi: 10.1016/j.jcis.2005.08.037.
57. Ojamäe L, Pisani C. Theoretical characterization of divacancies at the surface and in bulk MgO. J Chem Phys. 1998;109:10984–10995.
58. Ostrick B, Mühlsteff J, Fleischer M, Meixner H, Doll T, Kohl C-D. Adsorbed water as key to room temperature gas-sensitive reactions in work function type sensors: the carbonate – carbon dioxide system. Sens Actuators B. 1999.57:115–119.
59. Ostrick B, Pohle R, Fleischer M, Meixner H. TiN in work function type sensors: a stable ammonia sensitive material for room temperature operation with low humidity cross sensitivity. Sens Actuators B. 2000;68:234–239.
60. Over H, Kim YD, Seitsonen AP, Wendt S, Lundgren E, Schmid M, Varga P, Morgante A, Ertl G. Atomic-Scale Structure and Catalytic Reactivity of the RuO_2(110) Surface. Science. 2000;287:1474–1476. doi: 10.1126/science.287.5457.1474.
61. Persson P, Lunell S, Ojamäe L. Quantum-chemical prediction of the adsorption conformations and dynamics at the HCOOH-covered ZnO(10-10) surfaces. Int J Quantum Chem. 2002;89:172–180. doi: 10.1002/qua.10316.
62. Persson P, Ojamäe L. Periodic Hartree-Fock study of the adsorption of formic acid on ZnO(10-10). Chem Phys Lett. 2000;321:302–308.
63. Petrini D, Larsson K. Electron transfer from a diamond (100) surface to an atmospheric water adlayer: A quantum mechanical study. J Phys Chem C. 2007;111:13804–13812. doi: 10.1021/jp070565i.
64. Pillay D, Hwang GS. O_2-coverage-dependent CO oxidation on reduced TiO2(110): A first principles study. J Chem Phys. 2006;125:144706. dio: 10.1063/1.12354083.
65. Pisani C. Ab-initio approaches to the quantum-mechanical treatment of periodic systems. In: C Pisani, editors. Lecture Notes in Chemistry. Springer-Verlag, ISBN 3-540-61645-4. vol. 67, 1996. pp. 47–75.
66. Reetz MT, Helbig W. Size-selective synthesis of nanostructured transition metal clusters. J Am Chem Soc. 1994;116:7401–7402.
67. Reetz MT, Helbig W, Quaiser SA, Stimming U, Breuer N, Vogel R. Visualization of Surfactants on nanostructured Palladium Clusters by a Combination of STM and High-resolution TEM. Science. 1995;267:367–369.

68. Salamonsson A, Eriksson M, Dannetun H. Hydrogen interaction with platinum and palladium metal – insulator – semiconductor devices. J Appl Phys. 2005;98:014505-1–014505-9.
69. Salomonsson A, Petoral Jr RM, Uvdal K, Aulin C, Käll P-O, Ojamäe L, Strand M, Sanati M, Lloyd Spetz A. Nanocrystalline ruthenium oxide and ruthenium in sensing applications – an experimental and theoretical study. J Nanoparticle Res. 2006;8:899–910. doi: 10.1007/s11051-005-9058-1.
70. Schalwig J, Kreisl P, Ahlers S, Müller G. Response mechanism of SiC-based MOS Field-Effect gas sensors. IEEE Sens J. 2002;2(5):394–402.
71. Scharnagel K, Eriksson M, Karthigeyan A, Burgmair M, Zimmer M, Eisele I. Hydrogen detection at high concentrations with stabilized palladium. Sens Actuators B. 2001;78:138–143.
72. SenSiC AB, http://www.sensic.se/(2008-08-17).
73. Spetz A, Armgarth M, Lundström L. Hydrogen and ammonia response of metal-silicon dioxide- silicon structures with platinum gates. J Appl Phys. 1988:64:1274–1283.
74. Spetz A, Helmersson U, Enquist F, Armgarth M, Lundström L. Structure and ammonia sensitivity of thin platinum or iridium gates in metal-oxide-silicon capacitors. Thin Solid Films. 1989;177:77–93.
75. Steinhoff, G., Baur, B., Wrobel, G., Ingebrandt, S., Offenhäusser, A., Dadgar, A., Krost, A., Stutzmann M, Eickhoff, M. Recording of cell action potentials with AlGaNGaN field-effect transistors. Appl Phys Lett. 2005;86(3):033901-1–033901-3. see also Erratum. Appl Phys Lett. 2006;89:019901.
76. Sun Q, Reuter K, Scheffler M. Hydrogen adsorption on RuO_2(110): Density-functional calculations. Phys Rev B. 2004;70:235402. doi: 10.1103/PhysRevB.70.235402.
77. Tse J. Ab initio molecular dynamics with density functional theory. Annu. Rev Phys Chem. 2002;53:249–290. doi: 10.1146/annurev physchem 53.090401.105737.
78. Uemura Y, Taniike T, Tada M, Morikawa Y, Iwasawa Y. Switchover of reaction mechanism for the catalytic decomposition of HCOOH on a TiO_2(110) surface. J Phys Chem C. 2007;111:16379–16386. doi: 10.1021/jp074524y
79. Vuong DD, Sakai G, Shimanoe K, Yamazoe N. Hydrogen sulfide gas sensing properties on thin films derived from SnO_2 sols different in grain size. Sens Actuators B. 2005;105:437–442.
80. Wallin M, Grönbeck H, Lloyd Spetz A, Skoglundh M. Vibrational study of ammonia adsorption on Pt/SiO_2. Appl Surf Sci. 2004;235:487–500. doi: 10.1016/j.apsusc. 2004.03.225.
81. Wallin M, Grönbeck H, Lloyd Spetz A, Eriksson M, Skoglundh M. Vibrational analysis of H_2 and D_2 adsorption on Pt/SiO_2. J Phys Chem B. 2005;109:9581–9588.
82. Wallin M, Byberg M, Grönbeck H, Skoglundh M, Eriksson M, Lloyd Spetz A. Vibrational analysis of H_2 and NH_3 adsorption on Pt/SiO_2 and Ir/SiO_2 model sensors. In Proc IEEE Sensors 2007, Atlanta, USA, Oct 28–31, 2007. pp. 1315–1317.
83. Wang H, Schneider WF. Effects of coverage on the structures, energetics, and electronics of oxygen adsorption on RuO_2(110). J Chem Phys. 2007;127:064706 doi 10.1063/1. 2752501.
84. Weidemann O, Hermann M, Steinhoff G, Wingbrant H, Lloyd Spetz A, Stutzmann M, Eickhoff M. Influence of surface oxides on hydrogen-sensitive Pd:GaN Schottky diodes. Appl Phys Lett. 2003;83(4):773–775.
85. Wendt S, Schaub R, Matthiesen J, Vestergaard E K, Wahlström E, Rasmussen MD, Thostrup P, Molina LM, Lægsgaard E, Stensgaard I, Hammer B, Besenbacher F. Oxygen vacancies on TiO_2(110) and their interaction with H_2O and O_2: A combined high-resolution STM and DFT study. Surf Sci. 2005;598:226–245. doi: 10.1016/j.susc.2005.08.041.
86. Wingbrant H, Svenningstorp H, Kubinski DJ, Visser JH, Andersson M, Unéus L, Löfdahl M, Lloyd Spetz A. MISiC-FET NH_3 sensors for SCR control in exhaust and flue gases. In: CA Grimes, EC Dickey, editors. Stevenson Ranch, Ca, USA: Encyclopedia of Sensors, American Scientific Publishers; 2006. vol 6, pp. 205–218.

87. Wurtzinger O, Reinhardt G. CO-sensing properties of doped SnO_2 sensors in H_2-rich gases. Sens Actuators B. 2004;103:104–110.
88. Yoshimoto M, Takagi S, Umemura Y, Hada M, Nakatsuji H. Theoretical study on the decomposition of HCOOH on a ZnO(10-10) surface. J Catalysis. 1998;173:53–63.
89. Zangooie S, Arwin H, Lundström I and Lloyd Spetz A. Ozone Treatment of SiC for Improved Performance of Gas Sensitive Schottky Diodes. Mat Sci Forum. 2000;338–342:1085–1088
90. Zhanpeisov NU, Fukumura H. Oxygen vacancy formation on rutile TiO_2(110) and its interaction with molecular oxygen: A theoretical density functional theory study. J Phys Chem C. 2007;111:16941–16945. doi: 10.1021/jp074869g.
91. Zhou Z, Zhao J, Chen Y, von Ragué Schleyer P, Chen Z. Energetics and electronic structures of AlN nanotubes/wires and their potential application as ammonia sensors. Nanotechnology. 2007;18:424023 (7 pp). doi: 10.1088/0957-4484/18/42/424023.
92. Åbom AE, Haasch RT, Hellgren N, Finnegan N, Hultman L, Eriksson M. Characterization of the metal-insulator interface of field-effect chemical sensors. J Appl Phys. 2003;93(12):9760–9768.
93. Åbom A E, Persson P, Hultman L, Eriksson M. Influence of gate metal film growth parameters on the properties of gas sensitive field-effect devices. Thin Solid Films. 2002;409:233–242.

Chapter 5
Solid-State Electrochemical Gas Sensing

Norio Miura, Perumal Elumalai, Vladimir V. Plashnitsa, Taro Ueda, Ryotaro Wama, and Masahiro Utiyama

5.1 Introduction

Arduous efforts are required worldwide to tackle the issues of environmental pollutions, which govern a prime role in global warming (climate change). The importance of this issue has been also emphasized by the Nobel Prize winners (2007) awarded for the contribution to the climate change (http://nobelprize.org/nobel_prizes/peace/laureates/2007/index.html). Therefore, over the years, there has been a growing concern on the environmental pollution due to emissions from automotive vehicles and other industrial furnaces. As the automotive industries expand drastically every year even in developing countries, strict regulations have to be implemented to reduce the pollutions. The major causative hazardous gases are usually CO, NO_x (NO and NO_2) and hydrocarbons (HCs). Thus, severe norms are being enforced on what concerns on automobile's emission in Europe, United States and Japan. Figure 5.1 shows the regulation values for the pollutant gases emitted from gasoline passenger cars to be strictly followed by the automobile industries (http://en.wikipedia.org/wiki/European_emission_standards; www.dieselnet.com/standards; www.jama.or.jp/exhaust/table_01.html). For example, the regulation values according to Euro VI (2014) for CO, NO_x and HCs are set 1, 0.06 and 0.1 g/km, respectively. Such low limit values for car exhaust emission strongly require for the in situ monitoring of gas concentration in exhausts as well as the development of emission-free or low-emission automobiles.

The on-board diagnosis (OBD) system equipped with in-situ exhaust gas sensors including air/fuel sensor (λ sensor) which are located upstream or downstream of a three-way catalyst (TWC), will be realized in the future to be used inside vehicles to monitor all the gaseous components related to air pollution. Therefore the demand for reliable, compact and low-cost solid-state on-board sensors has been enhanced substantially. To monitor

N. Miura
Art, Science and Technology Center for Cooperative Research, Kyushu University, Kasuga-shi, Fukuoka 816-8580, Japan
e-mail: miurano@astec.kyushu-u.ac.jp

E. Comini et al. (eds.), *Solid State Gas Sensing*,
DOI: 10.1007/978-0-387-09665-0_5, © Springer Science+Business Media, LLC 2009

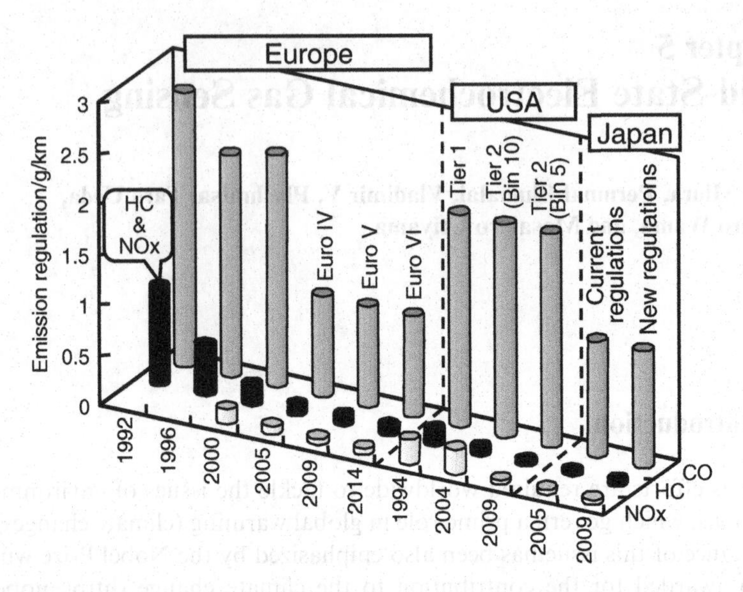

Fig. 5.1 The trend of the regulations for exhausts emission from gasoline passenger cars

automotive exhausts (e.g., NO_x concentration from *ca.* 10 ppm up to *ca.* 2000 ppm), the sensor should be able to work in very harsh environment where the temperature fluctuates from room temperature up to even 900°C. In addition, the necessity of high-performance solid-state gas sensors is not only limited to on-board monitoring but also outdoor applications such as monitoring of pollutant gases in atmosphere, air quality assessment in houses or work places, gas sensing for safety and security, industrial process control and so on.

Among the various electrochemical gas sensors based on solid electrolytes so far developed, the yttria-stabilized zirconia (YSZ)-based gas sensors seem to be most reliable and realistic for actual application, because the YSZ-based oxygen (air/fuel) sensor has been used for commercial on-board sensors for more than 30 years. Thus, the electrochemical gas sensors combining YSZ with oxide-based sensing-electrode (SE) have been extensively examined for these ten years and are now considered to be the most potential candidate of on-board sensors in terms of their special features such as high temperature operation, excellent sensing performances and good chemical and mechanical stability.

Table 5.1 shows the general overview of the solid-state electrochemical gas sensors reported so far mostly aiming at monitoring of combustion exhausts. It is seen from this table that significant progress has been recently made on the YSZ-based gas sensors operating in several modes (mixed-potential-type, amperometric and impedancmetric). We have also been working on the YSZ-based sensors focusing on the fundamental issues as well as application point of

Table 5.1 An overview of the solid-state electrochemical gas sensors operating in different modes aiming mainly at detection of combustion exhausts

Structure gas, RE | solid electrolyte | SE, gas	Gas	Conc. (ppm)	Temp(°C)	Year	Ref.
(Mixed-potential-type)					
base, Pt | YSZ | $CdMn_2O_4$, NOx(+base)	NOx	5–4000	500–600	1996	[1, 2]
base, Pt | YSZ | $CdCr_2O_4$, Pt, NOx(+base)	NOx	20–600	500–600	1997	[3]
base, Pt | YSZ | WO_3, NOx(+base)	NOx	5–200	500–700	2000	[4]
base, Pt | YSZ | $NiCr_2O_4$, NOx(+base)	NOx	25–436	550–650	2001	[5]
base, Pt | YSZ | $ZnCr_2O_4$, NOx(+base)	NOx	50–436	600–700	2002	[6]
base, Pt | YSZ | $ZnFe_2O_4$, NOx(+base)	NOx	50–436	600–700	2002	[7]
base, Pt | YSZ | Cr_2O_3, NOx(+base)	NOx	10–200	600	2003	[8]
base, Pt | YSZ | ZnO, NOx(+base)	NOx	40–450	600–700	2004	[9]
NOx(+base), Pt | YSZ | $LaFeO_3$, NOx(+base)	NOx	20–1000	450–700	2004	[10, 11]
NOx(+base) | YSZ | NiO, NOx(+base)	NOx	10–400	700–900	2005	[12–21]
NOx(+base) | YSZ | Cr_2O_3, NOx(+base)	NOx	50–400	700	2006	[22]
NOx(+base) Pt | YSZ | $La_{0.8}Sr_{0.2}CrO_3$, NOx(+base)	NOx	0–500	450–650	2006	[23]
base, Pt | $Sn_{0.9}In_{0.1}P_2O_7$ | Pt-Rh, NOx(+base)	NOx	0–1000	250	2006	[24]
Nox(+base) | YSZ | NiO+Au, NOx(+base)	NOx	50–400	600–800	2007	[25]
base, Pt | YSZ | Pt, CO(+base)	CO	100–160	260–450	1980	[26]
base, Pt | YSZ | Pt, CO(+base)	CO	32–800	500–650	1995	[27]
base, SnO_2 | YSZ | CdO, CO(+base)	CO	20–4000	600	1998	[28]
base, Au | YSZ | $LaCoO_3$, Au, CO(+base)	CO	0–4000	600–700	2000	[29]
base, Pt | YSZ | Au, HCs(+base)	HCs	0–500	600	1998	[30, 31]
base, Au | YSZ | Pd, HCs(+base)	HCs	200–1000	700	1999	[32]
base, Pt | YSZ | In_2O_3(+0.1 wt.% Mn_2O_3), HCs(+base)	HCs	0–500	600	1999	[33, 34]
base, Pt | YSZ | CdO, HCs(+base)	HCs	50–800	600	2000	[35]
base, Pt | YSZ | $SrCe_{0.95}Yb_{0.05}O_{3-\alpha}$, HCs(+base)	HCs	0–1000	550–750	2001	[36]
HCs(+base), Pt | YSZ | Au + Nb_2O_5, HCs(+base)	HCs	0.50–8000	700	2002	[37, 38]
HCs(+base), Pt | YSZ | $La_{0.7}Sr_{0.3}CrO_3$, HCs(+base)	HCs	0–500	550	2003	[39]
base, Pt | YSZ | Pr_6O_{11}, HCs(+base)	HCs	0–500	750–800	2005	[40]
HCs(+base), Pt | YSZ | Au + NiO, HCs(+base)	HCs	0–790	650	2007	[41]

Table 5.1 (continued)

Structure gas, RE | solid electrolyte | SE, gas	Gas	Conc. (ppm)	Temp(°C)	Year	Ref.
base, Pt | YSZ | In_2O_3, HCs(+base)	HCs	50–800	790	2008	[42]
(Amperometric)					
base, Pt | YSZ | $CdCr_2O_4$, NOx(+base)	NOx	20–800	500–600	1998	[43]
NOx(+base), Pt | YSZ | Rh-Pt-Au, NOx(+base)	NOx	200–1200	550–600	2005	[44]
NOx(+base), $La_{0.5}Sr_{0.5}MnO_3$ | $La_{0.8}Sr_{0.2}Ga_{0.8}Mg_{0.1}Ni_{0.1}O_3$ | $La_{0.4}Sr_{0.6}Mn_{0.8}Ni_{0.2}$, NOx(+base)	NOx	500–2000	500–700	2006	[45]
NOx, Pt+zeolite | YSZ | Pt+zeolite, NOx	NOx	100–800	500	2007	[46]
CO(+base), Pt | ScSZ | ITO, CO(+base)	CO	0–500	600	2007	[47]
base, Pt | YSZ | Pt, C_3H_6(base)	HCs	200–100	440–560	1998	[48]
base, Pt | $Ce_{0.8}Gd_{0.2}O_{1.9}$ | Au, HCs(+base)	HCs	0–500	600	1999	[49]
O_2(+Ar), Pt | YSZ | Au, HCs(+Ar)	HCs	100–700	700–750	1999	[50, 51]
base, Pt | YSZ | Au, Pd, HCs	HCs	0–2000	700	2000	[52]
base, Au, Pt | $SrCe_{0.95}Y_{0.05}O_{3-\alpha}$ | Au, Pt, HCs(+base)	HCs	0–1000	600	2001	[53]
HCs(+base), $La_{0.6}Sr_{0.4}Co_{0.3}O_3$ | $La_{0.8}Sr_{0.1}Ga_{0.8}Mg_{0.1}Ni_{0.1}O_3$| Pt, HCs(+base)	HCs	250–1000	300	2003	[54]
HCs(+base), $LaSr_{0.5}Mn_{0.5}O_3$ | $La_{0.8}Sr_{0.2}Ga_{0.8}Mg_{0.1}Ni_{0.1}O_3$| Pt, HCs(+base)	HCs	500–2000	350	2004	[55, 56]
HCs(+base), $La_{0.5}Sr_{0.5}MnO_3$ | $La_{0.7}Sr_{0.3}Ga_{0.7}Fe_{0.3}O_3$ | $(Pt_{0.9}Co_{0.1})_{0.9}(LDC)_{0.1}$, HCs(+base)	HCs	500–2000	250	2005	[57]
base, Pt | YSZ | ZnO+Pt, HCs(+base)	HCs	10–800	600	2007	[58, 59]
(Impedancemetric)					
base, Pt | YSZ | $ZnCr_2O_4$, NOx(+base)	NOx	50–400	600–700	2002	[60, 61]
CO(+base), Pt | YSZ | Au-Ga_2O_3, CO(+base)	CO	0–800	550	2005	[62]
base, Pt | YSZ | ZnO+Pt, HCs(+base)	HCs	10–800	600	2006	[63]
base, Pt | YSZ | ZnO, HCs(+base)	HCs	0.05–0.8	600	2007	[64]
base, Pt | YSZ | In_2O_3, H_2O(+base)	H_2O	70–30000	500–900	2004	[65, 66]

view for over two decades. The following sections address the recent developments that have been done in our group for the YSZ-based electrochemical gas sensors in various sensing modes.

5.2 Mixed-Potential-Type Sensors

Mixed-potential-type gas sensors based on YSZ are considered to be a potential device for in situ measurements of various pollutant gases emitted from the different combustion processes. They can be performed either in tubular or in planar configuration in combination with an oxide-SE. When NO_2 is coexisting with O_2 for example, the following anodic reaction of O_2 and the cathodic reaction of NO_2 proceed simultaneously at the SE/YSZ interface [1].

$$\text{(anodic)} \quad 2O^{2-} \rightarrow O_2 + 4e^- \tag{5.1}$$

$$\text{(cathodic)} \quad 2NO_2 + 4e^- \rightarrow 2NO + 2O^{2-} \tag{5.2}$$

When the rate of the anodic reaction is equal to that of the cathodic reaction, the mixed-potential arises at SE. Since the reaction kinetics depends strongly on the SE material, the SE which favors Eq. (5.2) rather than Eq. (5.1) will give more intense non-Nernstian response. Besides the charge-transfer reactions at the interface, the SE material usually has certain catalytic activity to the gas-phase reaction. Thus, in the case of NO_2 detection, the following decomposition reaction (5.3) of NO_2 to NO has always to be considered. Thus, the morphology, composition and thickness of the SE layer play a vital role for the establishment of actual concentration of the target gas at the SE/YSZ interface.

$$\text{(gas - phase)} \quad 2NO_2 \rightarrow 2NO + O_2 \tag{5.3}$$

Furthermore, since the state of SE/YSZ interface is considered to be responsible for the mixed potential [12], it is expected that the application of the nanostructured SEs permits to obtain highly sensitive and highly selective responses to target gases as well as to improve the response/recovery rates of sensor. In the following sections, we focus on the importance of the selection of SE materials and improvement methods of sensing performances for various mixed-potential-type gas sensors based on YSZ.

5.2.1 High-Temperature-Type NOx Sensors

Since the temperature of car engines sometimes reach even up to 900°C during high-speed cruising, an extensive search was made to find out a suitable SE material for the detection of NO_x at high temperature in excess of 800°C. Figure 5.2 shows the comparison of sensitivity to 400 ppm NO_2 at 850°C (or 700°C) for the zirconia-based sensors attached with each of various SEs. It turned out that NiO was found to be the most suitable SE material which gave high sensitivity even at 850°C [13,14]. Thus, the NO_x sensing characteristics were examined in the wide temperature range of 700–900°C. As a result, the best operating temperature for the sensor

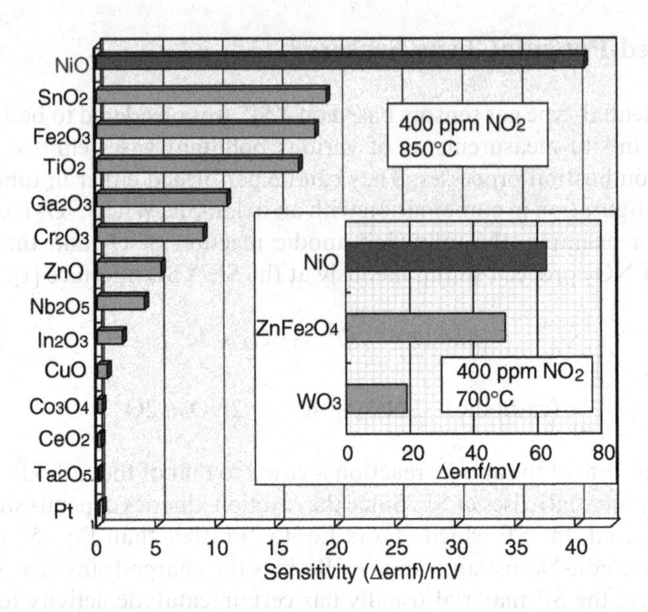

Fig. 5.2 Comparison of sensitivity to 400 ppm NO_2 at 850°C (or 700°C) for the sensors attached with each of various SEs

attached with NiO-SE was found to be 800°C, where the sensitivity was high and the response/recovery were good. It was also confirmed that the present sensor showed better NO_2 sensing characteristics; higher sensitivity and faster recovery, under the wet condition (5 vol. % H_2O) rather than under dry condition [14]. Such a performance is highly preferable to its actual application to automobile exhausts, which contain a lot of water vapor.

The sensing performances of mixed-potential-type sensors have been reported to depend strongly on morphology, sintering temperature, composition and thickness of SE layer. Thus, these key parameters are needed to be optimized in order to fabricate a stable and reliable NO_x sensor. The morphology of sensing layer can be controlled by various ways such as changing material synthesis method, use of sintering/pore-creating agents and varying sintering temperature of SE during processing. Quite recently, we have reported that the morphology (grain size and pore size) of NiO-SE can be well controlled by varying sintering temperature from 1100 to 1400°C [12]. With increasing sintering temperature, the average pore size was found to increase from *ca.* 0.5 μm to *ca.* 2 μm. As a result, the BET surface area of NiO-SE decreased from 0.5 to 0.1 m^2g^{-1}. This suggests that higher sintering temperature can give lower number of reaction site at gas/YSZ/NiO interface. The less surface area of NiO-SE will lead to lower heterogeneous catalysis for the decomposition of NO_2 to NO. Figure 5.3 shows variation of sensitivity to 200 ppm NO_2 and 70% recovery time at 800°C as a function of sintering temperature of NiO-SE. The inset shows response transients to 200 ppm NO_2 at 800°C. The sensitivity was strongly dependent on sintering

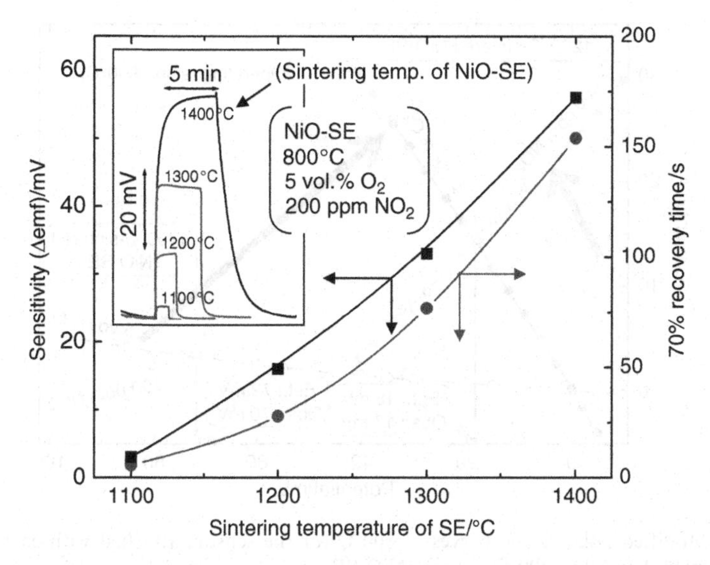

Fig. 5.3 Variation of sensitivity to 200 ppm NO_2 and 70% recovery time at 800°C for the sensor attached with NiO-SE sintered at different temperatures. Inset: Response transients to 200 ppm NO_2

temperature and high sensitivity was observed for the sensor attached with the 1400°C-sintered NiO-SE. However, the response/recovery rates became a bit lower for this sensor. Thus, the optimum sintering temperature of NiO-SE seemed to be 1300°C. In fact, the sensor attached with the 1300°C-sintered NiO-SE showed excellent selectivity to NO_2 even at 900°C with acceptable sensitivity [15].

Figure 5.4 shows the modified I-V curves obtained at 800°C for the sensors attached with each of the 1200°C-sintered and the 1400°C-sintered NiO-SEs. It is seen that the anodic polarization curve shifts upward as the sintering temperature is lowered. This means that the rate of anodic reaction of O_2 is higher at lower sintering temperature. Such a high rate of anodic reaction of O_2 leads to quicker recovery. On the other hand, the anodic current for the sensor attached with the 1400°C-sintered NiO-SE is low. Although such a low rate of anodic reaction of O_2 leads to poor recovery rate, it brings about higher NO_2 sensitivity because of the positive shift of mixed potential value which can be given by the intersection of both the anodic and the cathodic polarization curves, as shown in Fig. 5.4. The low rate of reaction is due to less number of reaction sites available at the NiO–YSZ interface. The sintering temperature of NiO-SE also affects strongly the gas-phase conversion of NO_2 into NO. The degree of NO_2 conversion tended to decrease as the sintering temperature was increased. Thus, it was confirmed that the catalytic activities to the gas-phase reaction of NO_2 as well as the anodic reaction of O_2 played an important role in deciding the NO_2 sensitivity of the present sensor.

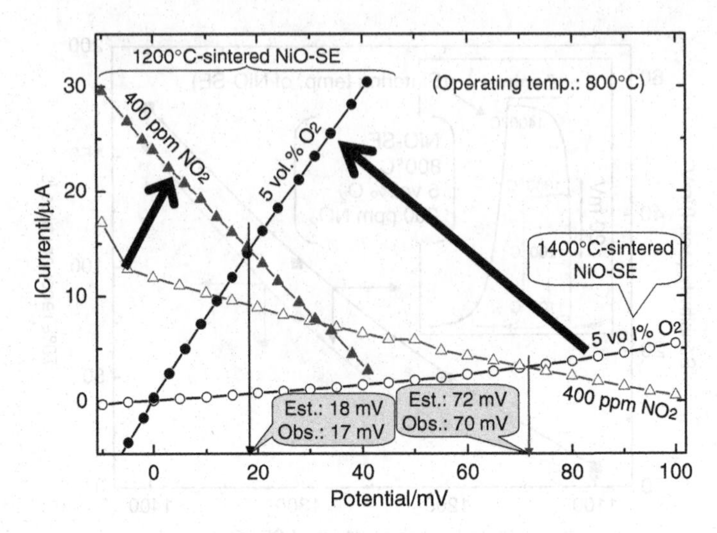

Fig. 5.4 Modified polarization curves at 800°C for the sensors attached with each of the 1200°C-sintered and the 1400°C-sintered NiO-SEs

In order to examine the effect of thickness of sensing layer on sensitivity, the Cr_2O_3 (or NiO)-SE layers having various thicknesses (in the range of 2.7–11 μm) were fabricated by means of slide-off-transfer printing technique. Figure 5.5 (a) shows the dependence of sensitivity to 200 ppm NO_2 on thickness of SE for the sensor attached with Cr_2O_3 (or NiO)-SE. Figure 5.5 (b) depicts response transients to 200 ppm NO_2 at 850°C for the sensor attached with each of NiO-SEs having various thicknesses. The NO_2 sensitivity depends strongly on the

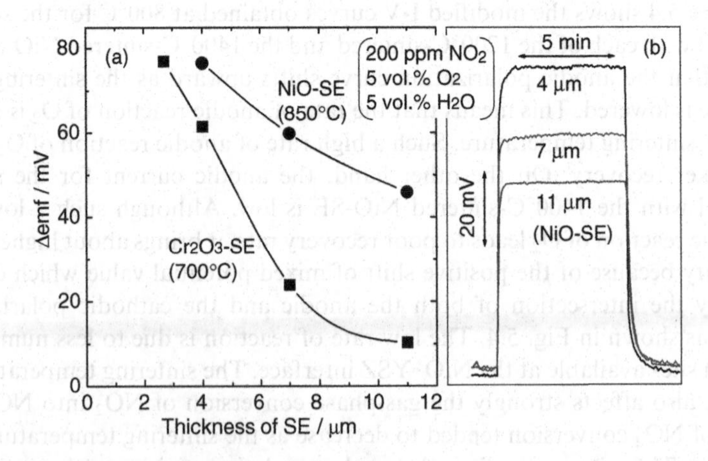

Fig. 5.5 (**a**) Dependence of sensitivity to 200 ppm NO_2 on the thickness of SE (NiO or Cr_2O_3); (**b**) response transients to 200 ppm NO_2 at 850°C for the sensor attached with each of NiO-SEs having various thicknesses

thickness of SE layer and higher sensitivity was obtained when thinner SE layer was used [12, 22]. Since the state of interface was confirmed to be same in each SE having different thickness, it was observed that the NO_2 sensitivity was altered by the degree of the gas-phase NO_2 decomposition reaction occurring in the oxide-SE matrix. The degree of NO_2 conversion decreased with decreasing thickness of SE layer and the lowest conversion was observed for the thinnest SE layer (4 μm for NiO, 2.7 μm for Cr_2O_3). Thus, the sintering temperature and thickness of sensing layer should be selected properly so as to obtain high sensitivity as well as quick response.

5.2.2 *Improvement in NO_2 Sensitivity by Additives*

The NO_2 sensitivity of mixed-potential-type sensor usually decreases drastically as the operating temperature is increased in excess of 700°C. Thus, improvement in sensing performance of NO_x sensor at such an extremely high temperature is still one of challenging issues. Figure 5.6 shows the comparison of sensitivity to 50 ppm NO_2 at 800°C for the sensor attached with each of NiO-SEs added with various oxides (10 wt.% each). Among the various additives examined, WO_3 was found to make highest improvement in NO_2 sensitivity [17]. We have also examined the effect of addition of noble metals such as Pd, Pt, Ru and Rh to NiO-SE on NO_2 sensitivity [18]. It was observed that only Rh

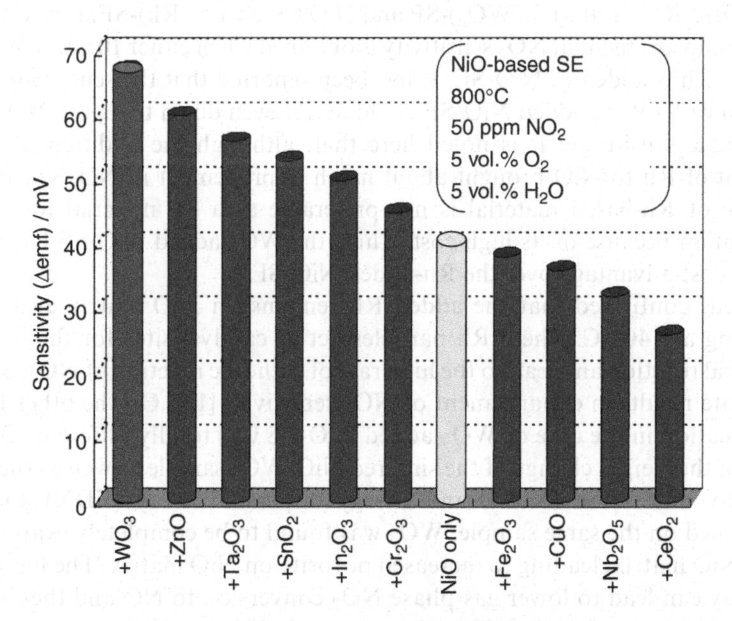

Fig. 5.6 Comparison of sensitivity to 50 ppm NO_2 at 800°C for the sensor attached with each of NiO-SEs added with various oxides (10 wt.% each)

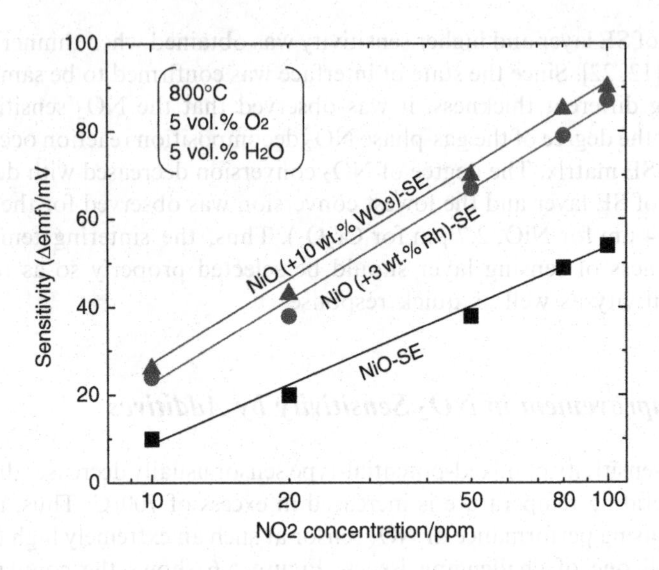

Fig. 5.7 Dependence of Δemf on concentration of NO_2 at 800°C for the sensor attached with each of NiO-SE, NiO (+ 10 wt.% WO_3)-SE and NiO (+ 3 wt.% Rh)-SE

(3 wt.%) improved the NO_2 sensitivity. Figure 5.7 shows the dependence of Δemf on concentration of NO_2 at 800°C for the sensors attached with each of NiO-SE, NiO (+ 10 wt.% WO_3)-SE and NiO (+ 3 wt.% Rh)-SE. It is seen that a large improvement in NO_2 sensitivity is obtained when either 10 wt.% WO_3 or 3 wt.% Rh is added to NiO-SE. It has been reported that the sensor attached with 10 wt.% WO_3-added NiO-SE could detect even down to 1 ppm NO_2 with acceptable sensitivity. It is noted here that, although the addition of small amount of Rh to NiO brought about much improvement in NO_2 sensitivity, the use of Rh-based material is not preferable as a SE material for actual application because of its high cost. Thus, the WO_3-added NiO-SE has much higher cost-advantage over the Rh-added NiO-SE.

It was confirmed that the added Rh remains on NiO matrix even after sintering at 1400°C. These Rh particles act as catalyst sites for the electrochemical reaction and lead to the high rate of cathodic reaction of NO_2. Such a high rate results in enhancement of NO_2 sensitivity [18]. On the other hand, the situation in the case of WO_3-added NiO-SE was totally different. On the basis of the weight change of the sintered NiO-WO_3 sample as well as the data of x-ray diffraction (XRD) and energy dispersive X-ray (EDX) analysis performed on the same sample, WO_3 was found to be completely evaporated from NiO matrix, leading to increased porosity on NiO matrix. The increased porosity can lead to lower gas-phase NO_2 conversion to NO and then to the increased NO_2 sensitivity. Thus, it was concluded that WO_3 acted as pore-creating agent in the sintering process at rather high temperature [17].

5.2.3 Hydrocarbon (C_3H_6 or CH_4) Sensors

Various hazardous gases such as NO_x, CO and HCs are exhausted to the environment from several sources. Among the various hazardous gases, some of HCs cause serious effects to atmospheric environment. For example, propene (C_3H_6) has strong photochemical reactivity which leads to urban photochemical smog and methane (CH_4) causes green-house effect. Thus, there is a strong demand for the detection and monitoring of HCs. Recently, on the process of searching a suitable material for detection of HCs, we have found that In_2O_3 could be the best SE material for selective detection of C_3H_6. The sensor attached with In_2O_3-SE could selectively detect C_3H_6 in the very low concentration range of 10–200 ppb at relatively low temperature of 450°C. Figure 5.8(a) shows the response transients to various concentration of C_3H_6 at 450°C. The sensor reaches the steady-state emf value within several minutes. The sensitivity to various concentration of C_3H_6 varies linearly on a logarithmic scale, as shown in Fig. 5.8(b). Such a linear variation is typical for a mixed-potential-type sensor. It is worthy to note that the present sensor can give rather good emf response as high as *ca.* 20 mV even to 20 ppb C_3H_6 and such a high sensitivity to ppb level gas has never been reported so far by using mixed-

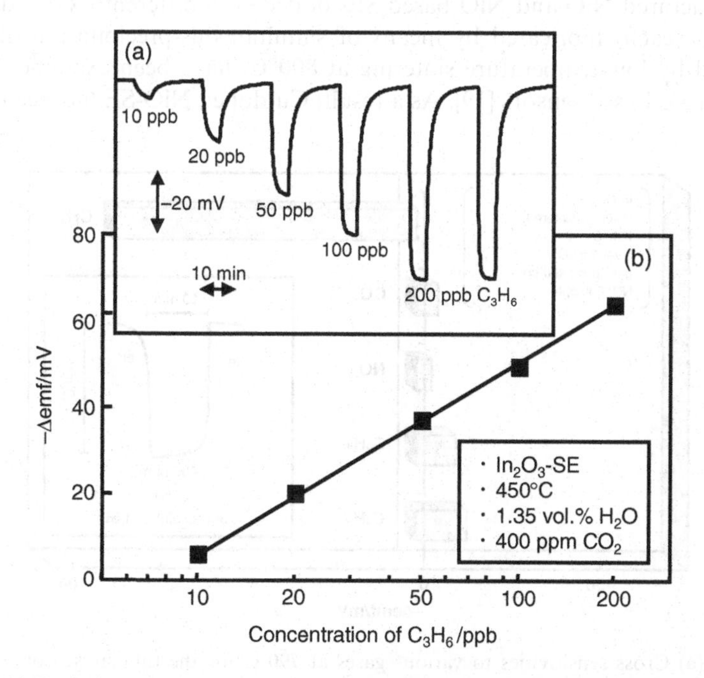

Fig. 5.8 (a) Variation of Δemf to different concentration of C_3H_6 at 450°C for the sensor attached with In_2O_3-SE, (b) response transients to various concentration of C_3H_6

potential-type gas sensors. Consequently, this sensor can be considered as one of potential candidates for environmental monitoring of C_3H_6.

The same sensor attached with In_2O_3-SE was examined for the detection of CH_4. The sensing characteristics were examined in the temperature range of 700–900°C. The sensor exhibited the highest sensitivity and selectivity to CH_4 at around 790°C, as shown in Fig. 5.9. It was found that the use of the 1300°C-sintered In_2O_3-SE exhibited much faster response/recovery, compared with that of the sensor attached with the 1400°C-sintered In_2O_3-SE [42]. Hence, it is clear that if the proper operating temperature is selected, the same SE material can be used for the selective and sensitive detection of CH_4 at 790°C or C_3H_6 at 450°C.

5.2.4 Use of Nanostructured NiO-Based Materials

For the last decade, a growing attention has been paid to the utilization of nanostructured materials in different chemical and electrochemical devices [19,20,21,25]. The application of nanostructured materials in YSZ-based sensors allows minimizing the size of sensor, decreasing the sintering temperatures for SEs and REs (lower power consumption) as well as improving the gas sensitivity and response/recovery rates of the sensor's signal. Recently, the nanostructured NiO and NiO-based SEs doped with different oxide additives (10 wt.% each) fabricated by means of simultaneous-precipitation method, followed by low-temperature sintering at 800°C, have been examined in the planar YSZ-based sensors [19]. As a result, Cu-doped NiO-SE has been found

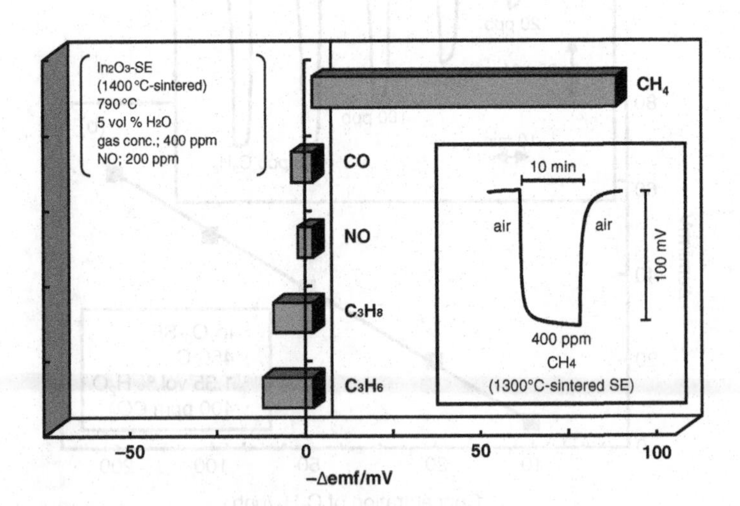

Fig. 5.9 (**a**) Cross sensitivities to various gases at 790°C for the tubular sensor using the 1400°C-sintered In_2O_3-SE under the wet condition, (**b**) response transient to 400 ppm CH_4 for the sensor attached with the 1300°C-sintered In_2O_3-SE

to show higher NO_2 sensitivity at 800°C under the wet condition (5 vol.% H_2O) in comparison with pure NiO-SEs. Simultaneously, the sensor using Cu-doped NiO-SE showed rather good NO_2 selectivity at the same operating temperature with negligible responses to CO, C_3H_8 and NO (400 ppm each). Unfortunately, however, the utilization of nanostructured SEs fabricated by screen-printing technique is limited probably due to their high catalytic activity towards the gas-phase decomposition of NO_2 to NO.

In order to examine the effect of the thickness of NiO-SE on the sensing characteristics, the sensors using thin-film NiO-SEs having thickness of 60–180 nm have been fabricated by using radio-frequency (r.f.) sputtering technique [21]. To improve the selectivity towards NO_2, the laminated-type (60-nm-thick NiO + 60-nm-thick Au)-SE (hereafter abbreviated as NiO-Au) has been further fabricated and examined [25]. Figure 5.10 depicts the

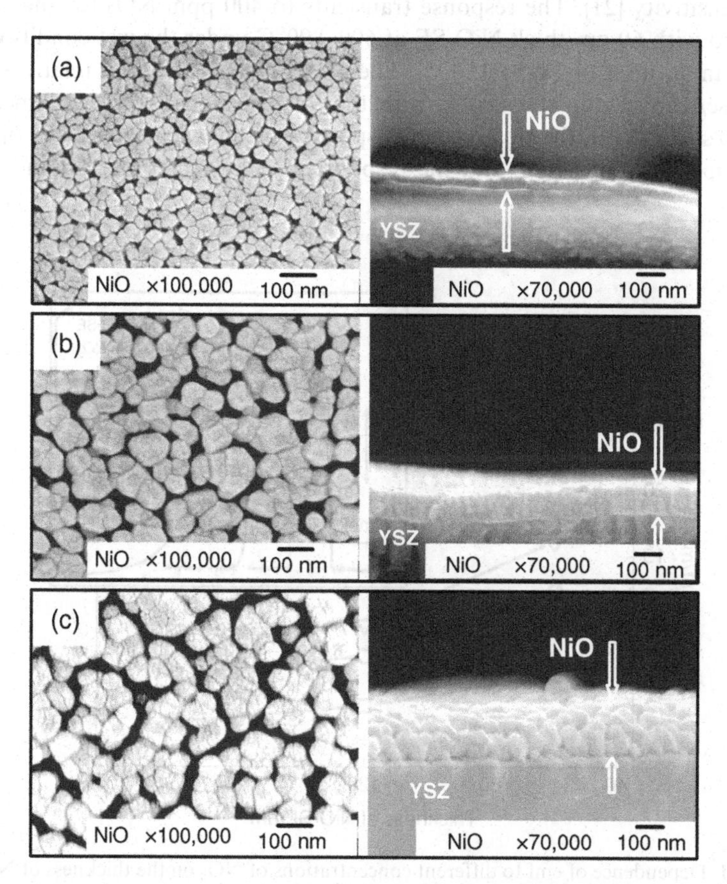

Fig. 5.10 Surface and cross-sectional SEM images of the thin NiO films formed on YSZ plate after sintering at 1,000°C. The sputtering times were **(a)** 20 min, **(b)** 40 min, and **(c)** 60 min [reprinted from Ref. [25] with permission from Springer]

representative surface and cross-sectional scanning electron microscope (SEM) images of the thin NiO films formed on YSZ plates by sputtering of Ni target for 20 min (a), 40 min (b) and 60 min (c), followed by oxidation and sintering at 1000°C. It is clearly seen that the thickness and grain size of the NiO film increased with increasing sputtering time. The thicknesses of the films obtained were about (a) 60, (b) 120 and (c) 180 nm [25]. For the laminated (NiO-Au)-SE, most of gold particles having the size of 40–70 nm were located on the surface of NiO thin-film. Moreover, based on the x-ray photoelectron spectroscopy (XPS) analysis, it has been shown that gold nanoparticles also permeate into the 60-nm-thick NiO layer forming the small gold nanoparticles in the bulk NiO and the additional hetero-junction NiO-Au–YSZ interface [25].

The dependence of NO_2 sensitivity on the thickness of NiO-SE at 800°C under the wet condition is shown in Fig. 5.11. It is seen that, at each NO_2 concentration, the sensor attached with 60 nm-thick NiO-SE gave the highest NO_2 sensitivity [21]. The response transients to 400 ppm NO_2 for the sensor attached with 60 nm-thick NiO-SE at 600–800°C under the wet condition are shown in an inset of Fig. 5.11 [25]. The present sensor showed relatively fast response/recovery at 700–800°C with the 90% response/recovery times less than 10 s. However, the steady-state response was only achieved at the operating temperature of 600°C and the response/recovery rates were found to be

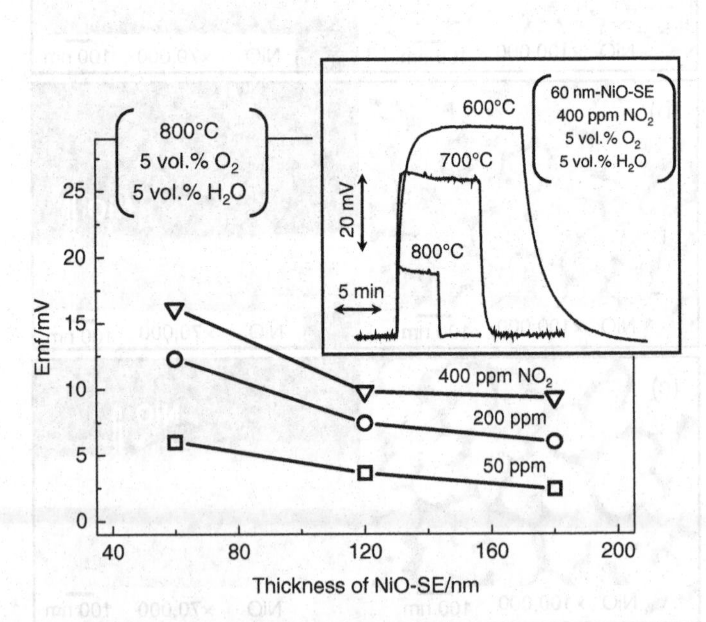

Fig. 5.11 Dependence of emf to different concentrations of NO_2 on the thickness of NiO-SE for the sensor operating at 800°C in the presence of 5 vol.% H_2O. Inset: response/recovery transients to 400 ppm NO_2 for the sensor using 60-nm-thick NiO-SE at different operating temperatures [reprinted from Ref. [25] with the permission from Springer]

lower [25]. It is noted that the sensor's response at high operating temperature can be stabilized by increasing the sintering temperature of thin-NiO-SE. However, simultaneously, the thicker NiO-SE (120 nm-thick) gave the highest NO_2 sensitivity [21].

Figure 5.12 shows the comparison of response transients to NO_2 at 600°C under the wet condition for the sensors attached with each of 60 nm-thick NiO-SE and the laminated (NiO-Au)-SE. It is seen that the sensor attached with the laminated (NiO-Au)-SE gave more stable and higher emf responses to NO_2. In addition, this sensor exhibited much faster recovery; the 90% recovery time was as short as 30 s. Higher NO_2 sensitivity can be attributed to rather strong preferential NO_2 adsorption on the surface of nanogold [67, 68] followed by spillover towards (TPB) [69] providing higher actual NO_2 concentration in the vicinity of SE–YSZ interface. The higher recovery rate is probably ascribed to higher oxygen coverage of nanogold supported by NiO layer as a result of additional dissociative oxygen chemisorption on the gold nanoparticles dispersed on the oxide [70] and the presence of oxygen ion species on nanogold [70,71,72]. The higher oxygen coverage affects also the selective properties and the sensor attached with the laminated (NiO-Au)-SE exhibited the excellent NO_2 selectivity at 600°C under the wet condition, while the sensor using pure NiO-SE showed poor selectivity [25]. The excellent selectivity in the case of the layered SE may be caused by the high catalytic activity of gold nanoparticles located on the NiO layer for the gas-phase oxidation reaction of all oxidizable gases [25, 70,71,72].

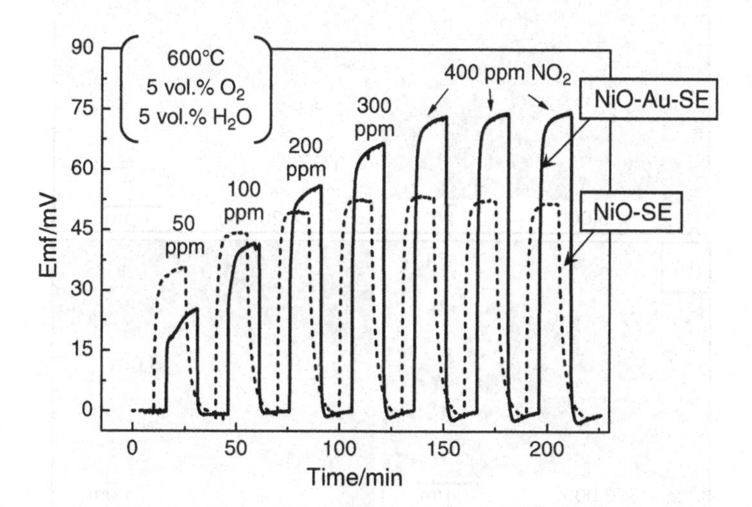

Fig. 5.12 Response transients to NO_2 at 600°C in the presence of 5 vol.% water vapor for the sensors attached with each of 60-nm thick NiO-SE and the laminated (NiO–Au)-SE [reprinted from Ref. [25] with the permission from Springer]

5.2.5 *Nanosized Au Thin-Layer for Sensing Electrode*

Based on the above mentioned results, we can speculate that the gold nanoparticles are presumably responsible for the highly sensitive and selective responses towards NO_2. To examine the ability of nanostructured gold towards the NO_2 sensitivity, the Au-SEs with various thicknesses have been fabricated by sputtering of Au for different times (30–720 s) onto a YSZ plate, followed by sintering at 950°C for 1 h. Then, they have been examined in YSZ-based planar sensor operated at 600°C. After sintering, the fabricated Au-SEs were found to consist of discrete gold particles with the rectangular-like structure where the length is longer than the height. Figure 5.13 shows the representative surface SEM images of the Au-SEs deposited for 120 s in comparison with the conventional Au-SE fabricated by screen-printing technique. It is seen that 120 s-sputtered Au-SE is composed of a combination of huge-Au agglomerates placed on the YSZ grains and the fine nano-Au particles positioned along the grain-boundaries of YSZ. The fabricated Au-SEs did not exhibit the usual metallic conductivity at any operating temperature. The complex-impedance plots of the sensor exhibited always one depressed semicircle, denoting that the conductivity is probably provided along the grain-boundaries of YSZ. It can be speculated that the small gold nanoparticles located along the grain-boundaries of YSZ form the low-conductive gold nanonetwork along the entire surface of YSZ. On the contrary, the Au-SE fabricated by screen-printing method showed always the perfect metallic conductivity.

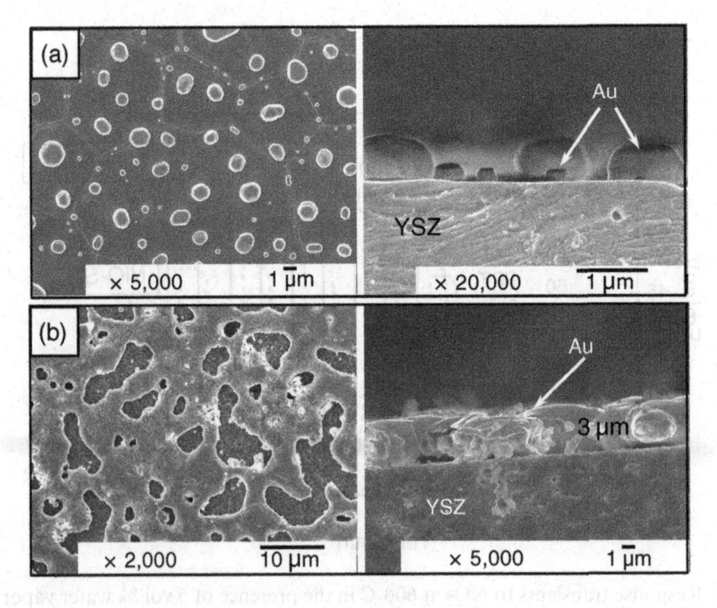

Fig. 5.13 Representative SEM images of Au-SEs fabricated by (**a**) sputtering for 120 s and (**b**) screen-printed techniques

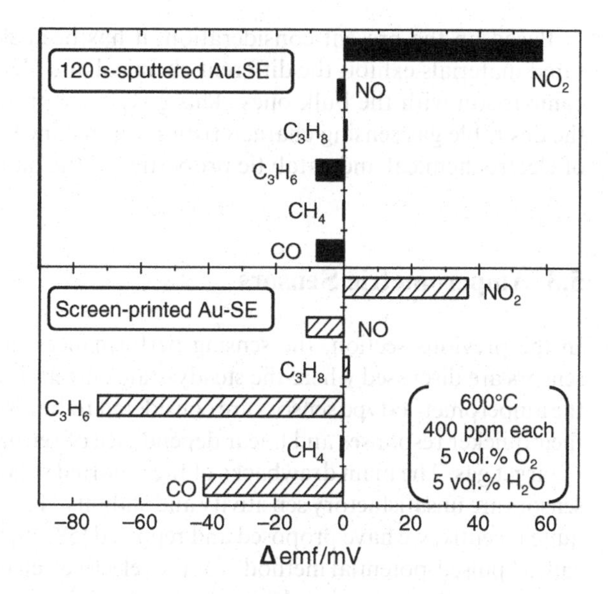

Fig. 5.14 Comparison of the cross sensitivities to different gases at 600°C under the wet condition (5 vol.% water vapor) for the sensors using each of the 120 s-sputtered Au-SE and the screen-printed 3 μm-thick Au-SE

All the fabricated sensors using thin Au-SEs showed rather selective response towards NO_2. Figure 5.14 shows the comparison of the cross sensitivities to different gases (CO, CH_4, C_3H_8, C_3H_6, NO, NO_2; 400 ppm each) at 600°C under the wet condition for the sensors using each of 120 s-sputtered Au-SE and screen-printed 3-μm-thick Au-SE. The sensor attached with 120 s-sputtered Au-SE was highly selective to NO_2, with minor responses towards both CO and propylene as well as negligible sensitivity to the saturated hydrocarbons and NO. The same reason for the NO_2 selectivity discussed in the case of the laminated (NiO-Au)-SE can be also adopted to the behavior of the present sensor. Furthermore, it should be also emphasized that the sensitivity to 400 ppm NO_2 depends strongly on the sputtering time of Au-SEs and gives the maximum value as high as 58 mV at the 120 s-sputtering time. The 90% response and recovery times were as short as 60 s and 15 s, respectively. For comparison, the recovery rate of the present sensor was about 10 times higher than that of the sensor attached with screen-printed 3-μm-thick Au-SE, because the massive bulk gold would be rather inert for the adsorption and activation of the oxygen species. It is believed that a combination of nanogold particles and large-gold agglomerates seems to be responsible for the reduction of the amount of active gold centers changing the catalytic activities of cathodic reaction of NO_2 and anodic reaction of O_2 at the interface. At the same time, it should be pointed out that all low-conductive Au-SEs were found to give very low polarization current (less than 50 nA), producing obviously very high impedance. Undoubtedly, such extremely low currents or high impedances are not applicable for the amperometric or impedancemetric sensors. On the other hand, it is noticeable merit for the mixed-potential-type sensor to observe the high enough emf values irrespective of the very low current or very high impedance.

Based on the present consideration, it has been shown that the nanostructured materials exhibit the different chemical and electrochemical properties in comparison with the bulk ones. This gives very promising way to accomplish the desirable gas sensing characteristics, for instance, utilizing the combination of electrochemical and catalytic properties of the nanostructured materials.

5.3 Amperometric Sensors

In the previous section, the sensing performances of the mixed-potential-type sensors are discussed where the steady-state current is equal to zero. Meanwhile, the amperometric-type sensors are considered to be also promising ones owing to their quicker responses and linear dependence of sensitivity on the concentration of target gas. The main drawbacks of the reported solid-electrolyte amperometric sensors are unsatisfactory sensitivity and stability. To overcome these drawbacks, quite recently, we have proposed and reported [58, 59] a novel measuring method called "pulsed-potential method" for the selective detection of propene (C_3H_6), in contrast to the conventional "constant-potential method". Figure 5.15 shows the schematic view of the representative applied-potential and current-response

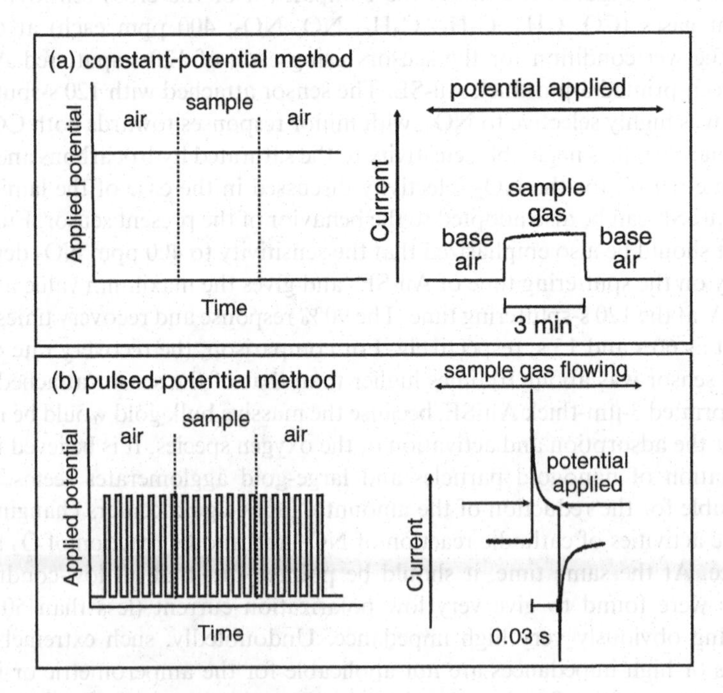

Fig. 5.15 Schematic views of applied potential and current response transients for the amperometric sensor operated by (**a**) "constant-potential method" and (**b**) "pulsed-potential method"

transients which can be obtained by the both methods. In the "constant-potential method", the current response of the sensor is measured at a fixed applied potential when the gas flow is changed from the base gas to the sample gas. On the contrary, in the "pulsed-potential method", the peak current is measured as a sensing signal when the potential is applied to the SE for a very short duration of 0.03 s under continuous flowing of the sample gas with the fixed concentration [58, 59].

Figure 5.16 shows the dependence of sensitivity (ΔI; defined as $I_{sample\ gas} - I_{base\ gas}$) on C_3H_6 concentration at 600°C under the wet condition (1.4 vol.% H_2O) for the sensor with ZnO (+8.5 wt.% Pt)-SE operated under the both measuring modes. According to the Fourier transform infra-red spectroscopy (FT-IR) measurements, ZnO was found to adsorb C_3H_6 selectively even at high temperature whereas the addition of Pt permitted to minimize the cross-sensitivity to NO_2. It is clearly seen that the C_3H_6 sensitivity obtained by the "pulsed-potential method" is much higher than that obtained by the "constant-potential method". For example, the ΔI value to 400 ppm C_3H_6 obtained by the new method is as high as 95 μA which is about twenty times higher than that obtained by the conventional method. In addition, the present sensor operated under both modes gave high selectivity to C_3H_6 with low cross sensitivities to other gases examined (including hydrogen and NO_2) [59]. Furthermore, the C_3H_6 sensitivity was found to be almost invariant with the change in water vapor (1.4–16 vol.%) and oxygen (3–20 vol.%) concentrations in the examined gas. Finally, the amperometric sensor operated under the pulsed-potential mode showed much better short-term stability exhibiting almost constant C_3H_6 sensitivity during the examined period of 35 days [59]. In the case of sensor operated under constant-potential mode, the sensitivity degraded rapidly down to *ca.* 25% of the initial value just after 5 days.

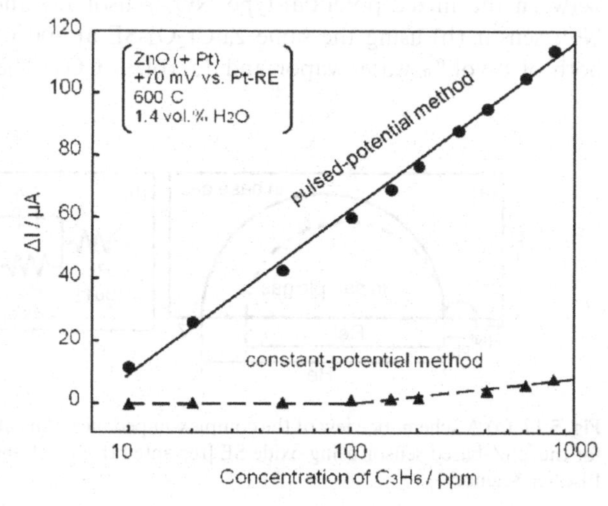

Fig. 5.16 Dependence of sensitivity on concentration of C_3H_6 at 600°C under the wet condition (1.4 vol.% H_2O) for the sensor using ZnO (+8.5 wt.% Pt)-SE operated by the "pulsed-potential method" and the "constant-potential method"; applied potential: +70 mV vs. Pt-RE [reprinted from Ref. [59] with the permission from MYU K.K., Tokyo]

5.4 Impedancemetric Sensors

Apart from the conventional mixed-potential and amperometric sensors, a new-type impedancemetric YSZ-based sensor has been proposed and examined for the detection of total NO_x [60, 61], water vapor [65] and different combustible HCs [66]. In this case, the change in the complex-impedance of device attached with the specific oxide-SE is measured as a sensing signal. In general, the equivalent circuit for the impedancemetric YSZ-based sensor attached with oxide SE can be expressed, as shown in Fig. 5.17. Here, R_b means the bulk resistance of YSZ, R_o and C_o are resistance and capacitance of oxide-SE, respectively, and R_e and C_e represent resistance and capacitance, respectively, due to the electrochemical reaction ($O_2 + 4e^- = 2O^{2-}$) occurring at the interface between YSZ and oxide-SE. When the sensor is exposed to the sample gas containing the certain concentration of examined gas, the resistance value (R_e) at the intersection of the large semicircle with the real axis at lower frequency (around 0.1 Hz) is shrinking, in comparison with that for the base gas. The difference between the impedance ($|Z_{air}|$) in the base gas and the impedance ($|Z_{gas}|$) in the sample gas at a fixed frequency is defined as the gas sensitivity ($|Z|$) [60].

5.4.1 *Sensing of Various Gases in ppm Level*

On the contrary to the mixed-potential-type sensor where the response direction to NO is opposite to that to NO_2, the impedancemetric sensor exhibited a completely different manner when the sensor is exposed to NO or NO_2. It has been reported that the impedance value decreased with an increase in the concentration of both NO and NO_2 [60, 61]. Figure 5.18 shows the comparison between the mixed-potential-type NO_x sensor (a) and the impedancemetric NO_x sensor (b) using the same $ZnCr_2O_4$-SE at 700°C in the presence of the both of 8 vol.% water vapor and 15 vol.% CO_2. The gas sensitivity for the

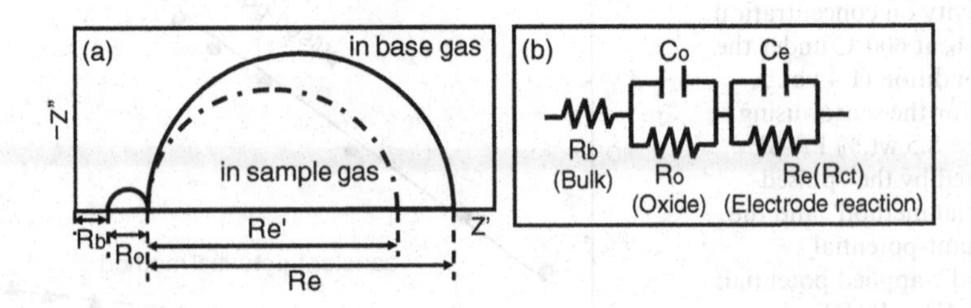

Fig. 5.17 (**a**) A schematic view of the complex-impedance plots and (**b**) the equivalent circuit, for the YSZ-based sensor using oxide SE [reprinted from Ref. [60] with the permission from Elsevier Science]

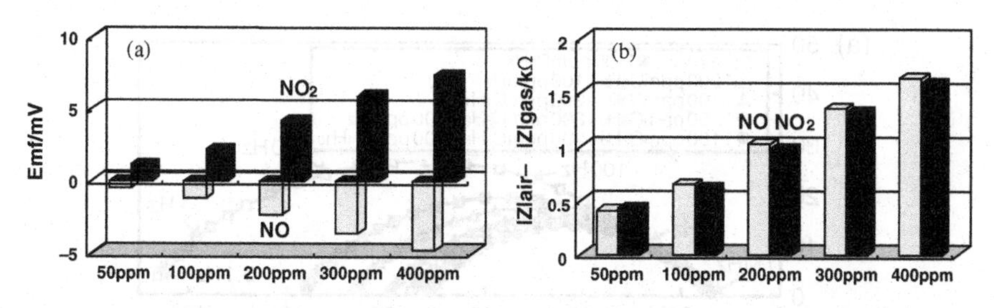

Fig. 5.18 Comparison of NO_x sensitivities between (**a**) the mixed-potential-type and (**b**) the impedancemetric NO_x sensors using the same $ZnCr_2O_4$-SE

impedancemetric sensor was measured at the fixed frequency of 1 Hz. As can be seen clearly, the NO response for the impedancemetric sensor shows the same direction as the NO_2 response does. On the other side, the mixed-potential-type NO_x sensor shows the output *emf* to NO which is opposite to that to NO_2. Furthermore, the impedancemetric sensor showed the near-identical behavior to NO and NO_2 at 700°C, which indicated that the present sensor can detect total NO_x (NO + NO_2) in the sample gas, regardless of NO/NO_2 ratio.

Another striking feature of the imperadncemetric sensors is the possibility to detect water vapor [65] and combustible hydrogen-containing gases [66]. The gas sensitivity of the impedancemetric YSZ-based sensor using In_2O_3-SE operated at 900°C varied linearly on a logarithmic scale of water-vapor concentration over a wide range of 70–30,000 ppm. Then, the same sensor has been also applied to the detection of various combustible hydrogen-containing gases, although it should be operated under dry condition [66]. Figure 5.19 (a) shows the Nyquist plots in the dry base and in the sample gas containing the mixture of various concentrations of hydrogen-containing gases for the sensor using In_2O_3-SE operated at 900°C. It is seen that the large semicircular arc is shrinking when the sample gas containing 100 ppm C_3H_6. The same shrinkage behavior has been also observed if the sensor was exposed to other HCs [66]. Furthermore, it is seen from Fig. 5.19 (a), even in the case of gas mixture, the similar shrinking-behavior has been also observed. The gas sensitivity at 1 Hz varied linearly with the logarithm of hydrogen number of the measuring mixed-gas, as shown in Fig. 5.19 (b). In addition, when the ZnO (+Pt)-SE was used as a SE for the impedancemetric sensor, the detection of C_3H_6 in the concentration range of 10–1000 ppm was achieved even under the wet condition (1 vol.% H_2O) [63].

5.4.2 Environmental Monitoring of C_3H_6 in ppb Level

It has been also shown that the impedancementric sensors are highly sensitive even to very low concentrations of the target gas and, therefore, it is expected that they can be suited not only for the detection of gases

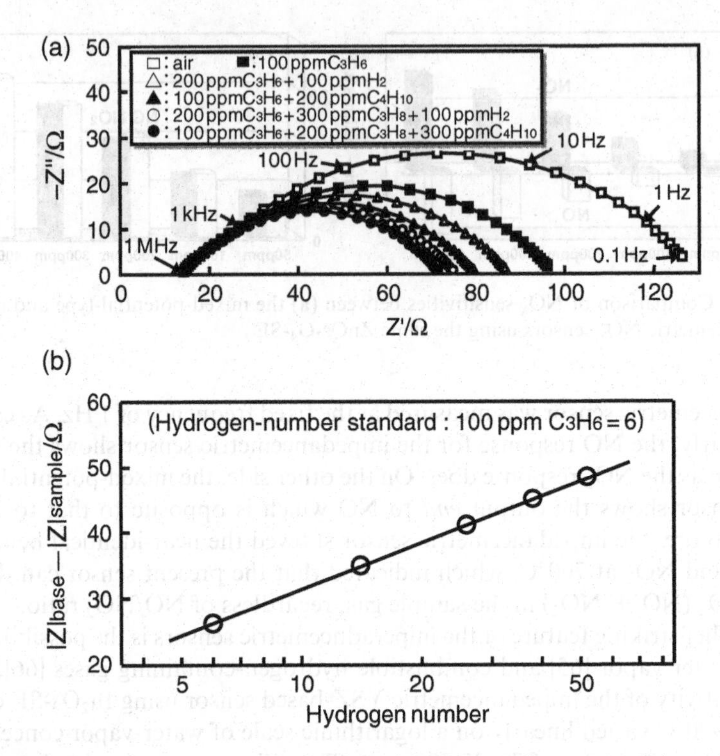

Fig. 5.19 (**a**) Nyquist plots in dry base air and in the sample gas containing mixture of hydrogen-containing gases and (**b**) dependence of gas sensitivity on hydrogen number of the sample gas, for YSZ-based impedancemetric sensor using In_2O_3-SE at 900°C under dry condition [reprinted from Ref. [66] with the permission from Elsevier Science]

from combustion exhausts, but also for the environmental monitoring. Quite recently, we proposed and reported the highly-sensitive impedancemetric YSZ-based sensor attached with pure ZnO-SE aiming at a selective detection of low concentration of C_3H_6 in the range of 0.05–0.8 ppm [64]. The representative response transients to C_3H_6 for the present sensor operating at 600°C and at the fixed frequency of 1 Hz are shown in Fig. 5.20 (a). It is noteworthy that the sensor can detect the low concentration of C_3H_6 down to 0.05 ppm (50 ppb) even under the wet condition (1.35 vol.% H_2O). Furthermore, the |Z| value changed quickly from the base level when the sample gas was introduced onto the sensor and the steady-state |Z| values were attained in due time in each case. It should be also pointed out that the present sensor showed a good repeatability of the response to C_3H_6. The dependence of sensitivity on the concentration of C_3H_6 showed the good linearity in the examined concentration range [64].

Another important parameter for the practical environmental HCs sensor is the selectivity to the desirable gas, because multitude of different gases exists

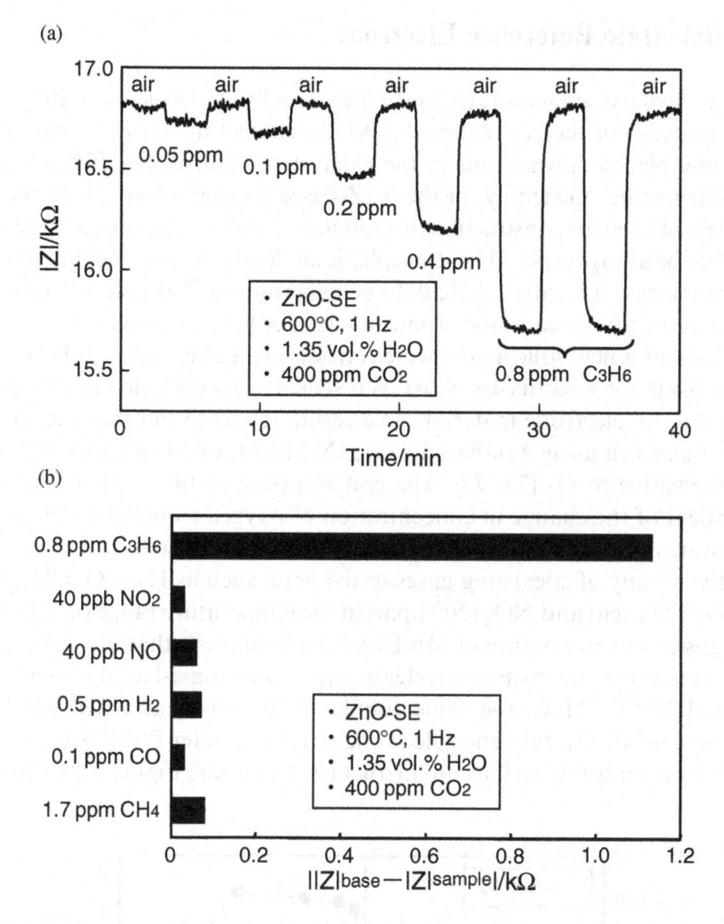

Fig. 5.20 **(a)** Response transients to C_3H_6 in the concentration range of 0.05–0.8 ppm [reprinted from Ref. [64] with the permission from Elsevier Science] and **(b)** cross-sensitivities to various gases (NO_2, NO, H_2, CO and CH_4 at 600°C in the presence of 1.35 vol.% H_2O, for the sensor attached with ZnO-SE at the fixed frequency of 1 Hz

simultaneously in surrounding atmosphere. Figure 5.20 (b) represents the cross sensitivities to various gases, such as NO_2, NO, H_2, CO and CH_4, at 600°C under the wet condition for the sensor using ZnO-SE. As shown in this figure, the sensitivity to 0.8 ppm C_3H_6 is remarkably higher than those to the other gases examined. Thus, it is confirmed that the present impedamcemetric sensor can give the selective response even to very low concentrations of C_3H_6 under the examined experimental conditions [64]. However, it is obvious that the impedancemetric sensors are just recently proposed and, no doubt, the additional investigations are necessary for the further improvement of the sensing performances for the different purposes.

5.5 Solid-State Reference Electrode

At present, two oxygen sensors (λ sensors) are usually used in gasoline engines; one placed upstream of the conventional TWC to control the air/fuel ratio and the second one placed downstream in the exhaust to control the efficiency of the catalytic converter. Generally, in the YSZ-based oxygen sensors, a Pt reference electrode is utilized for construction of a tubular- as well as a laminated-type sensor and should be always exposed to atmospheric air. Thus, the sensor configuration is not so simple and it is rather difficult to be miniaturized. The necessity of a Pt/air reference electrode increases the production cost of the oxygen sensors.

To find out a new solid-reference-electrode as an alternative to Pt/air reference electrode for zirconia-based oxygen sensor, various kinds of oxides were examined as an electrode material. As a result, it turned out that the zirconia-based tubular cell using Mn-based oxide ($NiMn_2O_4$ or Mn_2O_3)-SE was found to be insensitive to O_2 [73, 74]. The emf response of the tubular sensor was independent of the change in concentration of oxygen from 0.1 to 20 vol.% at temperatures below 600°C. In addition, the sensor was also found to be almost insensitive to any of coexisting gases tested here, such as H_2, CO, CH_4, C_3H_8, NO (400 ppm each) and NO_2 (200 ppm) in the temperature range of 400-900°C. Such a gas-insensitive nature of Mn-based oxide suggests that this material can be used as a stable reference electrode for a planar zirconia-based oxygen sensor at around 600°C. Thus, the same tubular YSZ sensor attached with outer Mn_2O_3 (or $NiMn_2O_4$) reference-electrode (RE) and outer Pt-SE was examined as an oxygen sensor as well as an air/fuel (A/F) sensor. Figure 5.21 shows the

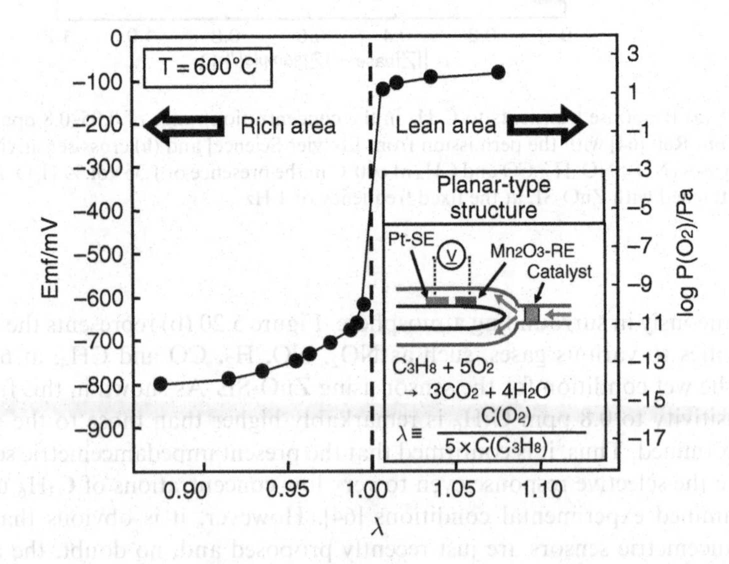

Fig. 5.21 Dependence of emf on air/fuel ratio for the planer-type structure sensor using Pt-SE and Mn_2O_3-RE operating at 600°C

dependence of emf on air/fuel ratio (λ). The emf change during the transition of λ value is as high as about 700 mV. This behavior was almost same as that of the tubular A/F sensor using a normal Pt/air-RE. This suggested that Mn_2O_3 (or $NiMn_2O_4$)-based RE can work stably even in the fuel-rich region where the partial pressure is usually extremely low such as around 10^{-13} Pa. Thus, Mn_2O_3 (or $NiMn_2O_4$) was found to be a promising material as a stable solid-reference-electrode for oxygen sensor as well as A/F sensor based on YSZ.

5.6 Conclusions and Future Prospective

We have developed a variety of high-performance solid-state electrochemical gas sensors for various applications. It has been shown that it is possible to optimize the sensing performances and receive the highly sensitive and highly selective responses to the target gases by controlling the thickness, sintering temperature and composition of SEs, and by altering the operating temperature and the measuring mode of the sensors. However, the main drawbacks for most of the solid-state electrochemical sensors are (i) a unique selectivity to the desirable gas in multi-component gas system (over about ten components) and (ii) the long-term stability of the sensor in a harsh surrounding atmosphere. For instance, the sensor for the automotive exhausts should be operated for more than ten years. No doubt, these drawbacks have to be overcome to a large extent by using all technologies and knowledge obtained up to now as well as applying new and original ideas of solid-state chemistry, electrochemistry, ceramics technology and so on.

The continuous large attention to the exhaust condition and the growing demands for the environmental monitoring require further improvement of the sensor's devices. The recent achievements in utilization of nanostructured SEs for the enhancement of gas sensing characteristics seem to be very attractive. It is believed that further investigations of nanostructured SEs will be able to promote significantly the sensing performances of the solid-state electrochemical sensor. It has been already shown in the semiconductive oxide sensors that the nanostructured materials can give the quite different and sometimes unexpected properties in comparison with bulk ones. To understand such distinctive features, the fundamental researches are obviously required. Furthermore, the utilization of nanostructured SEs allows miniaturizing the sensor's devices and minimizing the size and cost of sensors as well as power consumptions. Moreover, the portable sensors can be produced and commercialized for environmental monitoring by using the modern progress in MEMS technology. Anyhow, in the coming decade, we should mainly focus on miniaturization, unique sensing characteristics and long-life time stability which are the key parameters for the fabrication and commercialization of reliable solid-state electrochemical gas sensors.

References

1. Miura N, Kurosawa H, Hasei M, Lu G, Yamazoe N. Solid State Ionics. 1996;86–88:1069.
2. Miura N, Lu G, Yamazoe N, Kurosawa H, Hasei M. J Electrochem Soc. 1996;143(2):L33.
3. Lu G, Miura N, Yamazoe N. J Mater Chem. 1997;7:1445.
4. Lu G, Miura N, Yamazoe N. Sens Actuators B. 2000;65:125.
5. Zhuiykov S, Nakano T, Kunimoto A, Yamazoe N. Electrochem Commun. 2001;3:97.
6. Zhuiykov S, Ono T, Yamazoe N, Miura N. Solid State Ionics. 2002;152:801.
7. Miura N, Zhuiykov S, Ono T, Hasei M, Yamazoe N. Sens Actuators B. 2002;83:222.
8. Ono T, Hasei M, Kunimoto A, Miura N. Electrochemistry. 2003;6:405.
9. Miura N, Akisada K, Wang J, Zhuiykov S, Ono T. Ionics. 2004;10:1.
10. Bartolomeo E D, Luisa M, Traversa E. J Electrochem Soc. 2004;151:H133.
11. Bartolomeo E D, Kaabbuathong N, Grilli M L, Traversa E. Solid State Ionics. 2004;171:173.
12. Elumalai P, Wang J, Zhuiykov S, Terada D, Hasei M, Miura N. J Electrochem Soc. 2005;152:H95.
13. Miura N, Wang J, Nakatou M, Elumalai P, Hasei M. Electrochem Solid-State Lett. 2005;8:H9.
14. Miura N, Wang J, Nakatou M, Elumalai P, Zhuiykov S, Hasei M. Sens Actuators B. 2006;114:903.
15. Elumalai P, Miura N. Solid State Ionics. 2005;176:2517.
16. Elumalai P, Plashnitsa V V , Ueda T, Hasei M, Miura N. Ionics. 2006;12:331.
17. Miura N, Wang J, Elumalai, Terada D, Ueda T, Hasei M. J Electrochem Soc. 2007;154:J246.
18. Wang J, Elumalai P, Terada D, Hasei M, Miura N. Solid State Ionics. 2006;177:2305.
19. Plashnitsa V V, Ueda T, Miura N. Inter J Appl Ceram Tech. 2006;3:127.
20. Plashnitsa V V, Ueda T, Miura N. J Rare Metal Mat Eng. 2006;35:36.
21. Plashnitsa V V, Ueda T, Elumalai P, Miura N. Sens Actuators B. 2008; 130:231.
22. Elumalai P, Hasei M, Miura N. Electrochemistry. 2006;74:197.
23. Brosha EL, Mukundan R, Lujan R, Garzon FH. Sens Actuators B. 2006;119:398.
24. Nagao M, Namekata Y, Hibino T, Sano M, Tomita A. Electrochem Solid-State Lett. 2006;9:H48.
25. Plashnitsa V V, Ueda T, Elumalai P, Kawaguchi T, Miura N. Ionics. 2008;14:15.
26. Okamoto H, Obayashi H, Kudo T. Solid State Ionics. 1980;1:319.
27. Can Z Y, Narita H, Mizusaki J, Tagawa H. Solid State Ionics. 1995;79:344.
28. Miura N, Raisen T, Lu G, Yamazoe N. Sens Actuators B. 1998;47:84.
29. Brosha EL, Mukundan R, Brown DR, Garzon FH, Visser JH, Zanini M, Zhou Z, Logothetis EM. Sens Actuators B. 2000;69:171.
30. Hibino T, Kuwahara Y, Wang S, Kakimoto S, Sano M. Electrochem Solid-State Lett. 1998;1:197.
31. Hibino T, Wang S, Kakimoto S, Sano M. Sens Actuators B. 1998;50:149.
32. Hibino T, Wang S. Sens Actuators B. 1999;61:12.
33. Hibino T, Tanimoto S, Kakimoto S, Sano M. Electrochem Solid-State Lett. 1999;2:651.
34. Hibino T, Hashimoto A, Kakimoto S, Sano M. J Electrochem Soc. 2001;148:H1.
35. Miura N, Shiraishi T, Shimanoe K, Yamazoe N. Electrochem Commun. 2000;2:77.
36. Hashimoto A, Hibino T, Mori K, Sano M. Sens Actuators B. 2001;81:55.
37. Zosel J, Westphal D, Jakobs S, Muller R, Guth U. Solid State Ionics. 2002;152:525.
38. Zosel J, Ahlborn K, Muller R, Westphal D, Vashook V, Guth U. Solid State Ionics. 2004;169:115.
39. Mukundan R, Brosha EL, Garzon FH. J Electrochem Soc. 2003;150:H279.
40. Inaba T, Saji K, Sakata J. Sens Actuators B. 2005;108:374.
41. Zosel J, Tuchtenhagen D, Ahlborn K, Guth U. Sens Actuators B. 20008;130:326.

42. Ueda T, Elumalai P, Plashnitsa VV, Miura N. Chem Lett. 2008;37(1):120.
43. Miura N, Lu G, Yamazoe N. Sens Actuators B. 1998;52:169.
44. Schmidt-Zhang P, Zhang W, Gerlach F, Ahlborn K, Guth U. Sens Actuators B. 2005;108:797.
45. Dutta A, Ishihara T. Mat Manuf Processes. 2006;21:225.
46. Yang JC, Dutta PK. Sens Actuators B. 2007;123:929.
47. Li X, Kale GM. Sens Actuators B. 2007;123:254.
48. Somov SI, Guth U. Sens Actuators B. 1998;47:131.
49. Mukundan R, Brosha EL, Brown DR, Garzon FH. Electrochem Solid-State Lett. 1999;2:412.
50. Hibino T, Kuwahara Y, Wang S. J Electrochem Soc. 1999;146:382.
51. Hibino T, Wang S. Sens Actuators B. 1999;61:12.
52. Eguchi Y, Watanabe S, Kubota N, Takeuchi T, Ishihara T, Takita Y. Sens Actuators B. 2000;66:9.
53. Hibino T, Hashimoto A, Mori K, Sano M. Electrochem Solid-State Lett. 2001;4:H9.
54. Ishihara T, Fukuyama M, Dutta A, Kabemura K, Nishiguchi H, Takita Y. J Electrochem Soc. 2003;150:H241.
55. Dutta A, Ishihara T, Nishiguchi H, Takita Y. J Electrochem Soc. 2004;151:H122.
56. Ishihara T, Ishikawa S, Hosoi K, Nishiguchi H, Takita Y. Solid State Ionics. 2004;175:319.
57. Dutta A, Nishiguchi H, Takita Y, Ishihara T. Sens Actuators B. 2005;108:368.
58. Ueda T, Plashnitsa V V, Nakatou M, Miura N. Electrochem Commun. 2007;9:197.
59. Ueda T, Plashnitsa VV, Elumalai P, Miura N. Sens Materials. 2007;19:333.
60. Miura N, Nakatou M, Zhuiykov S. Electrochem Commun. 2002;4:284.
61. Miura N, Nakatou M, Zhuiykov S. Sens Actuators B. 2003;93:221.
62. Wu N, Chen Z, Xu J, Chyu M, Mao S X. Sens Actuators B. 2005;110:49.
63. Nakatou M, Miura N. Sens Actuators B. 2006;120:57.
64. Wama R, Utiyama M, Plashnitsa VV, Miura N. Electrochem Commun. 2007;9:2774.
65. Nakatou M, Miura N. Electrochem Commun. 2004;6:995.
66. Nakatou M, Miura N. Solid State Ionics. 2005;176:2511.
67. Wickham DT, Banse BA, Koel BE. Catal Lett. 1990;6:163.
68. Lu X, Xu X, Wang N, Zhang Q. J Phys Chem A. 1999;103:10969.
69. Morrison SR. Sens Actuators. 1987;12:425.
70. Schubert MM, Hackenberg S, van Veen AC, Muhler M, Plzak V, Behm RJ. J Catal. 2001;197:113.
71. Chester MA, Samorjai GA. Surf Sci. 1975;52:21.
72. Grzybowska-Swierkosz B. Catal Today. 2006;112:3.
73. Miura N, Nakakubo S, Shimamoto Y. Abs of the 15th Inter Conf on SSI. 2005;586.
74. Wama R, Nanakubo S, Miura N. Chem Sensors. 2007;23(Suppl A):145.

Chapter 6
Optical Gas Sensing

Ilaria Cacciari and Giancarlo C. Righini

6.1 Introduction

Optical sensors have a number of general advantages over corresponding chemical and electrical-electronic sensors in terms of contact-free measurements, higher resistance to corrosive or reactive environments, applicability in flammable atmospheres, usually fast response. Moreover, when using optical fibers, they show the capability of remote control monitoring, local, and distributed sensing, together with the intrinsic characteristics of being compact, flexible, and environment robust. In the case of gas sensing, an additional advantage is related to the capability of multigas detection using differences in the wavelength, intensity, polarization and phase of the probe light signal.

The spectroscopic techniques, with their numerous variations, are likely to remain the most common optical gas-sensing method. Several other measuring methods, however, are emerging, also thanks to the progresses in laser sources and optical components. Ellipsometry, surface plasmon resonance (SPR), hyperspectral imaging are examples of alternative approaches (even if they as well could be classified under a general "spectroscopy" label); on the instrumentation side, optical fibers and integrated optical circuits provide the tools for the implementation of efficient, lightweight, compact devices.

In very many cases the measurement is not direct, and is based on the interaction of the gaseous molecules with a thin film, which acts as a transducer by changing its optical properties (e.g., the refractive index, the reflection or absorption coefficient, the photoluminescence intensity, etc.) as a function of gas concentration.

In this review we attempt to provide an overview of the different optical techniques and of their applications to the sensing of gaseous species.

I. Cacciari
CNR, "Nello Carrara", Institute of Applied Physics (IFAC), 50019 Sesto Fiorentino, Italy
e-mail: g.c.righini@ifac.cnr.it

E. Comini et al. (eds.), *Solid State Gas Sensing*,
DOI: 10.1007/978-0-387-09665-0_6, © Springer Science+Business Media, LLC 2009

6.2 Spectroscopic Detection Schemes

The quantitative detection of gaseous substances and vapors is of great importance in several application areas such as toxic gas alert, industrial, and combustion processes control, human breath analysis for medical diagnosis, and environmental pollution monitoring.

When using optical detection methods, one fundamentally monitors the optical absorption/emission/scattering of a gas species at defined optical wavelength(s). The wavelength distribution of the chosen optical characteristic constitutes the feature that allows one to identify the presence of a specific gas species, while the intensity of the phenomenon (i.e., absorption/emission/scattering) gives information on the concentration. Thus, spectroscopic optical methods provide a fast, accurate and stable measurement of a gas species, provided that it has significant absorption or emission of optical radiation in a suitable wavelength band.

The advent of laser sources has permitted the development of versatile and powerful spectroscopic detection schemes, that in most cases are able to fulfil the challenging requirements for state-of-the-art gas sensors, including high gas sensitivity (part per billion – ppb – or better), high gas selectivity, real-time and continuous measurements, field suitability, multicomponent compatibility, automatic, or user-friendly operation. The simplest diode laser absorption method is based on the comparison of the absorption at a probe wavelength (where absorption by the gas species occurs) and at a reference wavelength (chosen to be free of any absorption by the target gas or any likely contaminant). Several more accurate detection schemes have been and are being exploited, based on multipass absorption spectroscopy (MPAS), cavity-enhanced spectroscopy (CES), cavity ringdown spectroscopy (CRS), tunable diode laser absorption spectroscopy (TDLAS), tunable infrared laser differential absorption spectroscopy (TILDAS), photoacustic spectroscopy (PAS), and others. The significant progresses of laser diode sources, especially in terms of emission bandwidth and tunability, linewidth, and optical power, have made these techniques very efficient and applicable to a very wide set of gas species.

For the fundamentals of optical spectroscopy, both theoretical and experimental, the interested reader is referred to one of the many textbooks and scientific papers [1, 2, 3, 4, 5, 6]. Here we report only a very few basic notes, and a few specific results. A separate paragraph is devoted to gas sensors using guided-wave devices (i.e., optical fibers or integrated optical circuits), even when the detection scheme is of the spectroscopic type.

The molecular absorption at a fixed wavelength (λ) is described by the Lambert-Beer law, which relates the transmitted power $P_T(\lambda)$ to the incident power $P_I(\lambda)$:

$$P_T(\lambda) = P_I(\lambda) \cdot \exp[-N\sigma(\lambda)L] \tag{6.1}$$

where N is the density of absorbing molecules, L the path length and $\sigma(\lambda)$ the absorption cross section. When N is very small, the detection sensitivity is limited mostly by the intensity fluctuations of the incident source. An improvement can be obtained by different methods, like modulating the wavelength or frequency of the incident light across a molecular absorption line or increasing the path length L in a multipass or intracavity arrangement [7].

Laser diodes are well suited for high-sensitivity absorption spectroscopy because of their much lower noise level than other laser sources. The distributed feedback (DFB) laser diodes first developed for optical communications may operate in the band from 1000 to 2000 nm, where the molecular overtone vibrational bands of gases like CO, CO_2, CH_4, HCl, HF, H_2S, NO, N_2O, NH_3, are located. Even if in principle sensitivities close to the quantum limit (shot-noise limit) may be achieved, in practice limitations arise from the optical noise due to interference fringes and from detector noise. An interesting discussion of the noise sources and of the advantage of frequency modulation spectroscopy is reported in the paper by Corsi et al. [8]. Table 6.1 shows some of the data they achieved by using a 1.5-m-long cell and DFB laser diodes. Three different detection schemes, namely direct absorption (DA), low wavelength modulation (LWM), and frequency modulation (FM) spectroscopy [9] were tested. It is evident that the combination of a DFB laser diode with two-tone frequency modulation spectroscopy makes possible to detect trace gases in air with a good sensitivity, at ppm (part per million) level.

The optical part of a multipass cell is commonly made of two mirrors located vis-à-vis on an optical axis: light is repeatedly reflected between either spherical [10] or astigmatic mirrors [11]. The latter arrangement allows reaching longer paths in a small volume and optimizes the separation of consecutive spots on the two mirrors minimizing the etalon effect. In both designs the laser beam is launched through a coupling hole in the front mirror at a certain angle with respect to the cell axis; the beam leaves the cell through the same coupling hole under an inverted angle.

Many strategies have been followed to develop and characterize new compact gas analyzers based on multipass cell in order to obtain high sensitivity and fast response. For instance, a minimum detectable concentration of 8 ppb for H_2CO (formaldehyde) has been achieved by using a pulsed difference-frequency mixing configuration that employs a broadly tunable cw external

Table 6.1 Sensitivity (in ppm × m) for a few gases obtained by using three different spectroscopic detection techniques. (DA = direct absorption; LWM = low wavelength modulation; FM = frequency modulation)

Gas species	λ (nm)	DA	LWM	FM
CO	1579.74	500	170	7
CO_2	1579.57	800	260	10
H_2S	1577.32	400	130	4

cavity diode laser and a diode-pumped Q-switched Nd:YAG [12]. Very good results have been also demonstrated by using different source schemes; thus, sensitivity of about 4 ppb has been obtained for HNO_3 [13] and of 0.1 ppb for many other smaller molecules in air [14] by using tunable diode-laser spectroscopy (TDLS). Very recently, new interests have risen not only in searching suitable sources but also in studying new approaches to detect simultaneously more gaseous species; a new design of a multipass cell based on two cylindrical mirrors with TDLS shows the possibility of detecting CO_2 and CH_4 with detection limit of 588 ppm and ~3 ppm, respectively [15].

In photoacoustic spectroscopy scheme [16] the modulated laser radiation absorbed by gas molecules releases heat to the gas sample by non-radiative relaxation of the excited states. The periodic heating of the gas generates an acoustic wave, which is detected in the photoacoustic cell. If the laser beam is modulated in the audio frequency range, the acoustic signal may be detected by a sensitive microphone. The recorded signal is directly proportional to the absorbed incident power $P_I(\lambda)$ and therefore to the gas concentration:

$$S(\lambda) = kN\sigma(\lambda)P_I(\lambda) \tag{6.2}$$

where k is a constant characteristic of the cell and it is derived from calibration measurements. With a not too weak incident power (> 1 mW), high detection sensitivity can be achieved. Other improvements are also possible using resonant photoacoustic cells with high Q factor or intracavity or multipass [17] arrangement; in the latter case it has been demonstrated that the detection sensitivity is enhanced by ~10 times for trace detection of water vapour near 1380 nm region [18]. Arrays of microphones can be used as well to enhance the signal-to-noise ratio. The detection limit therefore can reach sub ppb concentrations for many gaseous species opening up new application areas for ultrasensitive gas sensing.

In order to obtain a better understanding of the cell design that can influence the detection limit in the case of high background absorption noise, both experimental and theoretical studies have been recently performed [19]. The model proposed to understand the first longitudinal acoustic mode of a multipass photoacoustic cell has its analogue in an electric transmission line. With this simple model the best-suited number of passes for a given gaseous species and cell geometry can be determined.

The recent development of quantum cascade lasers (QCL), invented in 1994 [20], has offered an additional option for the development of high-resolution and high-sensitivity spectroscopic gas sensing, especially thanks to the tunability in the 3–20 μm wavelength region, where many gases exhibit strong rotational/vibrational absorption transitions, at least one order of magnitude larger than the overtone absorption lines in the near infrared. Photoacoustic spectroscopy combined with QCLs has proved to be effective in measuring ppb (and potentially sub-ppb) gas concentrations [21, 22, 23]. Room temperature quantum cascade lasers (QCL) are now available even in continuous wave mode, which is adding further potential for the development of robust and simple in situ gas detectors. A detection limit of 34 ppb of methane in flow and of 14 ppb for N_2O flux measurements was reached [24].

The cavity ringdown approach has been pioneered to detect very weak absorption bands in molecular oxygen [25], showing a very high sensitivity. In this approach [25, 26], using a picosecond laser source, the wavelength of the pulse is tuned to overlap absorption line of the gas before the laser pulse is optically injected into a passive resonator cavity made of two reflective mirrors in which the spacing is greater than the pulse coherence length. The laser pulse reflects between the mirrors and a small fraction of the radiation escapes out the cavity on each round trip. By measuring the decay time of pulse intensity, the concentration of the gas can be derived. Unfortunately this kind of sensors shows an intrinsically small dynamic range, and requires pulsed lasers and modulated continuous wave lasers of well-defined duration.

At one of the cavity ends a sensitive detector records an exponentially decaying pulse train. Comparing the decay time for an empty cavity with the corresponding time for a cavity filled by a specific gas, the sample absorbance per pass can be deduced. High sensitivities can be reached by this method because path lengths of kilometres can be easily achieved with highly reflective mirrors. These types of optical systems obviously need mirrors with very high reflectivity and precise alignment. Because of the litre-size volumes of the multipass cells, the gas exchange rates are relatively slow, in particular at low concentrations, causing a slow response (minutes-to-hours). Following the demand of portable sensors, a mobile spectrometer for in situ multi component monitoring of trace gases has been developed to measure sub-ppbv ($<10^{-9}$ per volume) concentrations; in particular, for ethane (C_2H_4) buffered in synthetic air a minimum detection limit of 70 ppt ($<70\ 10^{-12}$ per volume) has been achieved [17]. The multipass cell technique has been widely used to measure traces of ozone [7], ammonia [27], water vapour [28], CO_2 [29], NO_2 [30], and methane [31], exhibiting sensitivities of tens of ppb. Very recently, it has been also proved that a multipass cell can be employed to measure the isotopic composition of N_2O at trace level [32]. Combination of cavity ringdown spectroscopy with QCLs has again led to further advances in sensitive analytical techniques for trace gas analysis. As an example, a gas analyzer based on a pulsed mid-IR QCL operating near 970 cm^{-1} has been developed for the detection of ammonia levels in breath: a sensitivity of $\sim$50 ppb with a 20s time resolution has been demonstrated [33] for ammonia detection in breath with this system.

6.3 Ellipsometry

Ellipsometry is an optical, reflection-based, technique used since many years for thin film and surface characterization [34].

When a beam of light is incident on a sample with an arbitrary angle (θ_i), at the boundary of the medium, a part of the light will be reflected at θ_r and the

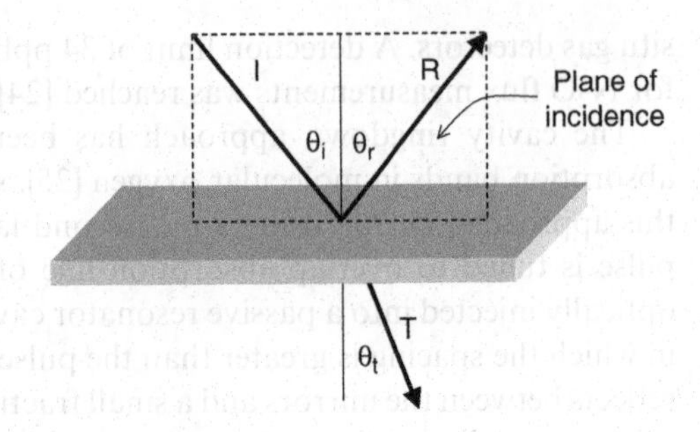

Fig. 6.1 Schematic illustration of incident, refracted and reflected beams

other part is transmitted through the sample at θ_t angle. Snell's law requires that all the three beams be in the incidence plane (Fig. 6.1). From the reflection and transmission measurements one can obtain the intensity ratios, R and T respectively, over a given range of wavelengths; R (T) is defined as the ratio of the light intensity being reflected (transmitted) over the light intensity I incident on the sample.

Ellipsometry is based on the measurement of the polarization changes arising upon reflection at oblique incidence of a polarized monochromatic plane wave. Typically the quantity measured is the ratio

$$\rho = \frac{p_r}{p_i} \tag{6.3}$$

where p_r and p_i are the state of polarization of the reflected and incident beam, respectively. With an optically isotropic sample there is no coupling between the two orthogonal polarizations with electric fields perpendicular (s direction) and parallel (p direction) to the incidence plane (Fig. 6.2). For such materials Eq. (6.3) becomes

$$\rho = \frac{R_p}{R_s} = \tan\psi \exp(i\Delta) \tag{6.4}$$

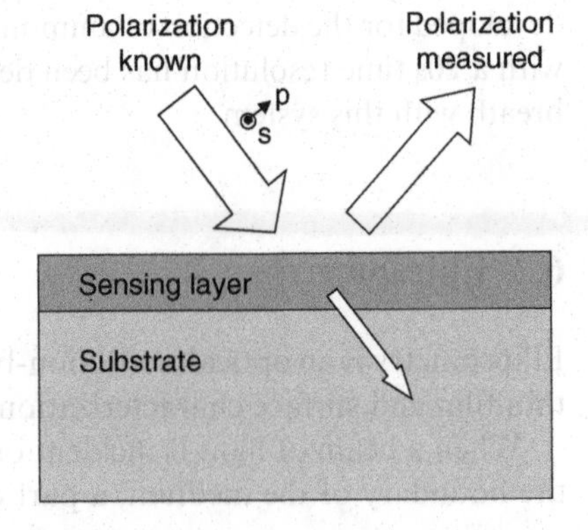

Fig. 6.2 Schematic illustration of ellipsometry-based sensing

R_p and R_p are the complex reflection coefficients for light polarized in p- and s-direction, respectively, and the ellipsometric angles ψ and Δ are typically the measured quantities.

Since ellipsometry measures the ratio of two values, it can be highly accurate and very reproducible. Different polarizations exhibit different reflectivity and different phase shifts on reflection at different monolayer or multilayer structures, and those values depend on the optical constant of each film. By using as a sensing layer a film which shows a reversible refractive index and/or thickness change when in contact with a given gas species, the measurement of the intensity changes of polarized light reflected at oblique incidence on the film gives us information about gas concentration. Typical sensing layers are thin (0.01–10 μm thickness) and are deposited on silicon or metal substrates.

Being ellipsometry an indirect sensing technique, the useful information has to be obtained by analysing the behaviour of ρ with a suitable optical model (see Fig.6.3) that describes the sample structure using as much information about the sample as possible. Generating theoretical data from the optical model that correspond to the experimental data and comparing generated data to experimental data, the unknown parameters in the optical model, such as thin film thickness or optical constants or both, are varied in agreement with a "best fit" procedure. By using regression algorithms to vary unknown parameters, the difference between the generated and experimental data can be in fact minimized.

In summary, the principle of an ellipsometric gas sensor is based on monitoring the changes of the polarization state, i.e., the ellipsometric parameters ψ and Δ for a thin film sample exposed to several gases. The advantages of this technique are primarily the rapid response, high selectivity and good sensitivity. Porous silicon is often employed as sensing layer because it exhibits a quite low detection limit threshold: about 10 ppm in the case of acetone vapours [35].

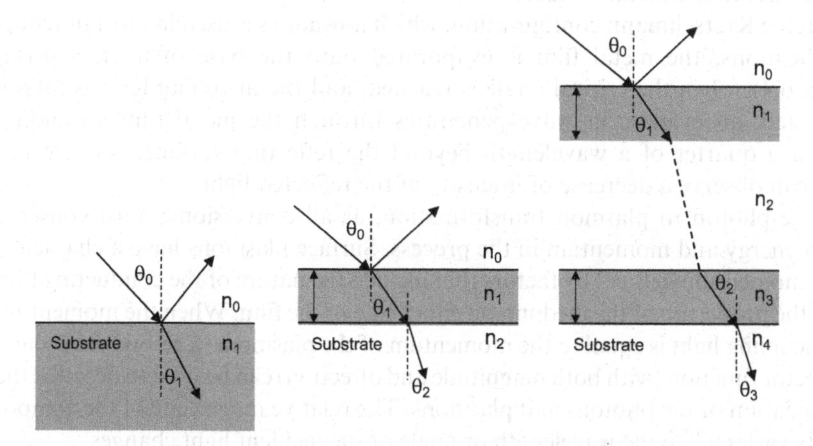

Fig. 6.3 Three different optical models for modelling ellipsometry data: on the *left* an ambient-substrate model, in the middle a two-phase model, on the *right* a multilayer model

Experimental results have proved that deposition of copper on a sensing layer made of porous silicon can improve sensitivity to low concentrations of alcoholic gases such as methanol, ethanol and 2-propanol [36]. Finally, sensitivity in the ppm range [37] can be achieved by optimizing the porous silicon layer in terms of thickness and porosity [38].

6.4 Surface Plasmon Resonance

Surface plasmon resonance (SPR) is a physical process that can take place when plane-polarized light hits a metal film under total internal reflection (TIR) conditions. The theoretical basis of the phenomenon started to be laid at the beginning of the last century, but a real progress was made past the mid of the century when Ritchie theoretically demonstrated the existence of surface plasma excitations at a metal surface [39] and Otto [40] and Kretschmann [41] devised the attenuated total reflection (ATR) prism coupling scheme to make practical the excitation of the surface electromagnetic waves by a light beam.

Light impinging at the interface between a metal and a dielectric material can, under the right circumstances, give rise to a resonant interaction between the waves and the mobile electrons (the outer shell and conduction-band electrons) at the surface of the metal. Surface plasmons, also known as surface plasmon polaritons, are the surface electromagnetic waves that propagate parallel along the metal–dielectric interface. If the metal is constituted by a thin film (with thickness of tens of nm), these waves are also on the boundary of the metal and the external medium (usually air): their amplitude is therefore very sensitive to any change of this boundary, such as the adsorption of molecules to the metal surface.

In the Kretschmann configuration, which nowadays is used in most practical applications, the metal film is evaporated onto the base of a glass prism (Fig. 6.4); when the critical angle is reached, and the incoming light is totally reflected, an evanescent wave penetrates through the metal film, extending about a quarter of a wavelength beyond the reflecting surface. As a result, one can observe a decrease of intensity of the reflected light.

The photon to plasmon transformation, as all conversions, must conserve both energy and momentum in the process. Surface plasmons have a characteristic momentum defined by factors that include the nature of the conducting film and the properties of the medium on either side of the film. When the momentum of incoming light is equal to the momentum of the plasmons, a resonance occurs. A vector function (with both magnitude and direction) can be used to describe the momentum of the photons and plasmons. The relative magnitude of the components varies when the wavelength or angle of the incident light changes.

The SPR angle mainly depends on the properties of the metal film, the wavelength of the incident light and the refractive index of the media on both

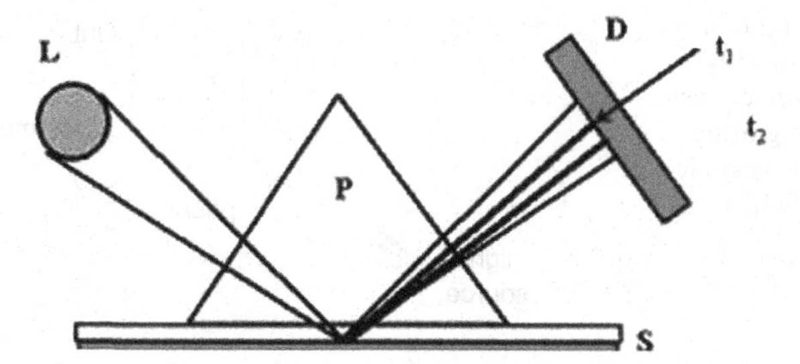

Fig. 6.4 Schematic illustration of surface plasmon resonance sensor. L: light source, D: photodiode array, P: prism, S: sensor surface. The two dark lines in the reflected beam projected on to the detector symbolise the light intensity drop following the resonance phenomenon at time t_1 and t_2, respectively. The line projected at t_1 corresponds to the starting time, while t_2 represents the position of resonance after interaction with the measurand

side of the film. The metal must have electrons in the conduction band that can resonate with the incoming light at a proper wavelength. Metals that satisfy this requirement are silver, gold, copper, aluminium, sodium, and indium. Among them, indium is quite expensive, sodium too reactive, copper and aluminium too broad in their SPR response. Gold and silver are therefore ideal candidates for the metal film in the visible light region; gold is preferable because, even though a silver film may yield a more distinct SPR spectrum with respect to gold, it is too susceptible to oxidation.

In general the thickness of the metal layer must be optimized: a thicker film corresponds to a shallower dip in reflective light intensity, a thinner film to a broader dip. The optimal thickness for gold films is around 50 nm. The light source should be monochromatic and p-polarized (polarized in the plane of the surface) to obtain a sharp dip. All the light not p-polarized will not contribute to the SPR and will increase the background intensity of the reflected light. A charge coupled device (CCD) is often used to record the dip profile: from it, the dip position is determined by numerical fitting. High resolution SPR sensors have been demonstrated by using a bicell or a quadrant cell photodetector [42].

The specificity of SPR sensors is usually achieved by modifying the metal surface with a functional layer to allow specific binding between the ligand molecules and the measurand. A good review of early SPR sensors was published by Homola et al. [43]. Another interesting, more recent, review of chemical and biologic sensors based on interferometry, luminescence and SPR was provided by Ince et al. [44].

Nowadays SPR technique is a well standing technique for optical gas-sensing applications [43, 45, 46, 47, 48]. Its compatibility with optical fibers offers the possibility of using this technique in remote sensing applications too, such as in-line process and environmental monitoring. The use of guided-wave configurations in conjunction with SPR is the subject of the following section.

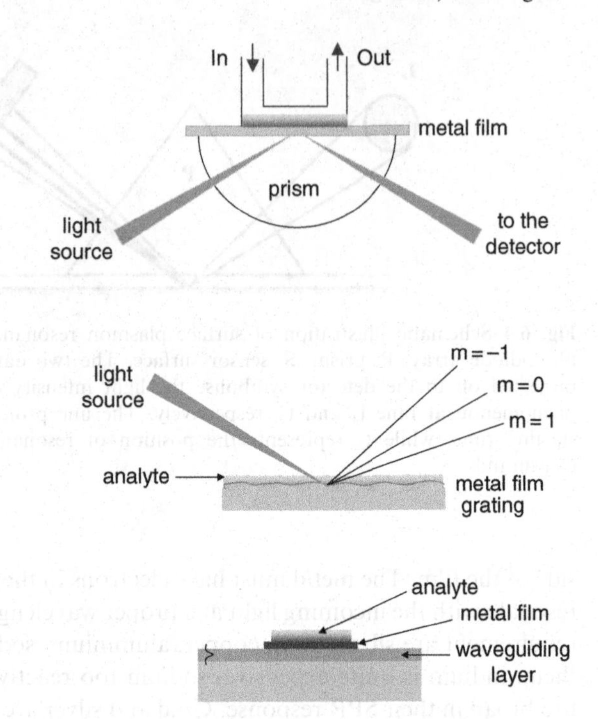

Fig. 6.5 Most common configurations of SPR sensors: from *top* to *down*, prism-coupling, grating coupling, and waveguide schemes [from [43]]

There are several configurations of SPR devices that are capable to generate and measure surface plasmon resonance: prism-coupled total reflection systems, grating coupled systems, and optical waveguide systems (Fig. 6.5).

In turn, inside the prism-coupled category different configurations are possible. In the Otto arrangement [40] the space between the metal and the TIR surface (Fig. 6.6a.) is filled with a lower refractive index medium. This configuration is commonly used for SPR in solid-phase media.

Since the distance between metal and TIR surface in the Otto configuration limits the SPR efficiency, when analyzing liquid solutions or gases it is preferable to use the Kretschmann configuration [41]. In this configuration

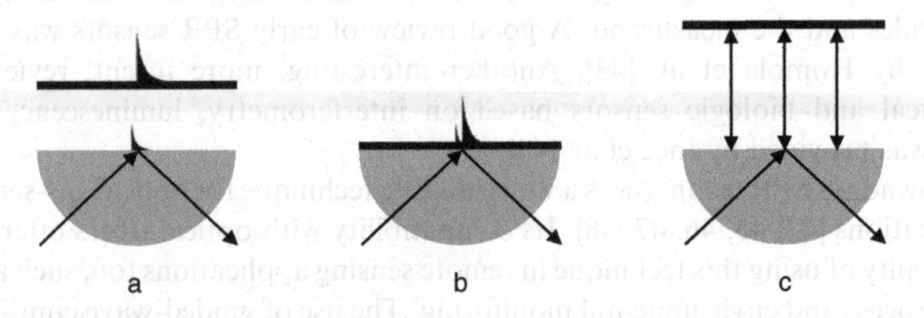

Fig. 6.6 Different prism coupling configuration

(Fig. 6.6b), the metal layer is directly on top of the TIR surface, thus enabling a more efficient plasmon generation. A third configuration (Fig. 6.6c) is similar to the Otto arrangement but uses a special layer to enhance TIR and a resonant mirror to couple the TIR light to plasmons. A layer of silica is deposited onto the prism base, and then a layer of titania on top of the silica layer. The silica layer is thin enough to allow the evanescent field generated by TIR in the prism to couple into the high refractive index titania: this frustrated total internal reflection (FTIR) allows the titania layer to act as an optical waveguide, so that an evanescent field is generated at the titania-functional layer interface. The exact angle of incident light at which the resonance between the guided-wave mode and the evanescently coupled light occurs is directly related to the refractive index of the surface adlayer.

Many efforts have been carried out to better understand the optical properties of different sensing layers, most researches being centred on conducting polymers; interesting results, in particular, have been obtained using polyimide films for humidity sensors [49, 50] or for selective sensors of different hydrocarbons gas molecules [51]. Very recently, this kind of films has been successful employed in combination with SPR for ethanol and methanol vapour sensors [52].Using a high density polyethylene thin film over gold layer, an SPR sensor has been developed for detecting *n*-dodecane vapor: preliminary results indicate that samples in the range of a few hundred ppmv of *n*-dodecane vapor in butane gas can be sensed [53].

Detection sensitivity of SPR sensors may be enhanced by embedding Au nanoclusters in a dielectric (silica) film [54]. It has been shown that the sensitivity of the biosensor can be improved simply by varying the volume fraction and the size of the embedded nanoclusters in order to control the surface plasmon effect (Fig. 6.7). In particular, the sensitivity of the device has been tested to detect pure nitrogen (N_2) and pure argon (Ar) that have very similar refractive index (at room temperature they differ by $\sim 1.5\ 10^{-5}$). The experiments revealed that the resolution of nanocluster SPR sensor is improved by a

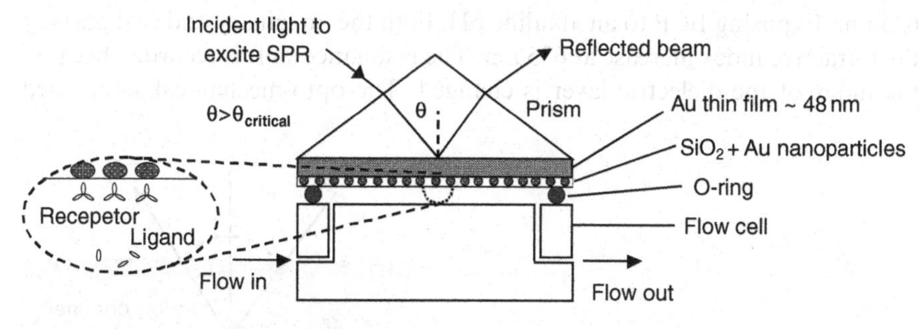

Fig. 6.7 Schematic illustration of SPR system employed for gas detection. In this prism coupling configuration [54], the Au:SiO_2 film is coated on an Au thin film by an RF magnetron co-sputtering process

factor of approximately eight times with respect to a standard SPR sensor, opening the way to achieve an ultrahigh-resolution detection system. It must also be remarked that there are other nanosized materials that can increase significantly the transduction mechanism; for instance, thin films of crystalline TiO_2 nanoparticles [55] appear to be extremely promising because of their large surface-to-volume ratio and of the particular surface properties of the nanocrystals.

In the SPR configuration, the glass prism can also be replaced by a diffraction grating covered with gold (Fig. 6.8). If the metal-dielectric interface is periodically distorted, the incident optical wave is diffracted into several beams directed away from the surface at different angles. The component of momentum of these diffracted beams along the interface differs from that of the incident wave by multiples of the grating wave vector. If the total component of momentum along the interface of a diffracted order is equal to that of the surface plasmon wave, a coupling between the two waves occurs.

From theoretical analysis and comparison of the sensitivity [56] it has been concluded that grating-based SPR sensors using wavelength interrogation are much less sensitive then their prism coupler-based counterparts. When using the angular interrogation mode, instead, the sensitivity of SPR sensors using diffraction gratings depends on the diffraction order and it is almost similar to that of SPR sensors based on prism couplers. As an example, for NO_2 detection, a sensor based on SPR in Si grating structures has shown a sensitivity of tens of ppb [57] comparable to the ppm-to-sub-ppm level achieved with a prism in Kretschmann configuration [46, 58].

Other optical arrangements have also been explored with the aim of achieving simpler and low-cost SPR gas sensors. A grating-based SPR sensor that uses relatively simple optics and low-cost electronics has been proposed to detect ammonia concentration down to few ppb [59]. The sensor consists of a rotating optical disk with a coated silver film (100 nm thick) and a transparent dielectric (yttria) film with thickness gradient. A uniform layer of 100 nm of bromocresol purple (BCP), as a pH indicator, is evaporated onto the yttria film. BCP shows in the neutral pH state a refractive index of 1.7 and a very little absorption at 635 nm. Exposing BCP to an alkaline pH, both the imaginary and real parts of the refractive index increase at 635 nm. The resonance shift is recorded because the index of the dielectric layer is changed. The opto-mechanical setup used

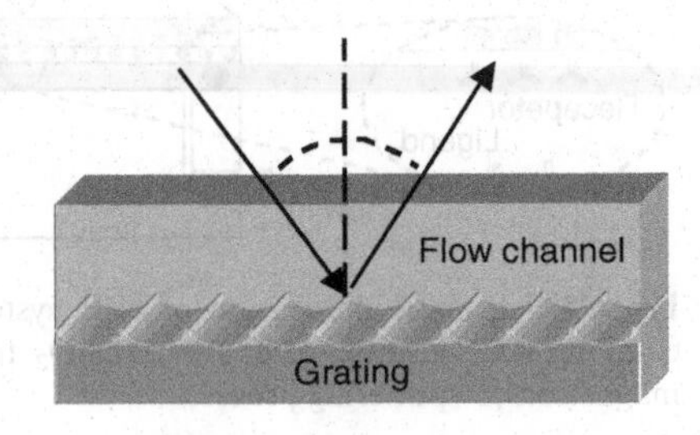

Fig. 6.8 Grating based SPR gas sensor

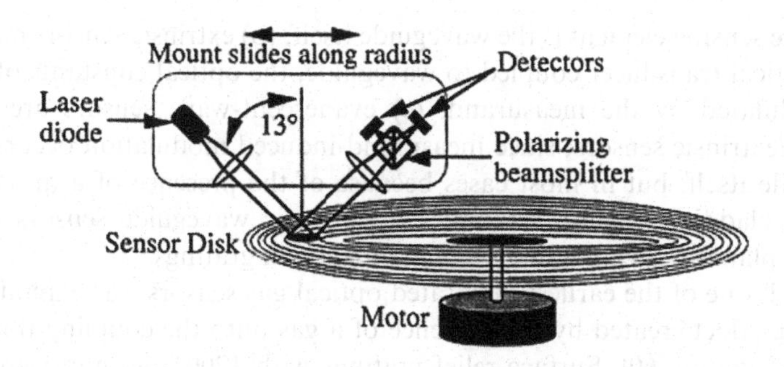

Fig. 6.9 Sensor disk based on SPR [59]

(Fig. 6.9) is made of a commercial CD player motor, used to rotate the disk at ~ 30 Hz, a 635 nm semiconductor laser pen with a focusing lens and a linear polarization output as source, and oriented in order that equal amplitudes of both polarizations were incident upon the disk at 13° within the plane of incidence perpendicular to the grooves. The spot diameter at the disk is very large compared to the grating period, but very small compared to the curvature radius of the grooves; therefore the spiral groove diffracts light as a rectangular grating with the same period. The reflected light was analyzed by a polarizing beam splitter and both s and p polarizations were detected with standard Si photodetectors.

6.5 Guided-Wave Configurations for Gas Sensing

Fiber optics and integrated optics (IO) technologies were primarily developed for telecommunications but the advances in the development of high quality and competitive price optoelectronic components and fibers have rapidly contributed to the development of guided wave sensors. While the detection principles in many cases remain the same described in previous paragraphs, one of the main advantages of guided-wave sensors is the possibility of fully exploiting the presence of the evanescent wave associated with the light confinement inside the waveguide. This permits a stronger interaction between the electromagnetic field and the gas species; besides absorption phenomena, the guided propagation is affected by the change in refractive index in the medium surrounding the waveguide.

The basic components of a complete guided wave sensor include an optical source, an optical interface for the source-to-waveguide light coupling, the waveguide itself (where the measurand-induced light modulation occurs), a photodetector and the electronics for amplification, signal processing and data display. Waveguide sensors may be classified as intrinsic, extrinsic, or evanescent-wave sensors: (a) intrinsic sensors are true waveguide sensors in which the sensing element is the waveguide itself; (b) extrinsic sensors make use of an optical transducer coupled to waveguide, the optical constants of which are modulated by the measurand; (c) evanescent-wave sensors are hybrid intrinsic/extrinsic sensors, since measurand-induced modulation occurs in the waveguide itself, but in most cases because of the presence of a measurand-sensitive cladding section. Examples of intrinsic waveguide sensors include fibers or planar waveguides containing diffraction gratings.

Indeed, one of the earliest integrated optical gas sensors was exploiting the switching effect created by the presence of a gas onto the coupling through a grating structure [60]. Surface-relief gratings with 1200 lines/mm were fabricated by an embossing technique on SiO_2-TiO_2 waveguiding films with refractive index $n_f \cong 1.8$ and thicknesses ranging from 120 to 150 nm, deposited on Pyrex ($n_s = 1.47$) substrates. By adsorption of H_2O (or other vapours, like acetone and ethanol) the optical thickness of the guiding film is increased, which makes the effective index of the guided mode to increase as well. This changes the coupling conditions, causing the propagating mode to switch off. The ultimate sensitivity limit for water vapour was deduced to be about 1/100 of one H_2O monolayer [60].

Another approach for the development of integrated optical gas sensors is based on interferometric measurements. Mach-Zehnder Interferometers (MZI) are easily fabricated in integrated optics, by means of standard photolithographic processes, and allow one a highly sensitive measurement of the phase shift induced by the presence of a perturbing agent above one of the arms of the interferometer. While the free-space configuration requires several optical components and a tight alignment, an IO circuit a few cm long represents a very stable and efficient solution. NO_2 sensing, for instance, has been demonstrated in a MZI geometry fabricated in a Ti:$LiNbO_3$ channel waveguide with a Langmuir-Blodgett film cladding over one arm [61]. Sensitivity in the range of ppb for NO_2 has been predicted for a similar device, fabricated in $LiNbO_3$ by ion implantation, having 60 mm length, 0.5×1 mm^2 cross section and a few grams of weight [62].

Further advances are being made possible by the use of nanocomposite films, which exhibit peculiar properties. Silica films doped with nickel or cobalt oxide nanocrystals, for instance, show a reversible variation of the optical Vis-NIR absorption when exposed to CO [63].

Absorption-based IO gas sensors can be quite sensitive: a composite optical waveguide realized by evaporating a tapered film of bromothymol blue (BTB) onto a potassium-ion-exchanged glass has proved to be effective for trace-ammonia detection [64]. In the presence of ammonia, the BTB film changed color from yellow to blue, inducing absorption of the 633-nm guided wave. The authors achieved a detection limit of one part in 10^9 of ammonia, so demonstrating that such a guided wave absorption–based ammonia-gas sensor was much more sensitive than one based on evanescent-wave absorption [65].

6.5.1 Integrated Optical SPR Sensors

An efficient way to exploit surface plasmon resonance takes advantage of the properties of optical waveguides. Because of the simpler way to control the optical path, small size, and ruggedness, this configuration provides numerous attractive features. By varying the angle of incidence of the light, a coupling condition is found and the beam, entering the region with a thin metal overlay, evanescently enters into the metal layer itself. If the surface plasmon wave is phase-matched with the guided mode, the light wave excites a surface plasmon wave at the outer interface of the metal. At the end of the waveguide the outcoming light can be detected by photodiodes.

Integrated optical waveguide SPR sensors appear promising for the development of multichannel sensing devices on a chip. Theoretically, the sensitivity of waveguide-based SPR devices is approximately the same as that of the corresponding prism coupled configurations. Despite increased design constraints compared to bulk prism-based SPR-sensing devices, all the main SPR detection approaches have been implemented in waveguide SPR sensors. Research of gas sensors based on SPR and integrated optical waveguide was pioneered at the beginning of nineties [66], and several devices were developed using either slab [67] or channel [68] single-mode IO waveguides.

By choosing a proper cladding layer, SPR can be employed to measure specific gases. A recent study proved the feasibility of a small and highly efficient sensor for hydrogen by using a thin palladium layer deposited on the top of a channel waveguide [69]. The Pd layer can act as a chemo-optical transducer and as a support for surface plasmon waves.

6.5.2 Fiber Optic SPR Sensors

Currently, the highest level of miniaturization of SPR devices can be reached with optical fibers, keeping the same detection limit ($\sim$ ppm) as in microoptical systems. The fiber-optic SPR (FO-SPR) sensors have numerous advantages over their prism-based counterparts: first of all, fibers are inexpensive and can easily be used to make disposable sensors for medical or otherwise sterile tasks. Fibers have no moving parts, and their mechanical flexibility and the ability to transmit optical signals over long distance give them a much broader operating range than Kretschmann sensors and make multiple sensor arrays a possibility. A further advantage with respect to most semiconductor gas sensor is their possibility of working at room temperature and the feasibility of safe operation in explosion endangered areas.

The first FO-SPR sensing structure [70] was made of a conventional polymer clad silica fiber, where a certain length of the fiber cladding had been stripped away and an active metal layer, such as silver, had been deposited around the exposed section of the fiber core. The length of the metallized section depends on the diameter of the fiber and determines the number of reflections occurring at the surface plasmon metal interface: if it is too short, not enough coupling will occur; if too long, the coupling will be very strong and it will be difficult to determine the minimum coupling intensity.

The sensing device may require a chromium mirror fixed at the fiber end and a gas sensitive polymer film deposited on the silver layer; the SPR section can be located at the tip (Fig. 6.10a) or in the middle of the optical fiber (Fig. 6.10b).

With respect to Kretschmann arrangement, in which a range of coupling angles is swept over the prism, the fiber allows sweeping through a number of coupling wavelengths. These wavelengths can be processed simultaneously using a broadband, multi wavelength source. By measuring the amount of each wavelength escaping from the fiber, a spectrophotometer can determine which wavelength had been coupled with the surface plasmon and one can derive the gas concentration. In another configuration, the fiber may be side-polished and then coated with a thin silver or gold metal layer.

Because the modal distribution of light in a fiber is sensitive to mechanical disturbances, and these phenomena occurring near the sensing area may produce intermodal coupling and modal noise, it has been suggested to use single mode optical fibers rather than large-diameter ($\sim$ 400 mm) multimode fibers. The operating range of a single mode fiber SPR is limited; it can be effectively tuned with a high refractive index dielectric overlayer, but it may cause a drop in the sensitivity of the sensor [73]. On the contrary, a side polished multimode fiber sensor using a halogen light source exhibited a resolution of the measurement based on wavelength interrogation equal to 3×10^{-6} refractive index units (RIUs).

To establish proper design parameters for the construction of highly sensitive fiber based devices for gas-sensing applications, several experimental and theoretical analysis have been conducted on fiber probes employing SPR. New

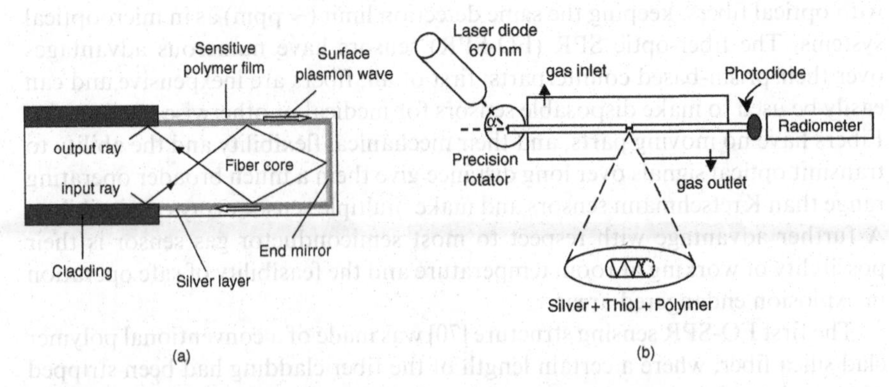

Fig. 6.10 Excitation of SPR (**a**) on the tip of an optical-fiber sensor [71], or (**b**) in the middle of the fiber [72]

devices for helium and propane [74, 75] sensing have been realized studying the diffraction of the guided mode out of a fiber core. After interaction with the metallized fiber tip, the spatial SPR has been observed as a dark strip within the light distribution of the diffracted beam pattern. Variations in the gaseous environment next to the fiber probe induce modifications on the SPR, with detection sensitivity down to 10^{-5} RIUs.

A comprehensive review of fiber optic sensors based on SPR has been published very recently [76].

6.5.3 Conventional and Microstructured Fibers for Gas Sensing

In the early optical-fiber sensor designs of the 1980s, optical fibers were used only to drive light from a light source to a sample cell and/or collect light emerging from the sample cell and possibly deliver it to a photodetector.

These sensors were simple in design and easy to make, but they were constrained by several factors such as a limited path length of the sample cell, and very low light-collection efficiency. A great improvement was represented by the utilization of the optical-fiber cladding as a transducer.

When a light beam propagates along an optical fiber, the electromagnetic field does not abruptly fall to zero at the core/cladding interface. Instead, the overlap of the input beam and the internally reflected beam leads to a field that enters into the cladding over a small distance (in a range comparable to the length of the light wavelength). This electromagnetic field, called evanescent field, tails but does not propagate into the cladding. Its intensity $I(z)$ shows an exponential decay with the distance z perpendicular to the interface:

$$I(z) = I_0 \exp\left(-\frac{z}{d_p}\right) \tag{6.5}$$

where I_0 is the intensity of the incident radiation. The depth of penetration d_p of the evanescent field is related to the angle of incidence θ at the interface, refractive indexes of core n_1 and cladding n_2, and the wavelength of the radiation λ as follows:

$$d_p = \frac{\lambda}{2\pi\sqrt{n_1^2 \sin^2\theta - n_2^2}} \tag{6.6}$$

Molecules to be sensed in the environment surrounding the fiber diffuse into the coating and interact with the evanescent field at the reflection point [77, 78]. This phenomenon has been used to design a number of sensors for monitoring volatile organic compounds, ozone etc. [79, 80, 81]. Moreover, the fiber cladding can be doped by a chemical reagent, so that gaseous molecules diffusing into the cladding react with the reagent to give a product that in turn interacts with the evanescent field. The parameter to be measured is the light intensity carried by the fiber, that is modulated by the change in the absorption spectrum of the transducer agent.

Ammonia (NH_3) is a toxic gas even at low concentrations. Environmental and health safety restrictions or recommendations determine a very low allowable concentration. Ammonia gas detection is widely used, from air conditioning to searching for life on Mars, and there is an increasing demand for cheap and reliable NH_3 sensors. Several kinds of fiber optic sensors for ammonia have therefore been developed as well. Low detection limits were reached with optical fibers coated with a PVC film containing one [82] or two [83] dye materials that display near infrared absorption band. A detection limit of about 10 ppm was obtained using a dye (bromocresol purple BCP) immobilized in the substitutional cladding by a sol-gel process [84].

A recent work [85] reported that for conducting polymers, such as polyaniline and polypyrrole, which undergo a reversible change in conductivity upon exposure to chemical vapors, the conductivity change is accompanied by optical property change in the material. Therefore, polyaniline and polypyrrole were selected as the modified cladding material for the detection of hydrochloride (HCl), ammonia (NH_3), hydrazine (H_4N_2), and dimethyl-methl-phosphonate (DMMP), respectively. The measured sensitivities, in terms of the percentage optical intensity change, were: 51% for Ammonia, 91% for Hydrazin, 30% for HCl and 15.75% for DMMP. Polyaniline was used in another proposed ammonia sensor, based on variations of reflected light spectra resulting from interactions of ammonia with the sensing layer. The sensor was therefore an optrode-type optical-fiber sensor, where the light reflected from the face of the fiber is monitored [86]. The sensor head consisted of polyaniline and Nafion® layers deposited on the face of the fiber. DuPont Nafion is a copolymer of tetrafluoroethylene and perfluoro-3,6-dioxa-4-methyl-7-octenesulfonyl fluoride: it is highly resistant to chemical effects, but the presence of its exposed sulfonic acid groups allows it to absorb 22% by weight of water. The sensor response to ammonia concentration changing from 78 ppm to 5000 ppm was measured.

Two more fiber sensors for ammonia were proposed by a same research group: the first is an evanescent wave sensor, where sol-gel deposition process is employed for immobilizing a reversible NH_3-sensitive dye on the middle unclad region of a multimode fiber [87], while the other one additionally exploits the properties of a long-period grating [88].

A specially designed light guiding fused silica capillary was used to implement long-path evanescent wave optical sensor, showing the capability of detecting trace of ammonia down to 10 ppm level [89].

Following the demand of compact and high sensitive devices, an optode sensor was developed for the detection of NO_2 by immobilizing a selective chemical transducer, constituted by $[Ti(Pc)_2]$, a sandwich-type molecule [metal diphthalocyanine, bis(pthalocyaninato)titanium(IV)] on an alumina disc and using a single optical fiber for the interrogation [90]. The interaction with NO_2 leads to a spectral change of this compound in the red and green components of the visible spectrum. A detection limit of 0.6 ppm was obtained.

A clear advantage of fiber optic based sensors is due to their miniaturization and flexibility, that are key factors in medicine for *in vivo* measurements. An example is represented by a sensor for the continuous monitoring of a clinically-relevant parameter, the partial pressure of gastric carbon dioxide [91]. The tip of the fiber probe consists of a piece of black plastic that contains the CO_2-sensitive membrane; the latter consists basically of a dye/quaternary ammonium ion pair, dissolved in a thin layer of ethylcellulose. The absorption of the layer at 590 nm wavelength depends on the CO_2 concentration; the quantity to be measured is therefore the ratio between the detected intensities at 590 and 850 nm, with the wavelength at 850 nm used as a reference. The obtained resolution varies between 0.2 h Pa (at 0 h Pa) and 0.6 h Pa (at 150 h Pa); the response time is less than 1 minute.

One limitation to the use of telecommunication fibers, which are easily available and relatively cheap, is represented by the fact that the fundamental vibrational absorbance features of gases are generally located in the infrared region (IR), outside the transmission window of silica fibers. Happily, many gases of interest (methane, carbon dioxide, hydrogen sulphide, carbon monoxide) also have overtone and combination absorption lines in the near infrared ($\lambda \approx$ 1-2 μm). Thus, remote, multi-point monitoring of hazardous and environmentally important gases is well feasible; the use of fiber lasers and high sensitivity (cavity enhanced) spectroscopic techniques may increase the potential of these techniques [92].

Methane, for instance, is the main component of natural gas and marsh gas: early detection of methane has an utmost impact on houses' and industrial mines' safety. This explains why several fiber based sensors have been proposed and developed for the measurement of methane concentration: from early devices, demonstrating a minimum detectable concentration for methane and acetylene in air of 5 ppm and 3 ppm respectively [93], to various more recent configurations [94, 95, 96, 97, 98].

It is well known that Bragg gratings written in the core of an optical fiber (also called FBGs – fiber Bragg gratings) [99] are widely used as direct sensing elements for strain and temperature (because these two factors directly change the period of the grating itself) [100]. They can also be used as transduction elements, converting the output of another sensor where a strain or temperature change is induced by the measurand; as an example, a fiber Bragg grating gas sensors may use a selective absorbing coating, which in the presence of a gas expands, so generating a strain which is measurable by the grating itself. Technically, the absorbing material is the sensing element, converting the concentration of gas into a given strain or temperature increase, while the Bragg grating translates that effect into a wavelength/intensity change. An example is provided by a very recent paper where a multipoint in-fiber sensor, capable of hydrogen leak detection in air as low as 1% concentration with a response time smaller than a few seconds, has been presented [101]. This sensor makes use of FBGs covered by a catalytic sensitive layer made of a ceramic doped with noble metal which, in turn, induces a temperature elevation around the FBGs in the presence of hydrogen in air. The same principle has been used in a FBG-based methane sensor, which employs a special catalyst: the temperature of the catalyst rises rapidly as the concentration of methane increases, and hence it can be detected by an FBG-type sensor through monitoring the wavelength shift of the reflection spectrum. The experimental results show the good performances of our system, such as the high resolution and temperature stability [98].

In analogy with a traditional gas cell, an optical fiber can be employed as transmission medium for delivering light to a microcapillary tubing: light passes through the sample contained in the microcapillary tubing to another optical fiber, which in turn delivers the radiation to a detector. In the case of acetylene the minimum detectable concentration is less than 10 ppm [102].

Different fiber designs including fibers with a small hole placed in the core center [103] and D-shaped optical fibers [104] have also been proposed. However, a poor overlap between the gas volume and the mode field of the propagating light results in weak absorption and therefore in the requirement of long-length fibers.

For measurements in the IR region hollow optical waveguides have been suggested [105], but they are usually multimode and exhibit rather high losses, which, besides limiting the practical waveguide length to a few meters, makes the sensor not so efficient.

The architecture of a new kind of optical fiber, the so called microstructured optical fiber (MOF) or photonic crystal fiber (PCF) [106], overcomes some drawbacks of conventional fibers and suggests a variety of new strategies for optical sensing. A PCF is an optical fiber which obtains its waveguide properties not from a spatially varying material composition, but from an arrangement of tiny and closely spaced air holes, which go through the whole length of fiber. PCFs are generally divided into two categories: high -and low-index guiding fibers. As the traditional fibers, high-index guiding fibers are guiding light in a solid core by the total internal reflection (TIR) principle, fulfilled by the lower effective index in the microstructured air-filled region. In sensing devices based on these fibers, the evanescent field of the light interacts with the gas present in the holes [107], and molecular species can be detected using absorption spectroscopy [108]. The gas detection sensitivity is mainly determined by the fractional power carried by the evanescent wave in the air holes. In general, these fibers exhibit less than 1% of the modal power in the holes but with an optimized fiber design the overlap at a particular wavelength can reach 30%. Another limitation of this type of fiber in gas sensing arises from the small-sized air holes, which limit gas exchange and, thus, response times of the fibers.

Low-index guiding fibers guide light by the photonic bandgap (PBG) effect. In this type of fiber, the light is confined within the air core by a two-dimensional PBG formed by the periodic structure of the cladding allowing transmission over a limited wavelength range [109]. In these fibers the overlap between the gas and the mode field can be significantly improved with respect to a traditional optical fiber by carefully designing the microstructured cladding (such as small core size and large air fill fraction) [110]. It has been demonstrated that in PBFs more than 98% of the guided mode field energy can propagate in the air regions of the fiber [111, 112]. This leads to a strong reduction of the length of fiber needed for gas-sensing applications [108, 113]. PBG fibers are also bending insensitive, and that makes easier the realization of portable and compact devices [114, 115].

Sensing PCFs are usually manufactured in silica glass, but attention has been devoted to the fabrication of PCFs in soft glass [116] and polymers [117] too.

Hollow-core photonic band-gap fibers filled with gases have been proved to be feasible as direct absorption gas sensors [118]. Saturated absorption spectroscopy has also been demonstrated using PBG fibers [119]. As in conventional fibers, selectivity may be achieved by special coatings: as an example, a toluene sensor with a detection limit of about 0.007 vol.% has been realized by applying porous xerogel layers onto the walls of air holes (with diameters in the range 10–50 μm) of three different microstructured fibers prepared on purpose [120].

6.6 Conclusions

Gas detection has been recognized since many years as a very important technology because it has a strong impact on many industrial and social objectives, like ensuring a sustainable development through the monitoring of industrial processes, providing early detection for personal and homeland security, monitoring pollution in the environment, allowing early and non-invasive diagnostics in healthcare [121].

Optical methods for gas detection, in particular, may offer a number of advantages over other techniques, including a high degree of gas specificity, non-contact or remote measurements, potential of absolute measurements (not requiring calibration mixtures), intrinsically safe systems even in hazardous and explosive environments thanks to the use of optical fibers.

The bulk optical systems for gas sensing may present disadvantages of high cost and low level of field robustness; both these factors are related to the constraints on the accurate alignment of optical components, which makes their manufacture expensive. The use of fiber optic and integrated optic based systems, however, has largely overcome this problem. A remaining drawback has to do with fouling (dirt, dust etc) and/or condensation of moisture on the optical surfaces exposed to the measurand.

Optical absorption spectroscopy still is the most widely used optics-based gas detection technique.

In fact, many gases of interest (like CH_4, CO, CO_2, HF, HCl, H_2S, NO, N_2O, NH_3, SF_6, etc.) exhibit absorption in the infrared (fundamental) and near-infrared (overtone) regions of the spectrum, the strength of the absorption at a particular wavelength corresponding to the concentration of the gas.

Since the optical response, for a given (low) gas concentration, is proportional to the optical path length of the gas cell, various techniques (intracavity or multipass arrangements) have been developed to increase the equivalent optical path length. Optical fibers too offer a convenient solution to this problem.

An increase of sensitivity and/or selectivity may be achieved by the use of transducer layers, which change their optical properties depending on the interaction with the measurand gas, and can be interrogated with a variety of optical methods by employing cheap optoelectronic components.

Tunable laser diodes in the 1.3–1.6 μm region, first developed for telecommunication applications, represent a good example of not so expensive and easily available sources; their narrow emission lines allow a high degree of gas specificity with high signal to noise ratio. Very interesting perspectives come from the more recent development of other laser sources like the quantum cascade lasers (QCLs); in particular, single mode distributed feedback QCLs exhibit excellent properties in terms of narrow linewidth, average power and room temperature operation. Their emission wavelength, based on intersub-band transition in a multiple-quantum well heterostructure, can be tailored over a wide spectral range in the mid IR region, so allowing us the possibility of direct measurement of fundamental absorption lines in that spectral region.

Some devices use the principle of optical interferometry to detect the refractive index changes with high sensitivity. Others exploit different interrogation methods of the sensing component, including surface plasmon resonance in a thin metal film, interaction with an evanescent optical field, Bragg or long period gratings on integrated waveguide structures or in optical fibers.

Finally, the combination of optical fibers with micro- and nanotechnologies is receiving a growing attention, as it offers great potential for the realization of novel sensors. Microstructured and photonic crystal fibers, with their characteristic air-hole structure running along the length of the fiber, which can be filled with gases, are very good candidates for distributed sensing based on direct absorption or evanescent wave absorption or fluorescence spectroscopy. Sub-micron tapered fibers and nanowires, with the largest part of the guided power being transmitted through the evanescent field, produce an increased sensitivity to the optical properties of the surrounding media.

Selected examples of optical gas sensors fabricated according to the various methods have been presented here, without any claim of being exhaustive. Different applications have demonstrated the effectiveness of optical methods for sensitive, selective, real time trace gas concentration measurements. Since more than a decade, significant advances in materials and processes have facilitated the miniaturization of a wide variety of sensing devices; a further exploitation of nanocomposite materials and nanotechnologies in photonics will open the way to even more compact and efficient devices as well to new sensing concepts.

Acknowledgments Useful discussions with IFAC colleagues of the Optoelectronics and Photonics section are gratefully acknowledged. Particular thanks are due to Francesco Baldini, Massimo Brenci, Riccardo Falciai, Anna Grazia Mignani, Stefano Pelli.

References

1. Demtröder W. Laser Spectroscopy, Basic Concepts and Instrumentation. Berlin: Springer-Verlag; 1996.
2. Menzel R. Photonics Linear and Nonlinear Interactions of Laser Light and Matter. Berlin: Springer-Verlag; 2001.
3. Mukamel S. Principles of Nonlinear Optical Spectroscopy. New York: Oxford University Press; 1999.
4. Tkachenko NV. Optical spectroscopy: methods and instrumentations. Amsterdam: Elsevier; 2006.
5. Abramczyk H. Introduction to laser spectroscopy. Amsterdam: Elsevier; 2005.
6. Sigrist MW. Editor, Air Monitoring by Spectroscopic Techniques. Somerset N.J: John Wiley & Sons; 1994.
7. Wenz H, Demtröder W, Flaud J. Highly sensitive absorption spectroscopy of the ozone molecule around 1.5 μm. J Mol Spectrosc. 2001:209:267–277.
8. Corsi C, Inguscio M. High sensitivity trace gas monitoring using semiconductor diode lasers. In: S Martellucci et al. editors. Optical Sensors and Microsystems: New Concepts, Materials, Technologies. New York: Kluwer Academic/Plenum Press; 2000.
9. Pavone FS, Inguscio M. Frequency- and wavelength-modulation spectroscopies: comparison of experimental methods using an AlGaAs diode laser. Appl Phys B. 1993;56:118–122.
10. Herriott DR, Kogelnik H, Kompfner R. Off-axis paths in spherical mirror interferometers. Appl Opt. 1964;3:523–526.
11. McManus JB, Kebabian PL, Zahniser MS. Astigmatic mirror multipass absorption cells for long-path-length spectroscopy. App Opt. 1995;34:3336–3348.
12. Seiter M, Sigrist MW. Trace-gas sensor based on mid-IR difference-frequency generation in PPLN with saturated output power. Infrared Phys Technol. 2000;41:259–269.
13. Toci G, Mazzinghi P, Miele B, Stefanutti L. An airborne diode laser spectrometer for the simultaneous measurement of H_2O and HNO_3 content of the stratospheric cirrus clouds. Optics and Laser Eng. 2002;37:459–480.
14. Werle PW, Mazzinghi P, D'Amato F, De Rosa M, Maurer K, Slemr F. Signal processing and calibration procedures for in situ diode-laser absorption spectroscopy. Spectrochim Acta Mol Biomol Spectros. 2004;60:1685–1705.
15. Kasyutich VL, Martin PA. Multipass optical cell based upon two cylindrical mirrors for tunable diode laser absorption spectroscopy. Appl Phys B. 2007;88:125–130.
16. Sigrist MW. Air Monitoring by Laser Photoacoustic Spectroscopy. In Ref. [6], 1994.
17. Nägele M, Sigrist MW. Mobile laser spectrometer with novel resonant multipass photoacoustic cell for trace-gas sensing. Appl Phys B. 2000;70:895–901.
18. Song K, Oh S, Jung EC, Kim D, Cha H. Application of laser photoacoustic spectroscopy for the detection of water vapor near 1.38 μm. Microchem J. 2005;80:113–119.
19. Rey JM, Marinov D, Vogler DE, Sigrist MW. Investigation and optimisation of a multipass resonant photoacoustic cell at high absorption levels. Appl Phys B. 2005;80: 261–266.

20. Faist J, Capasso F, Sivco DL, Sirtori C, Hutchinson AL, Cho AY. Quantum cascade laser. Science. 1994;264:553–556.
21. Elia A, Di Franco C, Lugarà PM, Scamarcio G. Photoacoustic spectroscopy with quantum cascade lasers for trace gas detection. Sensors. 2006;6:1411–1419.
22. Grossel A, Zéninari V, Joly L, Parvitte B, Durry G, Courtois D. Photoacoustic detection of nitric oxide with a Helmholtz resonant quantum cascade laser sensor. Infrared Phys Techn. 2007;51:95–101.
23. Lendl B, Ritter W, Harasek M, Niessner R, Haisch C. Photoacoustic monitoring of CO_2 in biogas matrix using a quantum cascade laser. Proc 5th IEEE Conf On Sensors. 2007; 338–341. doi: 10.1109/ICSENS.2007.355475.
24. Grossel A, Zéninari V, Parvitte B, Joly L, Courtois D. Optimization of a compact photoacoustic quantum cascade laser spectrometer for atmospheric flux measurements: application to the detection of methane and nitrous oxide. Appl Phys B: Laser and Optics. 2007;88:483–492.
25. O'Keefe A, Deacon DAG. Cavity ring-down optical spectrometer for absorption measurements using pulsed laser sources. Rev Sci Instrum. 1998;59:2544–2551.
26. Engeln R, Berden G, Peeters R, Meijer G. Cavity enhanced absorption and cavity enhanced magnetic rotation spectroscopy. Rev Sci Instrum. 1998;69:3763–3769.
27. Leleux D, Claps R, Englich FV, Tittel FK. Novel laser-based gas sensors for trace gas detection in a spacecraft habitat. Proc. of Bioastronautics Investigators' Workshop, Galveston, Texas. 2001.
28. Podobedov VB, Plusquellic DF, Fraser GT. Investigation of the water-vapor continuum in the THz region using a multipass cell. J Quant Spectrosc Radiat Transf. 2005;91: 287–295.
29. Mazzotti D, Giusfredi G, Cancio P, De Natale P. High-sensitivity spectroscopy of CO_2 around 4.25 μm with difference-frequency radiation. Optic Laser Eng. 2002;37: 143–158.
30. Nikolaev IV, Ochkin VN, Spiridonov MV, Tskhai SN. Diode ring-down spectroscopy without intensity modulation in an off-axis multipass cavity. Spectrochim Acta Mol Biomol Spectros. 2007;66:832–835.
31. Nadezhdinskii A, Berezin A, Chernin S, Ershov O, Kutnyak V. High sensitivity methane analyser based on tuned near infrared diode laser. Spectrochim. Acta Mol Biomol Spectros. 1999;55:2083–2089.
32. Waechter H, Sigrist MW. Mid-infrared laser spectroscopic determination of isotope ratios of N2O at trace levels using wavelength modulation and balanced path length detection. Appl Phys B. 2007;87:539–546.
33. Manne J, Sukhorukov O, Jäger W, Tulip J. Pulsed quantum cascade laser-based cavity ring-down spectroscopy for ammonia detection in breath. Appl Opt. 2006;45: 9230–9237.
34. Azzam RMA, Bashara NM. Ellipsometry and Polarized Light. North-Holland, Amsterdam; 1977.
35. Zangooie S, Bjorklund R, Arwin H. Vapor sensitivity of thin porous silicon layers. Sensor Actuator B Chem. 1997;43:168–174.
36. Wang G, Arwin H. Modification of vapor sensitivity in ellipsometric gas sensing by copper deposition in porous silicon. Sensor Actuator B Chem. 2002;85:95–103.
37. Wang G, Arwin H, Jansson R. An optical gas sensor based on ellipsometric readout. IEEE Sensor J. 2003;3:739–743.
38. Zangooie S, Bjorklund R, Arwin H. Water interaction with thermally oxidized porous layers. J Electrochem Soc. 1997;144:4027–4035.
39. Ritchie RH. Plasma losses by fast electrons in thin films. Phys Rev. 1957;106:874–881.
40. Otto A. Excitation of nonradiative surface plasma waves in silver by the method of frustrated total reflection. Z Physik. 1968;216:398–410.

41. Kretschmann E, Raether H. Radiative decay of non-radiative surface plasmon excited by light. Z Naturforsch. 1968;23:2135–2136.
42. Zhang HQ, Boussaad S, Tao NJ. High-performance differential surface plasmon resonance sensor using quadrant cell photodetector. Rev Sci Instrum. 2003;74:150–153.
43. Homola J, Yee SS, Gauglitz G. Surface plasmon resonance sensors: review. Sensor Actuator B Chem. 1999;54:3–15.
44. Ince R, Narayanaswamy R. Analysis of the performance of interferometry, surface plasmon resonance, and luminescence as biosensors and chemosensors. Anal Chim Acta. 2006;569:1–20.
45. Conoci S, Palumbo M, Pignataro B, Rella R, Valli L, Vasapollo G. Optical recognition of organic vapours through ultrathin calix[4]pyrrole films. Colloid Surface Physicochem Eng Aspect. 2002;198–200:869–873.
46. Kato K, Dooling CM, Shinbo K, Richardson TH, Kaneko F, Tregonning R, Vysotsky MO, Hunter CA. Surface plasmon resonance properties and gas response in porphyrin Langmuir-Blodgett films. Colloid Surface Physicochem Eng Aspect. 2002;198–200: 811–816.
47. Manera MG, Leo G, Curri ML, Cozzoli PD, Rella R, Siciliano P, Agostiano A, Vasanelli L. Investigation on alcohol vapours/TiO_2 nanocrystal thin films interaction by SPR technique for sensing application. Sensors and Actuators B Chem. 2004;100:75–80.
48. Manera MG, Cozzoli PD, Curri ML, Leo G, Rella R, Agostiano A, Vasanelli L. TiO_2 nanocrystal films for sensing applications based on surface plasmon resonance. Synthetic Met. 2005;148:25–29.
49. Story PR, Galipeau DW, Mileham RD. A study of low cost sensors for measuring low relative humidity. Sensor Actuator B Chem. 1995;25:681–685.
50. Gu L, Huang QQ, Quin M. A novel capacitive-type humidity sensor using CMOS fabrication technology. Sensor Actuator B Chem. 2004;99:491–498.
51. Podgersek RP, Franke H, Feger C. Selective optical detection of n-heptane/iso-octane vapors by polyimide lightguides. Opt Lett. 1995;20:501–503.
52. Manera MG, De Julián Fernández C, Maggioni G, Mattei G, Carturan S, Quaranta A, Della Mea G, Rella R, Vasanelli L, Mazzoldi P. Surface plasmon resonance study on the optical sensing properties of nanometric polyimide film to volatile organic vapors. Sensor Actuator B Chem. 2007;120:712–718.
53. Muñoz Aguirre N, Martínez Pérez L, Colín JA, Buenrostro-Gonzalez E. Development of a Surface Plasmon Resonance n-dodecane vapor sensor. Sensors. 2007;7:1954–1961.
54. Hu WP, Chen SJ, Huang KY, Hsu JH, Chen WY, Chang GL, Lai KA. A novel ultrahigh-resolution surface plasmon resonance biosensor with an Au nanocluster-embedded dielectric film. Biosens. and Bioelectron. 2004;19:1465–1471.
55. Manera MG, Cozzoli PD, Curri ML, Leo G, Rella R, Agostano A, Vasanelli L. TiO_2 nanocrystal films for sensing applications based on surface plasmon resonance. Synthetic Met. 2005;148:25–29.
56. Homola J, Koudela I, Yee SS. Surface plasmon resonance sensors based on diffraction gratings and prism couplers: sensitivity comparison. Sensor Actuator B Chem. 1996;54:16–24.
57. Nikitin PI, Anokhin PM, Beloglazov AA. Chemical Sensors Based on Surface Plasmon Resonance in Si Grating Structures. International Conference on Solid-State Sensors and Actuators Transducers. 1997;2:1359–1362.
58. Nagamura T, Yamamoto M, Terasawa M, Stiratori K. Highly performance sensing of nitrogen oxides by surface plasmon resonance excited fluorescence of dye-doped deoxyribonucleic acid films. Appl Phys Lett. 2003;83:803–805.
59. Challener WA, Ollmann RR, Kam KK. A surface plasmon resonance gas sensor in a 'compact disc' format". Sensor Actuator B Chem. 1999;56:254–258.

60. Tiefenthaler K, Lukosz W. Integrated optical switches and gas sensors. Opt. Lett. 1984;10:137–139.
61. Fushen C, Qu L, Yunqi L, Yu X. Integrated optical interferometer gas sensor. Microwave & Opt Techn Lett. 1996;11:213–215.
62. Chiarini M, Bentini GG, Bianconi M, Cerutti A, Pennestri G, Wang P, She L, Mazzoldi P, Sada C. Integrated Mach-Zehnder micro-interferometer for gas trace remote sensing. Proc. of SPIE. 2006;6031:603106.
63. Martucci A, Buso D, Guglielmi M, Zbroniec L, Koshizaki N, Post M. Optical gas sensing properties of silica film doped with cobalt oxide nanocrystals. J Sol-Gel Sci Technol. 2004;32:243–246.
64. Qi Z-M, Yimit A, Itoh K, Murabayashi M, Matsuda N, Takatsu A, Kato K. Composite optical waveguide composed of a tapered film of bromothymol blue evaporated onto a potassium ion–exchanged waveguide and its application as a guided wave absorption–based ammonia-gas sensor. Opt Lett. 2001:26:629–631.
65. Chen X-M, Itoh K, Murabayashi M, Igarashi C. A highly sensitive ammonia gas sensor based on an Ag+/K+ composite ion doped glass optical waveguide system. Chemistry Lett. 1996;26:103–104.
66. Lambeck PV. Integrated opto-chemical sensors. Sensor Actuator B Chem. 1992;8: 103–116.
67. Lavers CR, Wilkinson JS. A waveguide-coupled surface-plasmon sensor for acqueous environment. Sensor Actuator B Chem. 1994;22:75–81.
68. Harris RD, Wilkinson JS. Waveguide surface plasmon resonance sensors. Sensor Actuator B Chem. 1995;29:261–267.
69. Tobiŝka P, Hugon O, Trouillet A, Gagnaire H. An integrated optic hydrogen sensor based on SPR on palladium. Sensors and Actuators B. 2001;74:168–172.
70. Jorgenson RC, Yee SS. A fiber-optic chemical sensor based on surface plasmon resonance. Sensor Actuator B Chem. 1993;12:213–220.
71. Niggermann M, Katerkamp A, Pellmann M, Bolsmann P, Reinbld J, Cammann K. Remote sensing of tetrachlorethene with a micro-fiber optical gas sensor based on surface plasmon resonance spectroscopy. Sensor Actuator B Chem. 1996;34:328–333.
72. Abdelghani A, Chevelon JM, Jaffrezic-Renault N, Veilla C, Cagnaire H. Chemical vapour sensing by surface plasmon resonance optical fiber sensor coated with fluoropolymer. Anal Chim Acta. 1997;337:225–232
73. Slavik R, Homola J, Čtyroký J. Optical fiber surface plasmon resonance sensor for an acqueous environment. 12th Intl. Conf. on Optical Fiber Sensor, Williamsburg, USA, Tech. Digest Series. 1997;16:436–439
74. Fontana E, Dulman HD, Doggett DE, Pantell RH. Surface plasmon resonance on a single mode fiber. Instrum. and Meas. Techn. Conf. IMTC/97. IEE Proc. 'Sensing, Processing, Networking' 1997;1:611–616.
75. Fontana E, Dulman HD, Doggett DE, Pantell RH. Surface plasmon resonance on a single mode optical fiber. IEEE Trans. on Instrumentation and Measurement. 1998;47:168–173.
76. Sharma AK, Jha R, Gupta BD. Fiber-optic sensors based on Surface Plasmon Resonance: a comprehensive review. IEEE Sensor J. 2007;7:1118–1129.
77. Zaatar Y, Zaouk D, Bechara J, Khoury A, Llinaress C, Charles JP. Fabrication and characterization of an evanescent wave fiber optic sensor for air pollution control. Mater Sci Eng B: Solid State Mater Adv Technol. 2000;74:296–298.
78. Tao SQ, Winstead CB, Jindal R, Singh JP. Optical-fiber sensor using tailored porous sol-gel fiber core. IEEE Sensor J. 2004;4:322–328.
79. Bürck J, Conzen JP, Beckhaus B, Ache HJ. Fiber-optic evanescent wave sensor for in situ determination of non-polar organic compounds in water. Sensor Actuator B Chem. 1994;18:291–295.

80. Potyrailo RA, Hobbs SE, Hieftje GM. Near-ultraviolet evanescent-wave absorption sensor based on a multimode optical fiber. Anal Chem. 1998;70:1639–1645.
81. Culshaw B. Fiber-optic sensors: application and advances. Optics Photon News. 2005;16:24–29.
82. Grady T, Butler T, MacCraith B, Diamond D, Mc Kervey MA. Optical Sensor for Gaseous Ammonia With Tuneable Sensitivity. Analyst. 1997;122:803–806.
83. Malins C, Landl M, Šimon P, MacCraith BD. Fiber optic ammonia sensing employing novel near infrared dyes. Sensor Actuator B Chem. 1998;51:359–367.
84. Cao W, Duan Y. Optical fiber-based evanescent ammonia sensor. Sensor Actuator B Chem. 2005;110:252–259.
85. El-Sherif M, Bansal L., Yuan J. Fiber optic sensors for detection of toxic and biological threats. Sensors. 2007;7:3100–3118.
86. Opilski Z, Pustelny T, Maciak E, Bednorz M, Stolarczyk A, Jadamiec M. Investigations of optical interferometric structures applied in toxic gas sensors. Bull Polish Acad Sci – Techn Sci. 2005;53:151–156.
87. Suresh Kumar P, Abraham VS, Vallabhan CPG, Nampoori VPN, Radhakrishnan P. Fiber optic evanescent wave sensor for ammonia gas. Proc. SPIE. 2004;5280: 617–621.
88. Suresh Kumar P, Abraham VS, Vallabhan CPG, Nampoori VPN, Radhakrishnan P. Long-period grating in multimode fiber for ammonia gas detection. Proc. SPIE. 2004;5279:331–335.
89. Tao S, Gong S, Fanguy JC, Hu X. The application of a light guiding flexible tubular waveguide in evanescent wave absorption optical sensing. Sensor Actuator B Chem. 2007;120:724–731.
90. Baldini F, Capobianchi A, Falai A, Mencaglia AA, Pennesi G. Reversibile and selective detection of NO_2 by means of optical fibers. SensorActuator B Chem. 2001;74:12–17.
91. Baldini F, Falai A, De Gaudio AR, Landi D, Lueger A, Mencaglia A, Scherr D, Trettnak W. Continuous monitoring of gastric carbon dioxide with optical fibers. Sensor Actuator B Chem. 2003;90:132–138.
92. Stewart G, Whitenett G, Shields P, Marshall J, Culshaw B. Design of fiber laser and sensor systems for gas spectroscopy in the near-IR. Proc. SPIE. 2004;5272:172–180.
93. Tai H, Yamamoto K, Uchida M, Osawa S, Uehara K. Long distance simultaneous detection of methane and acetilene by using diode lasers-coupled with optical fibers. IEEE Photon Tech Lett. 1992;4:804–807.
94. Stewart G, Mencaglia A, Philp W, Jin W. Interferometric signals in fiber optic methane sensors with wavelength modulation of the DFB laser source. J Lightwave Technol. 1998;16:43–48.
95. Hodgkinson J, Pride R, Tandy C, Moodie DG, Stewart G. Field evaluation of multi-point fiber optic sensor array for methane detection (OMEGA). Proc. SPIE. 2000;4074: 90–98.
96. Cone OM, Garcia S, Mirapeix JM, Echevarria J, Madruga Saavedra FJ, Lopez-Higuera JM. New optical cell design for pollutant detection. Proc. SPIE. 2002;4578:283–290.
97. Wang S, Che R. Novel spectrum absorption fiber methane sensor with DFBLD. Proc. SPIE. 2005;5634:619–626.
98. Bin Z, Guo-rong L, Zu-guang G, Sai-ling H. An optical sensing system for the concentration of methane based on fiber Bragg gratings. Optoel Lett. 2007;3:410–412.
99. Hill KO, Meltz G. Fiber Bragg grating technology fundamentals and overview. J Lightwave Techn. 1997;15:1263–1276.
100. Othonos A, Kalli K. Fiber Bragg Gratings: Fundamentals and Applications in Telecommunications and Sensing. Norwood, MA, USA: Artech House; 1999.
101. Caucheteur C, Debliquy M, Lahem D, Megret P. Catalytic Fiber Bragg Grating sensor for hydrogen leak detection in air. IEEE PhotTechn Lett. 2008;20:96–98.

102. Alfeeli B, Pickrell G, Wang A. Sub-Nanoliter Spectroscopic Gas Sensor. Sensors. 2006;6:1308–1320.
103. Sudo S, Yokohama I, Yasaka H, Sakai Y, Ikegami T. Optical fiber with sharp optical absorptions by vibrational-rotational absorption of C2H2 molecules. IEEE Photon Tech Lett. 1990;2:128–131.
104. Stewart G, Jin W, Culshaw B. Prospects for fiber-optic evanescent-field gas sensors using absorption in the near-infrared. Sensor Actuator B Chem. 1997;38:42–47.
105. Harrington J. A review of IR transmitting hollow waveguides. Fiber Integrated Optic. 2000;19:211–227.
106. Russell PSJ. Photonic crystal fibers. Science. 2003;299:358–362.
107. Monro TM, Belardi W, Furusawa K, Baggett JC, Broderick NGR, Richardson DJ. Sensing with microstructured optical fibers. Meas Sci Tech. 2001;12:854–858.
108. Hoo YL, Jin W, Li C, Ho HL, Wang D, Windeler RS. Evanescent-wave gas sensing using microstructured fiber. Opt Eng. 2002;41:8–9.
109. Cregan RF, Mangan BJ, Knight JC, Birks TA, Russell PSJ, Roberts PJ, Allan DC. Singlemode photonic band gap guidance of light in air. Science. 1999;285:1537–1539.
110. Monro TM, Richardson DJ, Bennett PJ. Developing holey fibers for evanescent field devices. Electron Lett. 1999;35:1188–1189.
111. Lægsgaard J, Mortensen NA, Riishede J, Bjarklev A. Material effects in air-guiding photonic bandgap fibers. J Opt Soc Am B. 2003;20:2046–2051.
112. Humbert G, Knight J, Bouwmans G, Russell PSJ, Williams D, Roberts P, Mangan B. Hollow core photonic crystal fibers for beam delivery. Opt Express. 2004;12:1477–1484.
113. Pickrell G, Peng W, Wang A. Random-hole optical fiber evanescent-wave gas sensing. Optic Lett. 2004;29:1476–1478.
114. Hansen TP, Broeng J, Jakobsen C, Vienne G, Simonsen HR, Nielsen MD, Skovgaard PMW, Folkenberg JR, Bjarklev A. Air-guiding photonic bandgap fibers: Spectral properties, macrobending loss and practical handling. J Lightwave Tech. 2004;22:11–15.
115. Pawlat J, Sugiyama T, Matsuo T, Ueda T. Photonic Bandgap Fiber for a sensing Device. IEEJ Transactions on Sensors and Micromachines. 2007;127:160–164.
116. Konorov SO, Zheltikov AM, Scalora M. Photonic-crystal fiber as a multifunctional optical sensor and sample collector. Opt Express. 2005;13:3454–3459
117. Cordeiro CMB, Franco MAR, Chesini G, Barretto ECS, Lwin R, Brito Cruz CH, Large MCJ. Microstructured-core optical fibre for evanescent sensing applications. Opt Express. 2006;14:13056–13066.
118. Ritari T, Tuominen J, Ludvigsen H, Petersen JC, Sorensen T, Hansen TP, Simonsen HR. Gas sensing using air-guiding photonic bandgap fibers. Opt Express. 2004;12: 4080–4087.
119. Thapa R, Knabe K, Faheem M, Naweed A, Weaver OL, Corwin KL. Saturated absorption spectroscopy of acetylene gas inside large-core photonic bandgap fiber. Opt Lett. 2006;31:2489–2491.
120. Matejec V, Mrazek J, Podrazky O, Kanka J, Kasik I, Popisilova M. Microstructure fibers for sensing gaseous hydrocarbons. Proc. SPIE. 2007;6585:658511-1–658511-9.
121. Sberveglieri G, Editor. Gas Sensors: Principles, Operation and Developments. Springer. 1992.

Chapter 7
Thermometric Gas Sensing

István Bársony, Csaba Dücső and Péter Fürjes

The response of all kind of sensors to temperature changes has always to be considered as an important influencing factor of the stability of the reading. The principle of temperature sensing is one of the most reliable transduction methods. Temperature changes can be detected simply, sensitively and reliably, and thereby also rather inexpensively. Although most enabling technologies involve either new materials or material advances, the British MNT Roadmap for gas sensors, published in 2006 [1], considers among the enabling technologies, lower cost temperature measurements as well. Therefore, the field of *thermometric gas sensing*, which is often called *calorimetric gas sensing* too, remains the subject of much research and development.

Three main categories of thermometric gas sensing have to be discussed:

- Catalytic combustion;
- Thermal conductivity;
- Adsorption/desorption heat.

7.1 Detection of Combustible Gases

Thermometric gas sensing is widely used for the indirect detection of combustible gases. The operation is based upon the measurement of temperature changes generated during burning of the flammable gaseous components in the presence of oxygen of well-known concentration, which is typically normal air. The reading is calibrated in terms of concentration of the burned gas.

7.1.1 Combustion

Combustion or **burning** is a complex sequence of *exothermic chemical reactions* between a fuel and an oxidant accompanied by the production of heat (or both heat and light in the form of either a glow or flames). In an *ideal combustion reaction*, a combustible compound reacts with an oxidizing element resulting in

I. Bársony
Research Institute for Technical Physics and Materials Science – MFA, Hungarian Academy of Sciences, Budapest, Hungary

E. Comini et al. (eds.), *Solid State Gas Sensing*,
DOI: 10.1007/978-0-387-09665-0_7, © Springer Science+Business Media, LLC 2009

completely oxidized, chemically stable products. For example, combustion of hydrogen and oxygen results simply in water vapour:

$$2H_2 + O_2 \rightarrow 2H_2O + \text{heat} \tag{7.1}$$

The combustion of hydrocarbons is usually more complex:

$$CH_4 + 2O_2 \rightarrow CO_2 + 2H_2O + \text{heat} \tag{7.2}$$

In complete combustion, when a hydrocarbon burns in oxygen, the reaction will only yield carbon dioxide and water.

In most of the *real cases* oxygen *is obtained from the ambient air*. Therefore, the flue gas resulting from the combustion will contain, as by far the largest part, nitrogen and the excess oxygen:

$$CH_4 + 2O_2 + (N_2 + O_2) \rightarrow CO_2 + 2H_2O + (N_2 + O_2) + \text{heat} \tag{7.3}$$

It should be noted that a complete combustion is almost impossible to achieve –*combustion processes are never perfect*. In flue gases from combustion of carbon or carbon compounds both unburned carbon (as soot) and (e.g., incompletely oxidized) carbon compounds (CO and others) will be present. In all actual combustion reactions a wide variety of major and minor species will be present.

Air is a mixture of gases, typically:

Nitrogen	77.2 %
Oxygen	20.9 %
Water Vapour	0.9 %
Argon	0.9 %
Carbon Dioxide	0.03 %
Other Gases	0.07 %

Because its composition is reasonably constant, air is usually considered as a single gas, which simplifies the measurement of flammable gases for safety applications. Moreover, when air is the oxidant, some nitrogen or even sulfuric compounds will be oxidized to various oxides (NO_x, SO_x). The above combustion of methane in air will yield, in addition to the major products of carbon dioxide and water, the minor product carbon monoxide and nitrogen oxides, which are products of a side reaction (oxidation of nitrogen).

7.1.2 Thermal Considerations during Combustion

Direct combustion by atmospheric oxygen is a reaction mediated by radical intermediates. In chemistry, *radicals* are *atomic or molecular species with unpaired electrons on an otherwise open shell configuration*. They are usually highly reactive, so radicals are likely to take part in chemical reactions. It is necessary to maintain the high temperature required for radical production, the conditions for their production are naturally supported by the *heat* generated by combustion.

The total energy absorbed in bond breaking in an *exothermic reaction* is less than the total energy released in bond making. In other words, the energy needed for the (e.g.combustion) reaction to occur is less than the total energy provided. As a result, *the extra energy is released*, usually in the form of heat. When the medium in which the reaction is taking place gains heat, the reaction is exothermic. This is the case in combustion.

The absolute amount of energy in a chemical system is extremely difficult to measure or calculate, the enthalpy change ΔH, of a chemical reaction is, however, much simpler: The *enthalpy* or *heat content* (denoted *by ΔH*) is a quotient or description of thermodynamic potential of a system. It can be used under constant pressure to calculate the "useful" work obtainable from a closed thermodynamic system – often as a differential sum, describing the changes within exo- and endothermic reactions, which minimize at equilibrium.

$$\Delta H(S,P) = \Delta U + p\Delta V \tag{7.4}$$

The enthalpy of the ideal combustion reaction of methane

$$CH_4 + 2O_2 = CO_2 + 2H_2O \tag{7.5}$$

is –891 kJ/mol, obtained from the enthalpies of formation: CH_4 –75 kJ/mol, CO_2 -394, and $H_2O(l)$ –286 kJ/mol.

7.1.3 Catalysis

In chemistry and biology *the substance that increases the rate of a chemical reaction* (or generally accelerates a process) is called a "*catalyst*". The catalyst itself is not consumed by the overall reaction. It provides an alternative route to products, the catalytic route requiring lower activation energy than the uncatalyzed reaction. The lowered activation energy increases the reaction rate or provides lower temperature to accomplish the chemical reaction. Ideal catalysts may change in the course of a reaction but they regenerated, and re-used indefinitely.

Catalysts operate by *providing an (alternative) mechanism involving a different transition state and lower activation energy*. In consequence more molecular collisions possess the energy needed to reach the transition state. Hence, catalysts can enable reactions which, would not run without the presence of a catalyst (even if they are thermodynamically feasible), or perform them much faster, more specific, or at lower temperatures. This means that catalysts *reduce the amount of energy needed to start a chemical reaction*. Whether a catalyst is used or not, the net enthalpy change of a reaction remains the same; the catalyst just effects an easier activation. Catalysts have *no* effect on the chemical equilibrium of a reaction because the rate of both the forward and the reverse reaction are equally affected.

Enzymes are proteins that catalyse (*or* speed up) chemical reactions. Like all catalysts, enzymes work by lowering the activation energy (E_a) for a reaction. In enzymatic reactions, the molecules are converted into different products like in almost all processes in a biological cell. Most enzyme reaction rates are millions of times faster than those of comparable uncatalyzed reactions. *Enzymes do differ from most other catalysts by being much more specific, however. They are extremely selective* for their substrates and accelerate only a few reactions from among many possibilities. For example, the set of enzymes present in a cell determines which metabolic pathways occur in that cell.

Similar to other catalysts, enzymes *are not consumed by the reactions;* nor do they alter the equilibrium of these reactions. In the presence of an enzyme, the reaction runs usually in the same direction as it would without the enzyme, but much faster. If the equilibrium in a very exergonic reaction is displaced in one direction (during an exergonic process, energy is released *out of* the system, e.g., in the form of heat), the reaction is *effectively* irreversible. Under these conditions the enzyme will, only catalyze the reaction in the thermodynamically allowed direction.

7.1.4 Explosive Mixtures

In order to ignite an explosive mixture an ignition source typically a spark, or flame or hot surface is needed. The lowest temperature, which will cause a mixture to burn or explode, is called *the ignition temperature*. The power of the explosion depends on the "fuel" and its concentration in the atmosphere.

In the explosive mixture the concentration of combustible gas or vapour in air must be at a level such that the "fuel" and oxygen can react chemically. Consequently, not all concentrations of flammable gas or vapour in air will burn or explode.

- The lowest concentration of "fuel" in air which will burn is called the *lower explosive limit* (LEL) of the gas, and amounts to less than 5% by volume for most flammable gases and vapours. Therefore, there is a high risk of explosion even when relatively small concentrations of gas or vapour escape into the atmosphere. The LEL is different for each flammable gas, e.g., the *Institute of Gas Engineers and Managers* (IGEM) state levels of 4.9%vol in air for methane, compared with only 2.8%vol in air for ethane. Definitions of individual LELs vary but the IGEM figures are often accepted as the industry standard. [2]
- The *Upper Explosive Limit* (UEL) is the maximum concentration of "fuel" in air which will still burn. Concentrations above the UEL will not burn because there is insufficient atmospheric oxygen available for combustion

The combustable mixtures are mostly composed of organic compounds. The simplest organic compounds are those known as hydrocarbons being the main constituents of crude oil/gas. Methane is the first compound in the family known as *alkanes* or as *aliphatics*. *Alkenes* are similar but their molecular structure includes double bonds (e.g., ethylene and propylene). *Alkynes* contain triple bonds (e.g., acetylene). *Aromatic hydrocarbons* such as *benzene* have a ring molecular structure and burn with a smoky flame.

More complex organic compounds contain elements such as oxygen, nitrogen, sulphur, chlorine, bromine or fluorine. If these burn, the products of combustion will include – besides carbon dioxide and water – other compounds as well. For example, substances containing sulphur such as oil or coal will result in sulphur dioxide while those containing chlorine such as methyl chloride or polyvinyl chloride (PVC) will result in hydrogen chloride.

The lowest temperature at which vapour is given off by flammable liquids at sufficient rate to form an explosive mixture with air *is called flash point*. Liquids with flash points below normal ambient temperatures automatically release vapour in sufficient volume to provide an explosive mixture, so leakage of such liquids is potentially as dangerous as a flammable gas leak.

During thermometric detection of explosive gas concentrations, it is pivotal to eliminate the chance of explosion caused by the detection process and equipment itself by any means. These measures are strictly controlled by appropriate standards and norms, described and observed by the makers [3, 4, 5, 6, 7].

Catalytic combustion requires the presence of oxygen in stoichiometric or excess concentration. Therefore, parallel measurement of oxygen is also indispensable for reliable operation of pellistors.

7.2 Catalytic Sensing

The enthalpy generated by catalytic reaction can be detected in thermometric sensors and calibrated in terms of concentration.

The catalytic combustion of flammable gases has been the detection technology of choice for detection of explosive gases in mines and other high risk areas. An alarm is triggered by the sensor when a certain concentration threshold of flammable gas or vapour is exceeded. The mechanism governing the catalytic sensing of combustive gases in solid state gas sensors is called *heterogenous catalysis*. The overall process is a consecutive sequence of various physical and chemical steps:

- the diffusion of reactants to the sensor surface,
- exothermic adsorption, chemisorption and
- the chemical reaction;
- desorption and outdiffusion of gaseous products from the surface.

The gaseous products finally diffuse out of the sensor. The energy diagram of such a scheme is described in Fig. 7.1.

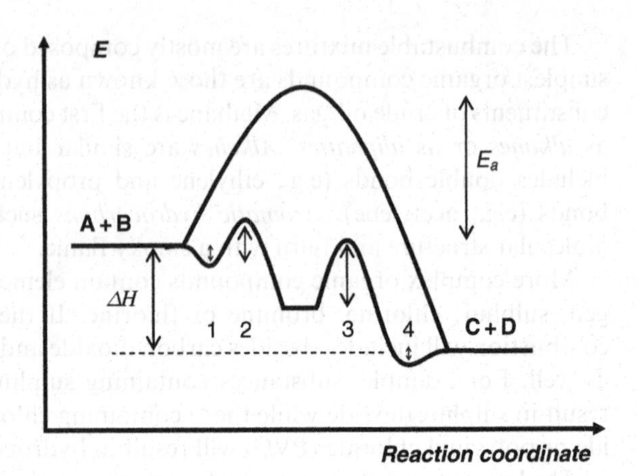

Fig. 7.1 Energetics of homogeneous gas phase reaction (*upper curve*) and heterogenous catalysis (*lower curve*) consisting of 1: adsorption, 2: chemisorption, 3: chemical reaction, 4: desorption. E_a: activation energy, ΔH: enthalpy of the exoterm reaction

Similarly, the enzymatic reactions were also used in the thermometric detection scheme *in microcalorimeters* [8], however, in connection with *liquid systems* rather than gases.

7.2.1 Pellistors

Pellet resistor sensors called *pellistors* are sometimes referred to as *catalytic bead sensors*. Pellistor detectors consist of two coils of fine platinum wire (Fig. 7.2) Both of them are embedded in a ceramic mass, (e.g., a bead of alumina) and connected electrically in a Wheatstone-bridge circuit. One of the ceramic beads is impregnated with a special catalyst which promotes oxidation, whilst the other without catalysts inhibits oxidation.

Current is passed through the coils so that they reach a temperature at which oxidation of the combustible gas readily occurs at the catalysed bead of the *active pellistor* (>450°C) in the presence of oxygen in an exothermic reaction. During operation, the *flammable vapour or gas* comes into contact with each of the pellistors. The generated heat of the heterogeneous catalysis causes an increase of the resistance of the conductive wire. The inactive bead (or compensator) is present to minimize the environmental effects such as temperature, heat conductivity variation of the ambient and humidity. As both elements

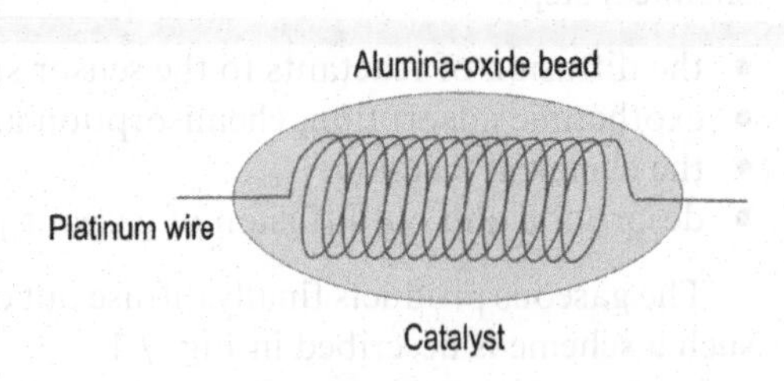

Fig 7.2 Structure of a conventional pellistor

behave similarly with temperature and humidity variation, no imbalance is seen in the bridge circuit, when no combustive gas is present and the heat transport between the two elements by gas flow is inhibited. The *change in resistance*, however, *causes an imbalance in the Wheatstone-bridge circuit* which is *within certain limitations proportional to the concentration of combustible gas* present (Fig. 7.3).

There are three alternatives to operate the sensor circuit:

- with constant bias voltage,
- with constant current bias or
- by keeping the resistance (temperature) constant.

Generally the response can be expressed as

$$\Delta V = k_1 k_2 \Delta H \tag{7.6}$$

where k_1 is the rate of chemical reaction,

k_2 is the characteristic coefficient of a given Wheatstone-bridge configuration describing its complex temperature dependence.

Due to the temperature dependence of k_1 and k_2 a quasi-linear response of the pellistor is expected for small temperature changes only, i.e., in a small concentration range of combustive gases both in constant voltage or current driven modes. Therefore, portable devices with a relatively simple driving circuitry can provide linear responses between 0 and 100% of LEL. At high operational temperature or at high concentration of the flue gas of large molecules, these systems suffer from loosing linearity.

In contrast, the constant resistance (temperature) method requires the measurement and control of heating power with a complex circuitry. The isothermal

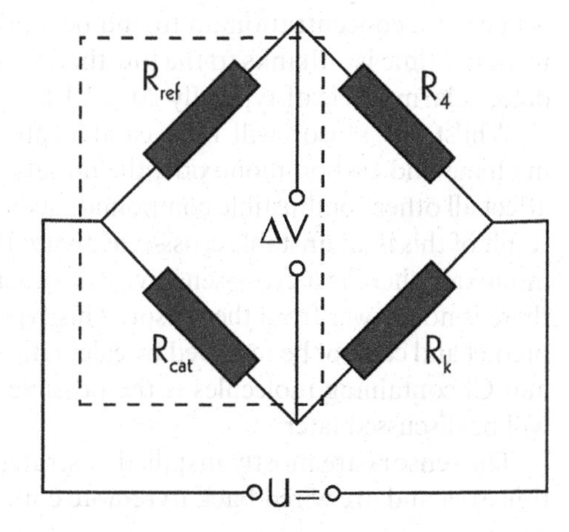

Fig. 7.3 Wheatstone-bridge arrangement of pellistors. The active (R_{cat}) and the reference (R_{ref}) elements are encapsulated by a diffusior head (dashed line) in order to eliminate false signals gas flow may result. The diffuser head also serves as flame arrestor

operation, however, provides linear response and makes this method attractive in calorimetry and oxidation kinetic studies.

Catalytic pellistor sensors are sensitive to most flammable gases and vapours; it is therefore possible to use *one single detector to monitor a wide range of flammable gases and vapours*. It is relatively straightforward to measure the proportion of gas or vapour present, using external circuitry. Gas detection in a pellistor occurs when *gas diffuses through a gas permeable membrane into the flameproof housing,* in which the detectors are packaged. The gas permeable membrane therefore plays a double role; it eliminates false signals possibly caused by gas flow, and absorbs the flame formed on the sensor head in case of malfunctioning.

7.2.1.1 Safe Detection of Explosive Mixtures

Mechanical safety and flameproof equipment are the basis of protection. Leaking of flames out of the device has to be prevented by the dense (usually bronze) diffuser-mesh in front of the pellistor, called *flame arrestor*. The equipment enclosure should be capable of containing an explosion within the sensor, thereby avoiding the possibility of the system becoming an external ignition source. The protection against unintended ignition of the explosive environment is provided by the mechanically stable, bulky, *explosion-proof housing*.

Intrinsic safety differs from other forms of protection by using the principle of electrically safe equipment. According to definition an intrinsically safe circuit does not use or release sufficient electrical energy, even under fault conditions (with two coincident faults applied to the system), to generate a spark or sufficient heat to ignite a combustible gas mixture.

The LEL ranges, e.g., from around 15% for anhydrous ammonia to around 0.5% for kerosene, and shows large variations for combustible gases and different organic vapours. However, mixtures of flammable gases may be oxidized even below the LEL by a suitable catalyst such as platinum or palladium. The change in the output voltage of the Wheatstone-bridge configuration is linear vs. concentration up to and beyond 100% LEL, for several gases. The response time is – thanks to the low thermal capacitance – only a few seconds to detect alarm levels of typically 20% LEL.

Whilst the sensor will only catalytically burn carbon compounds such as methane and carbon monoxide, the initiation of combustion will automatically affect all other combustible compounds such as H_2S which might be present. The result of this is an unstable cross-sensitivity: In the presence of methane or carbon monoxide there is a cross-sensitivity to other combustibles, without those gases there is no answer from the sensor. This type of cross-sensitivity is impossible to predict and can not be removed by calculation. The other unavoidable effect of S- and Cl-containing molecules is the possible poisoning of the sensor. This issue will be discussed later.

The sensors are mostly installed at strategic points in the hazardous application area and are wired back to remote control and monitoring equipment with three core armoured cables connected to the pellistor through a junction box. Pellistors are prone to a number of effects from the environment affecting their sensitivity and reliability.

7.2.1.2 Calibration of Pellistor Sensors

Detectors are usually calibrated for a certain type of gas. Most equipment is calibrated either for 0–100% LEL methane or for 0–100% LEL *n*-pentane. When a gas different from the calibrating gas is sensed, a "*correction factor*" should be applied.

Let us assume the *detector is calibrated for* n-*pentane* and the gas being measured is hydrogen. From the appropriate table *the correction factor* to be applied is determined to be 0.6. In case of a meter reading of 65, the corrected reading for hydrogen calculates to

$$65 \times 0.6 = 39\%\ \text{LEL}\ H_2$$

Since the reading is apparently higher with hydrogen, the alarm setting will operate at a lower level. For an alarm setting for n pentane of 20% LEL, the level at which the alarm will operate in hydrogen is $20 \times 0.6 = 12\%$ LEL H_2.

In some other instances the correction factor is >1. In those cases the detector reads low and the reading has to be increased to the proper value.

If a *methane calibrated detector* is used to measure acetone, the correction factor is 1.6; therefore a reading of 65% LEL would have to be corrected to $65 \times 1.6 = 104\%$ LEL acetone. An alarm setting of 20% LEL would therefore actually operate at $20 \times 1.6 = 32\%$ LEL acetone.

Correction factors for pellistors are provided by different makers. To obtain an approximate LEL value for the gases/vapours in Table 7.1, the meter reading on a pentane or methane calibrated detector has to be multiplied by the value int he table, respectively.

7.2.1.3 Reliability Issues

Pellistor technology has been used successfully in many industrial applications for more than three decades. The discrete pellistor sensor consists of two very thin coated wires exposed to the gas being measured. *Vibration* will cause fluctuations in the readings and failure is common after slight knocks. The pellistor sensor is designed for fixed installation for safety measurement, which it does very well. *The vibration sensitivity severely limits the use of pellistor in portable equipment.*

In *high gas concentrations* pellistors will provide *false and unsafe readings*. If a pellistor is exposed to a non-stoichiometric, oxygen deficient gas mixture, it may suffer a drop in sensitivity caused by "soothing". The degree of sensitivity reduction and whether the effect is permanent will depend on the

Table 7.1 Approximate LEL value for the gases/vapours (P = Pentane, M = Methane)

Detector Calibration	P	M	Detector Calibration	P	M
Acetaldehyde	0.8	1.6	Ethylene	0.7	1.2
Acetic Acid	0.8	1.6	Ethylene Dichloride	0.7	1.4
Acetic Anhydride	1.0	2.0	Ethylene Oxide	0.9	1.8
Acetone	0.9	1.6	*N* Heptane	1.3	2.3
Acetylene	0.8	1.7	*N* Hexane	1.3	2.0
Acrylonitrile	1.0	2.0	Hydrogen	0.6	1.2
Alkyl Alcohol	0.9	1.9	Kerosene	1.4	2.8
Ammonia	0.35	0.6	LPG	1.1	2.1
N Amyl Alcohol	1.4	2.8	Methane (LNG)	0.5	1.0
Aniline	1.2	2.5	Methanol	0.5	1.2
Benzene	1.1	1.9	Methyl Chloride	4.0	8.0
1.3 Butadiene	0.8	1.3	Methyl Cyclohexane	1.0	2.0
N Butane	0.8	1.6	Methylene Dichloride	0.5	1.0
Iso Butane	0.9	1.8	Dimethyl Ether	0.7	1.4
Butene 1	1.0	2.0	Methyl Ethyl Ether	1.0	2.0
N Butanol	1.4	2.9	Methyl Ethyl Ketone	1.1	2.2
I Butanol	0.9	1.9	Methyl *N* Propyl Ketone	1.6	3.2
Tert Butanol	0.6	1.3	Naphthalene	1.3	2.8
Butyl Acetate	1.5	3.0	*N* Nonane	1.4	2.8
N Butyl Benzene	1.4	3.0	*N* Octane	1.3	2.6
Iso Butyl Benzene	1.4	3.0	*N* Pentane	1.0	1.9
Carbon Monoxide	0.6	1.2	Iso Pentane	1.0	1.9
Carbon Disulphide	4.0	8.0	Petrol	1.0	1.9
Carbon Oxysulphide	0.5	1.0	Propane	0.65	1.3
Cyclohexane	0.9	2.0	*N* Propanol	1.0	2.0
Cyclopropane	0.8	1.6	*I* Propanol	0.9	1.8
N Decane	1.4	2.8	Propylene	0.9	1.8
Diethylamine	0.9	1.8	Propylene Oxide	1.0	2.0
Dimethylamine	0.8	1.6	Iso Propyl Ether	1.0	2.0
2.3 Dimethylpentane	1.1	2.2	Propyne	1.1	2.2
2.2 Dimethylpropane	1.1	2.2	Styrene Monomer	4.0	8.0
Dimethyl Sulphide	1.1	2.2	Tetra Hydra Furan	1.0	2.1
Dioxane	1.0	2.0	Toluene	1.1	1.9
Ethane	0.7	1.4	Trimethylbenzene	1.5	3.0
Ethyl Acetate	0.9	1.8	White Spirit	1.5	3.0
Ethanol	0.6	1.85	*o* Xylene	1.2	2.8
Ethyl Benzene	1.3	2.6	*m* Xylene	1.1	2.2
Ethyl Bromide	0.5	1.0	*p* Xylene	1.1	2.2
Ethyl Chloride	0.8	1.6			
Ethyl Cyclopentane	1.1	2.2			
Ethyl Ether	1.0	2.1			

After Crowcon – Gas Detection, Gas Monitoring, Gas Sampling [9] gas detected. Aromatic compounds having a high carbon content cause the worst problem as these compounds typically require excess oxygen for complete oxidation.

Even at normal operating conditions pellistors are *extremely susceptible to poisoning by chemicals such as sulphides and silicones*. These types of substances are frequently found in places such as coal mines and oil rigs. Poisonous chemicals react with the catalyst during the burning process thereby causing an inert layer to be built upon its surface. This build-up severely degrades the pellistor's ability to detect gas by blocking the catalyst, but is not evident at all during normal operation.

Catalytic sensors can be *poisoned temporarily* (e.g., by contaminating chemicals, compounds containing halogens, sulphur or chlorine) or *permanently* (e.g., by substances containing heavy metals, especially lead or silicones). The presence of sulphur dioxide will cause the sensor to fail in a very short time. Special tubing must also be used since the commonly used silicone tubing will also cause the sensor to lose sensitivity and fail prematurely. This has the effect of irreversibly reducing sensitivity.

Efficient *protection against poisoning* of fixed flammable gas detectors is provided by some vendors using *carbon filters* fitted on the outside of the *sintered flame arrestor*. It is suitable for hydrogen and methane monitoring but, unfortunately it reduces the sensitivity of the detector to propane (20% reduction), butane and pentane (50% reduction). The filter in the long-life pellistors resistant to halogens, silicones and lead must always be replaced whenever a detector is re-calibrated.

In the absence of adverse factors pellistor sensors will last for many years. However, they do *lose performance continuously by calibration drift*, typically by 20% per annum (e.g. by aging, changes in the heat-transfer, thermal and electrical contacts). Therefore *re-calibration of instruments* employing pellistors is recommended every 3–6 months. This high maintenance regime, when coupled with regular sensor-replacements and multiple sensor point arrays, increases the cost of ownership. Pellistor based detectors work satisfactorily in ambient temperatures of 40 to 60°C.

7.2.1.4 Limitations in Use of Pellistors

Methane is now recognized as one of the major contributors to the greenhouse effect, so a measurement of CH_4 is now essential.

Being a highly combustible gas and there is no reason to expect CH_4 to survive at high temperatures in the presence of oxygen. The fact is, it does, if only in small quantities. There was earlier no efficient way of measuring the concentration of methane in the flue gases. The only technology commonly available was the pellistor sensor, being an inaccurate and not particularly reliable method of measuring CH_4.

When considering the chemical and physical phenomena in sensing one must be aware of its limitations, i.e. the sensitivity of a pellistor is not comparable with other devices of various transduction principles (electrochemical, optical, resistivity type). Pellistors can typically detect combustives in a concentration range of c.a. 1000 ppm to a few tens of percents; however, the detection of 20–200 ppm volatile organic compounds (VOC) was also demonstrated by high sensitivity thermopiles. The lower limitation of commercial devices is the increasing noise/signal ratio while the upper limit is determined by stoichiometry.

The high cross-sensitivity of a single pellistor limits its reliable operation to known combustive gas component. Beside the strict requirement of stoichiometric gas composition or rather the presence of excess oxygen, the operation temperature of the device must be carefully selected and to be calibrated in order to avoid false signals. For example, the oxidation of alcohols and hydrogen takes place even on the surface of the passive, Al_2O_3 coated reference element at high temperatures, therefore, provides false readout by decreasing bridge voltage.

Another important issue in operation of pellistors is the *need for measurement of oxygen concentration*, as oxygen deficiency results in false signals and deteriorates sensor performance.

Methane can only accurately be measured using a dedicated infrared sensor. Care must be taken, however, to dry the gas efficiently, since the sensor will react slightly to the presence of water vapour as well as CH_4. Although it is generally believed that an infrared sensor is completely dedicated to one component, *all the alkanes have a wavelength similar to methane* and the sensor is so constructed such that it will react to their presence as well. Naturally, since it is calibrated for methane, the reaction to the longer chain alkanes will not be as accurate, but the correspondence is still better than that produced by a pellistor sensor.

7.2.2 Microcalorimeters in Enzymatic Reactions

The developed heat in the enzymatic reaction, e.g., the catalyzed conversion of glucose can be detected by sensitive thermometry. [2] used a non-perforated *silicon membrane microcalorimeter* to measure, e.g., glucose concentration in water.

For this purpose, the enzymatic layer (glucose-oxidase) was deposited on one side of the thin (<10 μm) Si membrane of a few mm diameter, which comes into contact with the fluid. The heat by the enzymatic reaction is sensed by means of an integrated thermopile (e.g., Si-Al) placed on the opposite side of the membrane. Typical thermal detectors used in such applications are thermistors, Bi-Sb thermopiles, integrated Si thermopiles, and polycrystalline Si thermopiles formed on SiN_x membranes (see in Section 7.5.2 too).

Microcalorimeters can also be incorporated as integrated elements in *microfluidic systems*. Microfluidics refers to the generic technology of manipulating fluids on a chip, including the integration of pumps, valves, mixers and reaction chambers that enable the fabrication of microreactors and lab-on-a-chip devices. They offer compactness, low power consumption, reduced reagent volumes (hence lower cost), faster response times, well-controlled reaction conditions and delivery of reagents at the cost of high electric fields required for electrophoretic pumping systems.

7.3 Thermal Conductivity Sensors

Because of the amount of power required to heat these sensors, they have to be mounted in a flameproof enclosure, in the same way as pellistors are. The precise mechanism, which occurs, is quite complex because the *thermal conductivity of gases varies with temperature and convection as well as conductivity plays a role.* Whilst most gases produce a linear output signal, this is not always the case.

Thermal conductivity (TC) *gas detectors, sometimes called katharometers,* operate by comparing the thermal conductivity of the sample with that of a reference gas. Measuring the TC of gases was one of the earliest forms of gas detection and it is suitable for volume levels of certain binary mixtures, i.e., two different gases, one of which can be air.

In the sensor a heated thermistor or platinum filament is mounted such that it is exposed to the sample. Another one, acting as a reference is enclosed in a sealed compartment. If the sample gas or vapour has a thermal conductivity higher than the reference, heat is lost from the exposed element and its temperature decreases, whilst if its thermal conductivity is lower than that of the reference the temperature of the exposed element increases. These temperature changes alter the electrical resistance, which is measured by means of a bridge circuit.

Power loss of a single filament thermistor by heat conduction via the ambient gas can be expressed as

$$P = k_{TC} \lambda \Delta T \quad (7.7)$$

where P is the power dissipation of the heater by thermal conduction of the gas,

λ the thermal conductivity of the given gas–gas mixture,
k_{TC} a constant, characteristic for a given geometry,
ΔT temperature difference between heater and ambient gas.

The presence of gases having a relative thermal conductivity to air of >1 leads to cooling of the exposed thermistor or filament. They are often measured by the TC technique, the higher their thermal conductivity the lower the concentration which can be measured (Table 7.2).

Gases with thermal conductivities of <1 are more difficult to measure, partly because water vapour may cause an interference problem (Table 7.3).

Gases with thermal conductivities close to 1 cannot be measured by this technique. These include ammonia, carbon monoxide, nitric oxide, oxygen and nitrogen.

Table 7.2 Thermal conductivities relative to air at 100°C

Helium	5.6
Hydrogen	6.9
Methane	1.4
Neon	1.8

Table 7.3 Thermal conductivity relative to air at 100°C

Argon	0.7
Butane	0.7
Carbon dioxide	0.7
Ethane	0.75
Freon/Halon	0.4
Hexane	0.5
Pentane	0.7
Propane	0.8
Water vapour	0.8
Xenon	0.2

In most cases the heat conductivity of gas mixtures can not be calculated by simple linear combination of the components' parameters. The relative heat conductivity of gas mixtures of polar and non-polar molecules shows extremum in the function of their concentration. (Fig. 7.4)

A few mixtures of two gases however, can be measured by TC techniques [10]. An example is methane in carbon dioxide mixtures found in sewage digester and coal gasification plants. An instrument can be scaled 0–100% methane in carbon dioxide, and such equipment works well, as long as precautions are taken to separate water from the gas stream. The change of the relative heat conductivity coefficient versus composition of ammonia in air is shown in Fig. 7.4.

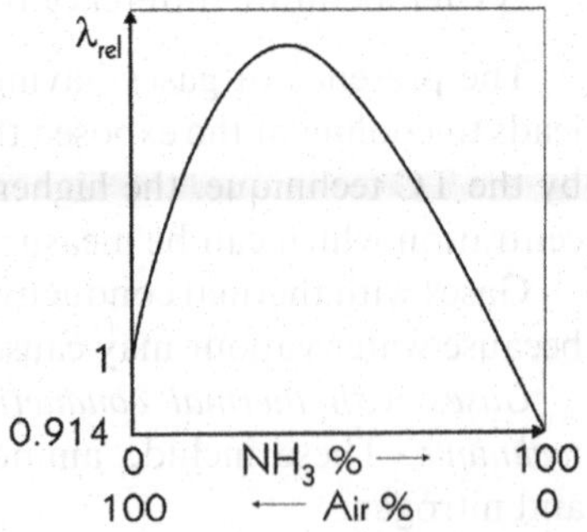

Fig. 7.4 Relative heat conductivity coefficients of NH_3-Air mixture

7.4 Calorimetric Sensors Measuring Adsorption/ Desorption Enthalpy

The low value of adsorption enthalpy (see Fig. 7.1) of known combustive gases can also be measured by high sensitivity thermopiles, provided the thermally isolated micro-hot-plate sensor structure is covered with a selective adsorbent. In the most attractive examples polymers are used for this purpose, however, due to their limited selectivity these sensors are mostly used as one of the elements of an integrated array [11] .

7.5 MEMS and Silicon Components

MEMS (micro-electro-mechanical system) is now widely applied to all miniaturized devices, generally three-dimensional microstructures. The term is somewhat misleading, in that not all so-called MEMS devices are "electromechanical" and few are "systems". For the fabrication of MEMS devices silicon is the most common material used due to the tremendous amount of knowledge, expertise and equipment available from the microelectronics industry. The fabrication techniques include isotropic and anisotropic etching (micromachining), thin film deposition, anodic bonding and the well known masking and doping techniques employed in IC manufacturing. Devices such as silicon microsensors, "lab on a chip" and micro-TAS (micro-total analytical systems) can all be viewed as MEMS units. High temperature applications (>400°C), however, may favour the use of alternative materials, such as SiC, particularly if integration of electronics with the sensor element is being considered.

Silicon MEMS technology has been very successful for sensing physical variables in mass produced devices (e.g., accelerometers, pressure sensors, and Hall effect position sensors). Its impact on gas sensing has been, however, with few products yet in high volume production, less dramatic so far. Nevertheless, research is widespread including among others: porous and nanoporous silicon layers; micro-hot-plates using Si and SiC (for high-temperature operation); integrated sensor arrays, electronic noses. The aim is to yield families of miniaturized devices that offer economic and operational improvements over conventional gas sensor types.

For decades the enthalpy generated by catalytic combustion of flammable gases (in the well-known pellistor) and the measurement of heat conductivity changes have been the preferred sensor technology for detection of explosive gases in mines and other high risk areas. In order to eliminate – at least partially – the shortcomings of the discrete sensors (in Section 7.2.1.4.), and to extend the capabilities of these type of thermometric gas sensing beyond the alarm signalling applications, by using silicon micromachining for dedicated MEMS structures successful attempts were made to reduce the size of the devices. This measurement technology is most cheaply employed in MEMS devices, similar to the structures used in digital mass-flow controllers.

7.5.1 Thermal Considerations

The most important issue in thermometric (i.e. calorimetric) sensor development is the fabrication of micro-hot-plates with minimum power dissipation, optimized for the targeted function. The maximum thermal isolation is required to provide inherently explosion proof operation and match the corresponding transmitter norms. As described in Chapter 1 the following physical mechanisms contribute to the loss of heat:

- conduction via the suspensions of the hot plate;
- conduction via the ambient gas;
- radiation from the hot surface.

Besides FEM modelling the dissipated power (P) of a given structure can be calculated using the following equations:
Conduction via suspensions:

$$P_{cond,solid} = \sum{}_i a_i \lambda_i (T_{hp} - T_e)/l_i \tag{7.8}$$

Conduction via ambient gas

$$P_{cond,gas} = A\lambda_g (T_{hp} - T_g) \tag{7.9}$$

Radiation

$$P_{rad} = 2\sigma A^2 \varepsilon (T_{hp}^4 - T_e^4) \tag{7.10}$$

where a_i is the cross section area of the suspension beam,
λ the thermal conductivity coefficient,
l the length of the suspension(s),
i denotes the material the suspension is made of,
T_{hp} and T_e are temperature of the hot plate and the environment, respectively,
A is the surface area of the hot plate,
λ_g is the thermal conductivity coefficient of the ambient gas/gas mixture,
T_g the temperature of the ambient gas,
σ the Stefan–Boltzmann constant,
ϵ the emission constant of the material covering the hot plate.

In a realistic MEMS structure with characteristic dimensions of a few tens or hundreds of micrometers, which operates at temperatures below 700–800°C, over 90% of the heat loss can be attributed to the conduction via the suspension beams. Therefore, it has to be minimized by a proper selection of geometry and construction materials (i.e., the lowest heat conductivity must be targeted). Nevertheless, the design is the result of a compromise between the optimum thermal isolation, thermo-mechanical stability as well as process compatibility of the applied materials. Obviously, when designing the hot plates one has to be aware of the targeted operation:

- in a pellistor structure all the heat loss paths must be minimized, while
- in heat conductivity sensors the device must be optimized, however, for maximum heat loss via the gas to be measured.

Micro-hot-plates operated at 550°C for catalytic combustion have been scaled to MEMS dimensions for many years, even in commercial devices with 150–200 mW power consumption of a pellistor pair [12, 13, 14, 15]. The silicon-oxynitride *membrane (non-perforated)* is formed by backside alkaline etching process. The same structure was also used in heat conductivity devices [14, 15]. The potentially low power consumption (<10 mW at 550°C targeted but not achieved yet) opens up prospective new markets in domestic, portable and sensor array gas detection. There are significant opportunities for low power devices, when combined with functionalized, high surface area sensing layers.

Inherently safe detection of combustible gases was demonstrated by a new type of integrated micropellistor and heat conductivity sensor [16]. The basic element in both types of thermometric devices is a *thermally isolated perforated micro-hot-plate* with a Pt heater, embedded in reduced stress non-stoichiometric silicon-nitride. The integrated pellistors are formed from a pair of an active and a passive resistor element in a Wheatstone-bridge arrangement. In the environment to be monitored they are heated up to 200–600°C with a power dissipation of 20–60 mW.

A novel, one-side porous silicon bulk micromachining process was developed [17] for the formation of the air gap around the devices. In a more advanced process a patented method [18] offering a thin single crystalline Si support for the heated hot-plate-membranes was demonstrated. Owing to the better mechanical stability obtained, the relatively large mass of porous bead, supporting the finely dispersed catalysts of Pt, Pd or Rh on the active device and for forming the chemically inert reference device, respectively, can be deposited on top of the micro-hot-plates without jeopardizing the heater's integrity. The obtained thermal response of the different hot plates is shown in Fig. 7.5.

The hot-plate structures, built with three different isolation schemes (marked in the figure as unsupported, supported bridge as well as bridge thermally isolated by porous silicon PS) [19, 20] were compared at similar driving conditions by means of fast thermometry.

The proposed simple thermal equivalent circuit allowed the extraction of thermal parameters. Temperature response to a square wave driving of 18-mW amplitude was simulated by the thermal equivalent circuit developed for the

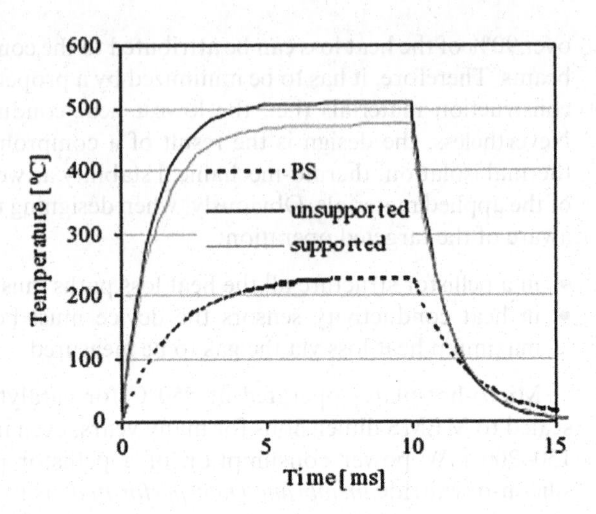

Fig. 7.5 Dynamics of perforated bare micro-hot-plates using different thermal isolation schemes

microheaters in [19]. The calculated performance showed a reasonable agreement with the empirically determined dynamic behaviour. While the porous silicon isolation leads to a severe deterioration of the thermal response, the use of virtually "intrinsic" thin crystalline Si support-pillars underneath the released membrane heaters did not significantly affect the speed of temperature settling. The enhanced thermo-mechanical stability of the loaded, suspended features was achieved at the cost of a mere 10–15% temperature loss via the additional heat sink.

7.5.2 Temperature Readout

Besides thermal isolation of the sensing element, the sensitivity and reliability of calorimetric sensors is determined by the sensitivity and the accuracy of the temperature measurement method. Several transduction principles can be considered:

- measuring the resistance of a high TC resistor;
- using the thermoelectric (Seebeck) effect;
- using pyroelectricity.

Due to the process compatibility with thin layer Si technology and the ability to work at elevated temperature needed for catalytic combustion, the overwhelming majority of calorimetric devices operate with one of the first two methods. When selecting between them, one has to consider both the targeted application and the limitations in processing.

Nevertheless, the high sensitivity of pyroelectic crystals (e.g., that of lithium tantalate is approximately 50–60 times higher than that of thin layer Pt resistors) makes their application for measuring the temperature changes generated by the adsorption/desorption process feasible. The heat generated by an exotherm catalytic reaction on the surface of the chemically sensitive material can be precisely measured by a *pyroelectric sensor*. They are extremely sensitive (detection limit of μW) heat detectors based upon the temperature dependent charge polarization of the pyroelectric crystal.

The most common sensor structure is made up of a pyroelectric crystal (e.g., of $LiTaO_3$) with two planar or interdigitated electrodes. The structure is coated by the catalyst and absorber layer or the reference material, respectively, on the top while the electrode is formed on the back. Advantages of this type of sensor are the enhanced sensitivity, low detection limit and simple fabrication technology.

Such a sensor operation was demonstrated for carbon monoxide and hydrogen detection using platinum and palladium catalysts, respectively.

The temperature change caused by the catalytic oxidation can also be detected through *thermoelectricity*, i.e., by the *Seebeck effect*. The measured voltage is generated by the temperature gradient established by the catalytic oxidation of the gas. The Seebeck effect sensors are fabricated using metal-oxide semiconductors (e.g., WO_3, SnO_2 and MnO) with and without catalyst (Pt or Pd). They read the potential difference established between two metals or different regions of the semiconductors applied. The measured voltage is proportional the temperature difference

$$\frac{dV}{dz} = \alpha \frac{dT}{dz} \tag{7.11}$$

where V is the voltage, $\frac{dT}{dz}$ is the temperature gradient and α the Seebeck coefficient.

In general the *thermopile readout* is more sensitive than the resistor TC, and therefore, it is better suited in almost all the applications, but preferred for detection processes accompanied by low temperature changes. As the thermopile measures the temperature difference between the two series of junctions formed on the active and reference sides of the hot plate, respectively, short pulse mode operation (a few seconds) is preferred [21, 22]. Due to the high thermal conductivity between the two sides of the hot plate, continuous operation would lead to reduced temperature difference and detriorate sensitivity. Readout by a thermopile requires independent heater element inserted in the thermopile meander or formed in an outer ring around the perimeter of the micro-hot-plate. Therefore, a double layer metallization of, e.g., polysilicon heater combined with Al/polysilicon thermopile is required.

The use of a simple single layer metallization processing and the requirements of detecting combustive gases at high operational temperature in the range of 20–100% LEL (as the most important application of calorimetric devices), however, keeps the *resistor temperature readout* the most viable alternative in long term. The temperature changes generated by combustion are high enough to measure with high TC metal or single crystalline Si resistors. Note that thin layers of the most widely used Pt resisitor have lower TC than the bulk phase material! With an appropriate technology a typical value of 2500 ppm/°C of Pt layer resistance variation can be achieved. In the temperature range of 300–600°C single crystalline Si shows higher, but non-linear TC. The active and the reference element is formed on two identical but thermally separated hot plates; therefore, continuous operation of the pellistor is allowed.

7.5.3 Integrated Calorimetric Sensors

Pellistor-type gas sensors detect the combustive gas concentrations by measuring the heat generated by their catalytic oxidation. Although a few attempts on the development of integrated calorimetric microdevices [12, 14, 15, 21, 23] had been published earlier, the demonstration of explosion-proof operation in explosive environment without any mechanical protection was only achieved recently [16]. The multisensor array of pellistors and heat conductivity sensor can even [16, 24] be used to measure extreme gas concentrations. The necessary catalytic temperature range of 200–600°C had to be achieved with minimum power consumption (of ca. 10–25 mW) in order to meet the requirements of system integration and first of all *to fulfil the transmitter norms.*

Microdevices with thin thermally isolated membranes holding the microheaters suffer from the difficulties of the combination of this “thin-film processing” with the essentially “thick-film fabrication” of the catalyst supporting bead of large area porous surface. The automation of the thick-film formation using some kind of a microdispenser technology is impaired by the frequent clogging of the dispenser for the highly viscous ceramic suspension. One has to resort to the less reproducible and less productive “dip-and-drop technique” [16], which was already successfully applied for conductivity type gas sensors [25].

The integrated sensors make use of the Wheatstone-bridge arrangement in determining the difference between the temperature of the *active and passive elements in the pellistor*. Here, as well as in case of the discrete device application, both the active and passive element are covered by a porous ceramic matrix. The active one of them contains the finely dispersed catalyst particles. In the device operation, it is absolutely required to control the thermal behaviour of the filament structure modified by the coating.

Sensors, utilizing the *heat conductivity principle*, are composed of a *pair of chemically passive elements*, having different heat exchange properties due to their dissimilar surface. Intrinsically safe, explosion-proof detection of hydrocarbons was demonstrated by measuring concentration levels up to LEL with no protective encapsulation, whatsoever.

Another attractive approach in low-power pellistor development is the high-temperature *nanohotspot or antifuse technique* reported recently [26]. The temperature of the 10–100 nm localized hot spot exceeds 1000 K at a cost of a few mWs. Controllability and reliable operation might be the bottlenecks in these potentially promising thermometric devices.

Several authors and patents describe microcalorimeter devices, of which we refer to the work of R.E. Cavicchi et al. [21]. The TMAH released micro-hot-plate measures the temperature differences between the catalyst coated and the reference side with an integrated thermopile. A polysilicon heater formed around the perimeter provides the uniform operatonal temperature up to 600°C. The device is capable of detecting VOCs and combustive gases in the range of 10–200 ppm.

7.6 Sensor Arrays and Electronic Noses

Arrays of chemical or biochemical sensors with varying degrees of response are utilized in artificial olfaction (electronic nose) systems. Artificial olfaction has high potential in numerous application areas, including food quality control, personal healthcare, and security as well as air quality monitoring in the automotive sector. Commercializing electronic noses started in the early 1990s, but most attempts failed because the target markets were not well selected and the sensor was not sufficiently repeatable.

The first report on integrated pellistor array was published in 1993 [26]. A new opportunity to extend the capabilities of miniaturized micropellistors or heat conductivity sensors [27] is their exploitation for olfactory imaging of explosive gas or vapour mixtures [16]. A four-wire transmitter system was constructed for detection and analysis of hydrocarbon mixtures up to their 100% LEL concentrations. Due to the reduced power consumption of $P < 50$ mW/individual sensor element in the integrated array and the adopted cyclic interrogation scheme, the driver and readout *electronics became inherently explosion-proof*. Consequently, *only the sensor chip must be explosion protected due to the high surface temperature of the calorimetric sensors* [29]. This leads to considerable reduction in production costs.

The *integrated calorimetric sensor array chip consists of three pellistors*, i.e. three microheaters covered with finely dispersed Pt containing porous Al_2O_3 ceramics on the active elements, and bare Al_2O_3 coating on their reference pairs. The porous layers were formed by dip and drop deposition of the appropriate suspensions, followed by a self-heating calcination step at 700°C. The maximum power dissipation of each sensor-pair is below 50 mW at 550°C. In order to enhance the selectivity of the system, the pellistors are operated at different temperatures in this approach. Application of different catalyst may of course also enhance selective gas detection.

Very low power catalytic detection of hydrocarbons is at the top of the wish list. Methane is an important target gas because of its high concentration in natural gas. Its catalytic burning requires, however, relatively high temperature with Pt catalyst (>540°C). Therefore, thermal conductivity measurement of

Fig. 7.6 Integrated micropellistors and heat conductivity sensor. **a** SEM view of a single micropellistor pair, **b** mounted array chips of three pellistors and a heat conductivity sensor (the reference element is encapsulated with a Si cap), **c** replaceable sensor head with the integrated sensor array and EPROM for calibration data (courtesy of WESZTA-T Ltd., Hungary)

methane is sometimes more feasible to be implemented in integrated gas-sensing. This is the case in the described array, too, where the fourth device on the chip is a *heat-conductivity sensor* constructed in the same porous Si micromachining sequence. The reference heater of this sensor is encapsulated with a Si cap. Figure 7.6 shows the microphotographs of the integrated micro-arrays for electronic nose application.

The system is capable to distinguish and measure methane, propane, butane and hexane mixtures. Half Wheatstone-bridges were formed from all of the sensor pairs in order to read their signal as bridge voltage. Principal component analysis (PCA) was used for classification of the explosive gas composition (Fig. 7.7)

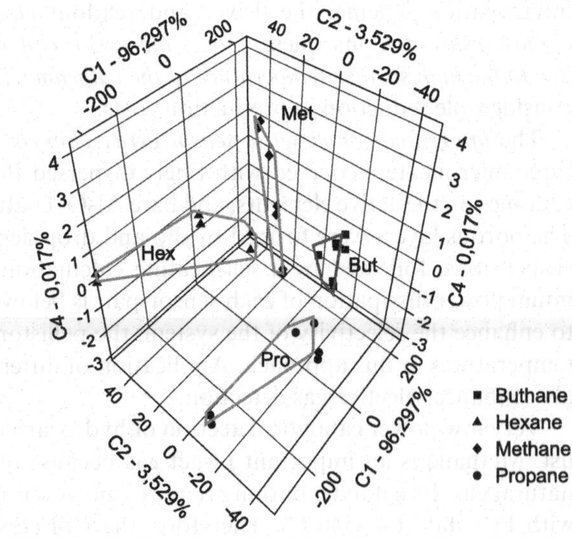

Fig. 7.7 Three-dimensional representation of the results of principal component analysis for classification of the explosive gas composition in mixtures of methane, propane, butane and hexane

The three-dimensional plot of the PCA results was constructed for concentration measurements of all the gases of interest. The representation reflects an appropriate distinction among the four gases with a good correlation between their concentration and the sensor responses, respectively.

An alternative method for the analysis of the set of sensor responses would be to employ an artificial neural network (ANN) [28, 29, 30]. In order to train the network for reliable predictions, however, a large set of training samples has to be prepared. The handling of explosive gas mixtures for training is rather cumbersome; therefore, in these applications PCA is preferred over ANN.

The most advanced integrated system presented so far is a Complementary Metal Oxide Semiconductor (CMOS) integrated polymer sensor array of *mass-sensitive cantilever resonant elements, capacitive type sensors and thermoelectric calorimeters*. The operation of the calorimeters is based on the Seebeck effect by measuring the enthalpy of adsorption and desorption of the combustive gases. In device fabrication after completing the CMOS circuitry Deep Reactive Ion Etching (DRIE) and alkaline etching process were used for the formation of the cantilever and the membrane, respectively [11].

References

1. MNT Gas Sensor Forum. MNT Gas Sensor Roadmap. 2006. www.technologyprogramme.org.uk/site/TechnologyReports. Accessed 2006.
2. Institution of Gas Engineers and Managers. Dealing with reported gas escapes, Publication SR/20. 1998. www.igem.org.uk/Technical/publications.asp. Accessed 1998
3. European Standard, EN 50014: Electrical apparatus for potentially explosive atmospheres, General requirements. 1998.
4. European Standard, EN 50018: Electrical apparatus for potentially explosive atmospheres, Flameproof enclosure “d”. 2000.
5. European Standard, EN 50019: Electrical apparatus for potentially explosive atmospheres, Increased safety “e”. 2000.
6. European Standard, EN 50020: Electrical apparatus for potentially explosive atmospheres, Intrinsic safety “i”. 2002.
7. European Standard, EN 50054: Electrical apparatus for the detection and measurements of combustible gases, General requirements and test methods. 1998.
8. Bataillard P, Steffgen E, Haemmerli S, Manz A, Widmer HM. An integrated silicon thermopile as biosensor for the thermal monitoring of glucose, Urea and Penicillin Biosens Bioelectron. 1993;8(2):89–93.
9. Crowcon Detection Instruments. Gas Detection, Gas Monitoring, Gas Sampling. 2007. www.crowcon.com. Accessed 2007.
10. Pollackdiener G, Obermeier E. Heat-conduction microsensor based on silicon technology for the analysis of 2-component and 3-component gas-mixtures. Sens Actuators B. 1993;13:345–347.
11. Hagleitner C, Hierleman A, Lange D, Kummer A, Kerness N, Brand O, Baltes H. Smart single-chip gas sensor microsystem. Nature. 2001;414:293–296.
12. Gall M. Si planar pellistor. A low-power pellistor sensor in Si thin-film technology. Sensors and Actuators, B: Chem. 1993;B4(3–4):533–538.
13. Krebs P, Grisel A. Low power integrated catalytic gas sensor. Sensors and Actuators, B: Chem. 1993;B13(1–3 pt 1):155–158.

14. Microsens SA. Integrated resistive gas sensors. 2007. www.microsens.ch/products/gas.htm. Accessed 2007.
15. Alphasense. Alphasense Sensor Data Sheets. 2007. www.alphasense.com. Accessed 2007
16. Kulinyi S, Brandszájsz D, Amine H, Ádám M, Fürjes P, Bársony I, Dücsõ Cs. Olfactory detection of methane, propane, butane and hexane using conventional transmitter norms. Sensors and Actuators, B: Chem. 2005;111–112 (SUPPL.):286–292.
17. Dücsõ Cs, Vázsonyi E, Ádám M, Szabó I, van den Berg A, Bársony I. Porous silicon bulk micromachining for thermally isolated membrane formation. Sens Actuators A. 1997;60: 228–234.
18. Dücsõ Cs, Ádám M, Fürjes P, Hirschfelder M, Kulinyi S, Bársony I. Explosion-proof monitoring of hydrocarbons by micropellistor. Sens Actuators B. 2003;95:188–193.
19. Bársony I, Fürjes P, ádám M, Dücsõ Cs, Vízváry Zs, Zettner J, Stam F. Thermal response of microfilament heaters in gas sensing, Sensors and Actuators, B: Chem. 2004;103(1–2): 442–447.
20. Kaltsas G, Nassiopoulou A. Bulk silicon micromachining using porous silicon sacrificial layer. Microelectron. Eng. 1997;35:397–400.
21. Cavicchi RE, Poirier GE, Tea NH, Afridi M, Berning D, Hefner A, Suehle J, Montgomery C. Micro-differential scanning calorimeter for combustible gas sensing. Sensors and Actuators, B: Chem. 2004;97(1):22–30
22. US Patent 6,079,873. Micron-scale differential scanning calorimeter on chip.
23. Aigner R, Dietl M, Katterloher R, Klee V. Si-planar-pellistor: Designs for temperature modulated operation. Sensors and Actuators, B: Chem. 1996;33(1–3):151–155.
24. Lee SM, Dyer DC, Gardner JW. Design and optimisation of a high-temperature silicon micro-hotplate for nanoporous palladium pellistors. Microelectron J. 2003;34(2):115–126.
25. Puigcorbe J, Vila A, Cerda J, Cirera A, Gracia J, Cane C, Morante JR. Thermo-mechanical analysis of micro-drop coated gas sensors. Sens Actuat A: Phys. 2002;97–98:379–385.
26. Kovalgin AY, Holleman J, Iordache G, Jenneboer T, Falke F, Zieren V, Goossens MJ. Low-power, antifuse-based silicon chemical sensor on a suspended membrane. J. Electrochem Soc. 2006;153(9):H181–H188.
27. Gall M. Si-Planar-Pellistor array, a detection unit for combustible gases. Sensors and Actuators, B: Chem. 1993;B16(1–3 pt 2):260–264.
28. Sommer V, Tobias P, Kohl D. Methane and butane concentrations in a mixture with air determined by microcalorimetric sensors and neural networks. Sensors and Actuators B: Chem. 1993;12:147–152.
29. Westa-T Ltd. Gas sensing and warning equipments for combustible and nosy gases. 2007. www.weszta-t.hu. Accessed 2007.
30. Debeda H, Rebiere D, Pistre J, Menil F. Thick film pellistor array with a neural network post-treatment. Sens Actuat B: Chem. 1995;B27(1–3 pt 2: 297–300.

Chapter 8
Acoustic Wave Gas and Vapor Sensors

Samuel J. Ippolito, Adrian Trinchi, David A. Powell, and Wojtek Wlodarski

8.1 Introduction

Acoustic wave devices are based on high-frequency mechanical vibrations. Originally developed for precision radio frequency (rf) signal-processing applications, they are widely utilized in mobile and wireless communications, and are routinely found in most modern day electronics [1]. As pointed out by Ballantine and Wohltjen [2], their inherent sensitivity to ambient environmental effects, which requires hermetic shielding or isolation in signal processing applications, has ironically become a windfall in the field of chemical and physical sensing.

Acoustic-wave based sensors offer a simple, direct and sensitive method for probing the chemical and physical properties of materials. The term *acoustic* is commonly used in the literature, even when referring to frequencies which are well above the audible range. Acoustic waves cover a frequency range of 14 orders of magnitude – from 10^{-2} Hz (seismic waves) and extending to 10^{12} Hz (thermo-elastic excited phonons) [3]. Acoustic wave devices such as those mentioned in this chapter operate in a narrow frequency range between $10^6 - 10^9$ Hz. In this chapter the discussion is concentrated on acoustic wave devices employed for measuring concentrations of gas- or vapor-phase analytes.

The utilization of acoustic wave devices for gas-phase sensing applications relies on their sensitivity towards small changes (perturbations) occurring at the 'active' surface. In order to monitor a specific gas or vapor, a sensitive layer is generally employed. In the presence of an analyte species, the waves' properties become perturbed in a measurable way that can be correlated to the analyte concentration.

Virtually all acoustic-wave-based devices use a piezoelectric material to generate the acoustic wave which propagates along the surface or throughout the bulk of the structure. Piezoelectricity is the ability of certain crystals to couple mechanical strain to electrical polarization, and will only occur in

S.J. Ippolito
RMIT University, Department of Applied Chemistry, Melbourne, Victoria, Australia
e-mail: sipp@ieee.org

E. Comini et al. (eds.), *Solid State Gas Sensing*,
DOI: 10.1007/978-0-387-09665-0_8,

crystals that lack a center of inversion symmetry [4]. By applying a time-varying electrical field, a synchronous mechanical deformation of the piezoelectric material will arise, resulting in the coincident generation of an acoustic wave in the material, and vice versa [5].

Acoustic wave devices come in a number of configurations, each with their own distinct acoustic and electrical characteristics. Two different groups of acoustic wave devices that are commonly employed for gas sensing will be discussed herein. The first are bulk acoustic wave (BAW) devices, which concern acoustic wave propagation through the bulk of the structure. This category of devices includes the quartz crystal microbalance (QCM) and thin-film resonators (TFRs), the latter encapsulating thin-film bulk acoustic resonator (TFBAR) and solidly mounted resonator (SMR) structures. The second type utilize acoustic waves confined to the surface of the piezoelectric material, and are known as surface acoustic wave (SAW) devices. Schematic diagrams of these can be seen in Fig. 8.1, and they will be discussed in detail in the following sections of this chapter.

Unlike the electrode structures found on QCM, TFBAR and SMR structures, SAW devices use patterned thin-film interdigital transducers (IDTs) to generate and detect the acoustic waves. Both QCM and TFRs are single port devices, whereas SAW devices can be configured as two-port delay line or as one-port resonator structures. All of these devices are mass sensitive, while SAW devices can be specifically designed to be highly sensitive towards sheet conductivity deviations at the active surface of the device. It should be noted that there are other members of the acoustic wave device family, such as thin-film flexural-plate-wave (FPW) delay lines and shear horizontal acoustic plate mode (SH-APM) devices [4], however they are not commonly used for gas or vapor-sensing applications.

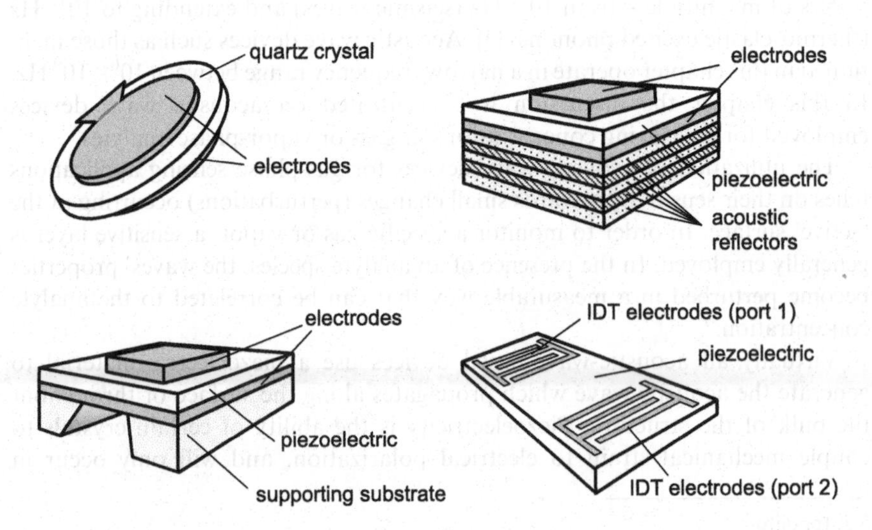

Fig. 8.1 *Top left*: QCM, *bottom left*: TFBAR, *top right*: SMR, and *bottom right*: two-port delay line SAW

The interactions between an analyte gas and the active surface of the device perturb the phase velocity of the propagating wave. The most commonly measured properties of acoustic modes are resonant frequency, phase shift or attenuation [6]. However, for single port devices such as QCMs and TFRs, direct measurements of impedance can be made. In any case, the measured change serves to quantify the analyte concentration. The perturbations affecting acoustic phase velocity can be attributed to by many factors, each of which represents a potential sensor response [7]:

Variations in any of these parameters alter the mechanical and/or electrical boundary conditions producing a measurable shift in the propagating acoustic wave phase velocity, v_0. Equation (8.1) illustrates the change in acoustic phase velocity, Δv, as a result of external perturbations, assuming that the perturbations are small and linearly combined:

$$\frac{\Delta v}{v_0} \cong \frac{1}{v_0}\left(\frac{\delta v}{\delta T}\Delta T + \frac{\delta v}{\delta \varepsilon}\Delta \varepsilon + \frac{\delta v}{\delta E}\Delta E + \frac{\delta v}{\delta \sigma}\Delta \sigma + \frac{\delta v}{\delta m}\Delta m + \frac{\delta v}{\delta \rho}\Delta \rho + \ldots\right). \quad (8.1)$$

However, the temperature dependence of each parameter, in addition to the overall temperature coefficient of the device structure must also be considered as the sensor response for many gas-phase-sensing applications is strongly dependent on operating temperature. Numerous physical, chemical, and biosensors could be developed based on monitoring the parameters in Table 8.1. In the case of gas and vapor sensors, parameters such as temperature and humidity will typically interfere with the desired response, and therefore care must be taken to limit their effect. Furthermore, the response of a sensor may be due to a combination of the parameters in Table 8.1, which also need to be considered in order to obtain a meaningful result.

8.1.1 Acoustic Waves in Elastic Media

Piezoelectricity, first described by the Curie brothers in 1880, is the ability of certain crystals to elastically deform by the application of a voltage, and vice-versa. This allows transduction between electrical and acoustic energy.

Table 8.1 Physical, electrical and thermal parameters to which acoustic waves are sensitive towards

•	T	–	temperature	•	ε	–	permittivity
•	E	–	electric field	•	σ	–	conductivity (electrical)
•	c	–	stiffness	•	η	–	viscosity
•	ρ	–	density	•	p	–	pressure
•	μ	–	shear elastic modulus	•	m	–	mass

Applying a periodic perturbation (electric field/voltage) to a piezoelectric material results in an elastic deformation (strain) that travels through the solid as a wave. These waves propagate in an elastic medium either as longitudinal (compression) or transverse (shear) lattice deformations, or as a combination of both. The propagation of the wave is often described or categorised in terms of the material lattice points' oscillatory pattern, which are termed as 'wave modes' and can become quite complex (see [1, 4, 8]). Table 8.2 lists several of the more commonly encountered acoustic modes used in sensing applications.

Figure 8.2 depicts the associated lattice point deformations for the three simplest cases, namely longitudinal, transverse and Rayleigh modes. It should be noted that the lattice point deformation of the Rayleigh mode is just a combination of both the longitudinal and transverse modes.

Quartz is a form of silicon dioxide (SiO_2) that has triclinic crystal symmetry, and is by far the most commonly utilized piezoelectric material for sensing applications. Like all piezoelectric materials, it is anisotropic, i.e. its material properties vary with different crystallographic orientations and there are no planes of material symmetry. Therefore, an acoustic wave device can exhibit different resonance frequencies and vibration modes depending on the chosen crystallographic orientation [8]. Hence, the acoustic mode, phase velocity, propagation direction, temperature coefficient etc. of the device are predominantly determined by the selected piezoelectric material crystallographic orientation. For instance, deviation in the crystal plane of quartz by as little as $0°5'$ can be advantageously used to engineer the temperature coefficient of the structure.

Table 8.2 Commonly encountered acoustic modes utilized in sensing applications

Wave types in solids	Wave propagation description
Bulk waves	
Longitudinal (compression)	Lattice point deformations parallel to wave direction.
Transverse (shear)	Lattice point deformations perpendicular to wave direction.
Surface waves	
Rayleigh	Displacement normal to the substrate surface with a component in the propagation direction forming an elliptical orbit – symmetrical mode.
Leaky SAW (or Pseudo SAW)	A high velocity surface wave that is imperfectly confined to the surface, leaking into the substrate.
Surface-skimming bulk wave, (SSBW)	Surface bulk mode propagating at a shallow angle to the surface.
Shear Horizontal Surface Acoustic Wave, (SH-SAW)	An SSBW mode, which is confined to the surface by a grating (array of electrodes or a periodically grooved surface).
Love Wave	An SSBW mode, which is confined to the surface by a guiding layer of lower acoustic shear velocity than the substrate.

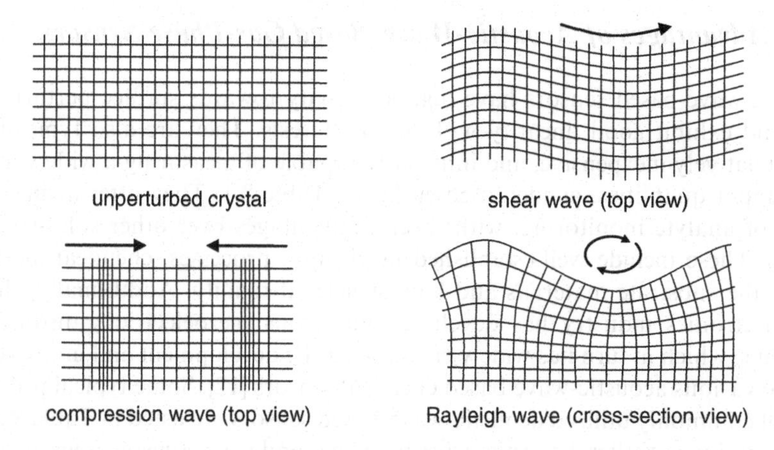

Fig. 8.2 Some propagation modes for acoustic wave devices

Figure 8.3 shows two commonly employed crystal cuts of quartz. In the case of an AT-cut QCM, the crystal is cut at 35°15′ with respect to the optical axis (z-axis), where the temperature coefficient of the crystal is at its smallest [3].

The propagation of acoustic waves in piezoelectric materials is governed by two coupled systems of equations: the mechanical equations of motion and Maxwell's equation for the electric behavior, the physics of which is well understood. These form the piezoelectric constitutive equations [9]. The fundamental equations describing mechanical wave propagation in piezoelectric media are described in detail elsewhere [1, 8, 9, 10, 11].

For each mode in a piezoelectric structure, an electromechanical coupling coefficient K^2 can be defined. It is a measure of the coupling between the electrical and mechanical fields, and thus has a very strong bearing on reflection, generation and detection of acoustic waves.

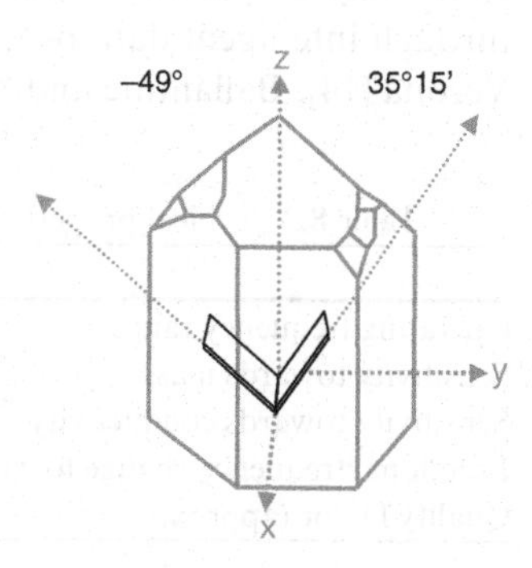

Fig. 8.3 Quartz crystal showing the AT-cut (35°15') and BT-cut (−49°) which belong to the Y-cut family

8.1.2 Advantages of Acoustic-Wave-Based Gas-Phase Sensors

Acoustic-wave-based sensors have high sensitivity towards surface perturbations and exhibit good linearity with low hysteresis. They are also typically small, relatively inexpensive, and inherently capable of measuring a wide variety of input quantities, as identified earlier in Table 8.1. They offer a simple means of analyte monitoring, with several advantages over other solid-state sensors. These include well-established fabrication processes, chemical inertness of the substrate materials and high structural rigidity. Additionally, the microelectronics industry has developed numerous piezoelectric transducer platforms, which in turn has widely encouraged the development and improvement of various acoustic-wave based chemical sensors [12]. Some typical properties of commonly utilized acoustic-wave-based sensors are listed in Table 8.3.

As mentioned earlier, acoustic-wave-based gas and vapor sensors operate on a change in the mass at the surface, or in the sheet conductivity of the sensitive layer. As mass sensitivity is defined per unit area, it implies that miniaturization will not result in a loss of sensitivity, making them highly suitable for lab-on-chip and other miniaturization applications. Another major advantage is that all acoustic-wave-based devices can be easily interfaced with external circuitry for direct frequency output signal monitoring. As a result, the frequency can be measured with high accuracy (<1 part in 10^7) [12]. Furthermore the associated electronics can be created with relatively simple and inexpensive discrete components.

It is inherently difficult to satisfy the requirements for all sensing applications with one single sensor design. This is because the operation specific requirements of the sensor determine the structural configuration, utilized materials, operating frequency and propagation mode of the acoustic-wave-based gas sensor employed. Therefore, in most cases, the selection of materials and structural parameters result in design tradeoffs between device sensitivity, long-term stability, selectivity and operating temperature.

With regards to measurements in multi-component gas mixtures, the use of sensor arrays, in which each element has its own signature response characteristic, has been widely investigated. Pattern recognition techniques have been commonly employed to interpret information obtained from a sensor array through intelligent data processing [13]. Early investigations by D'Amico and Verona [14], Ballantine and Wohltjen [2], and Grate et al. [15, 16, 17] reported

Table 8.3 Typical properties of commonly utilized acoustic-wave-based sensors

	QCM	TFR	SAW
Operating frequency range	5–30 MHz	500 MHz – 20 GHz	40 MHz – 1 GHz
Sensitivity towards mass	Yes	Yes	Yes
Sensitivity towards conductivity	No	No	Yes
Fractional frequency change to gas	~0.01%	~0.1%	~0.1%
Quality factor (approx.)	up to 10^5	up to 10^3	up to 10^4

that through the use of a sensor arrays the selectivity and overall performance of the sensing system may be enhanced. More recently Fernandez et al. [18] developed arrays for volatile organic compounds (VOC) and Joo et al. [19] for chemical warfare agents.

8.2 Thickness Shear Mode (TSM)-Based Gas Sensors

Thickness shear mode (TSM)-based resonators have acoustic vibrations that are transverse and shear to the surface and fall under the category of bulk acoustic wave (BAW) resonators. The most commonly employed TSM devices for gas and vapor-sensing applications are quartz-crystal microbalances (QCMs) and thin-film resonators (TFRs), which consist of a thin piezoelectric layer sandwiched between two electrodes as shown in Fig. 8.4. The application of an electric field between the two electrodes induces mechanical stresses that deform the piezoelectric material, thus launching acoustic waves. The crystal orientation of the piezoelectric is chosen such that it vibrates in a transverse shear mode when a voltage is applied to the metal electrodes. The primary factor in determining the acoustic wavelength of the structure is the thickness (h) of the piezoelectric and the acoustic velocity of the material. For instance a typical AT-cut QCM employing a 330-μm-thick substrate would operate at 5 MHz, whereas a 30-MHz device would use a significantly thinner 55-μm-thick substrate [3]. In the case of a QCM the thickness of the electrodes is much thinner than the piezoelectric material, however for TFRs this is not necessarily true.

When operated in thickness shear mode, the displacement maxima is at the crystal surface, making the device highly sensitive to small surface perturbations, such as mass accumulation [2, 4]. For sensing applications, masses that rigidly adhere to the surface will move synchronously with the surface, thus resulting in an effective increase in the devices thickness. Therefore, the mass loading of the surface effectively increases the wavelength, resulting in a fractional decrease in the structure's resonant frequency.

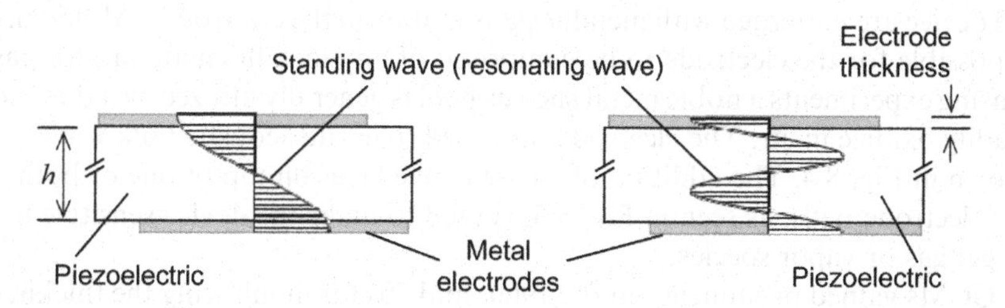

Fig. 8.4 Cross-section view of a TSM resonator showing the standing wave of the fundamental and 3rd harmonic

The sensitivity of a TSM device is the ratio of the fractional change in output signal per fractional change in gas or vapor concentration. Generally, the output signal in response to these small perturbations may be measured in two ways; the first being to monitor the change in resonant frequency upon exposure to the analyte, and the second involving monitoring the device impedance.

The readout of the resonant frequency, f_0, is routinely performed using an oscillator circuit connected to a frequency counter through a buffer or RF coupler as not to load the circuit. This mode of operation is termed 'active' since the crystal is used as the resonant frequency determining element in a closed loop feedback circuit [3]. More advanced systems use a phase-lock loop (PLL) and network analyzer techniques to determine the resonance frequency of the crystal, which are useful for extremely sensitive or low-temperature investigations [20].

8.2.1 Quartz Crystal Microbalance (QCM)-Based Gas Sensors

By far, the most commonly used acoustic-wave-based sensors are quartz crystal microbalances (QCMs). They have an extensive history of use for the quantification of physical and chemical adsorption in both commercial and fundamental research applications. Initially the devices were developed by Walter Cady in 1921 for stabilizing electronic oscillators [4], in which case they were labeled quartz crystal resonators (QCR). However, when employed for sensing, they are most commonly referred to as quartz crystal microbalances or QCMs, since the response is predominantly due to the mass loading of the analyte on the electrode surface. Although a large number of crystals exhibit piezoelectricity, quartz possesses a unique combination of mechanical, electrical, chemical and thermal properties. These properties have led to its commercial significance [3].

QCMs attained importance as analytical devices following the discovery of their linear relationship between mass loading and the frequency response, as described in 1959 by Sauerbrey [21]. A QCM typically consists of a thin circular AT-cut quartz crystal, which support thickness shear modes [12]. Both sides of the device are patterned with metallic pads that form the electrodes. Although it is possible for the electrode to be comprised of any metallic material, for gas-sensing experiments a noble metal such as gold is generally utilized as it does not readily oxidize in air. The electrodes are used to excite acoustic bulk waves, as shown in Fig. 8.4. The addition of gas-sensitive layer on top of one or both of the electrode pads, as seen in Fig. 8.5, is used to make the device sensitive to a target gas or vapor species.

QCMs gained prominence in the 1960s and 70s for monitoring the thickness of layers during their deposition in vacuum [22], and are still routinely used as in-situ thickness monitors [3]. They are well suited to such applications as they

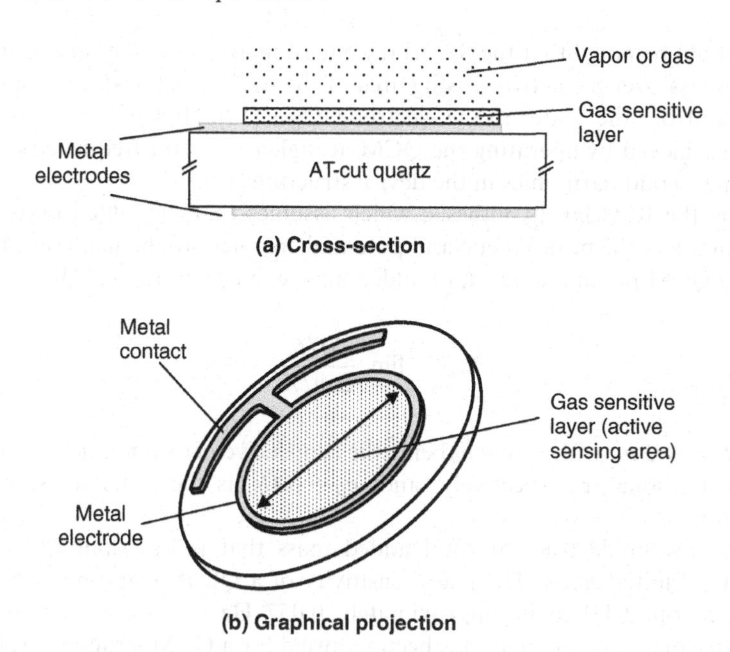

Fig. 8.5 (**a**) Cross section and (**b**) graphical projection of a QCM, illustrating a gas sensitive layer on one electrode

are capable of measuring small mass changes equivalent to a layer several nanometers thick. Subsequently, they have been modified to operate as gas-phase sensors by adsorption of target gas molecules onto the active sensing area. They are commercially available at operating frequencies up to 100 MHz, however for gas-sensing applications generally 5–30 MHz structures are used because higher frequency devices require thinner quartz layers, which are extremely fragile.

8.2.1.1 Gas and Vapor Sensitivity

The change in resonant frequency (Δf) can be directly related to the quantity of added mass (Δm) at the working electrode/s as defined by the Sauerbrey equation [21, 23]:

$$\Delta f = -\frac{2f_0^2}{A\sqrt{\rho\mu}}\Delta m, \tag{8.2}$$

where ρ is the density of the crystal, μ is the shear modulus of the crystal, A is the active area of the metal electrode and f_0 is the operating frequency of the device. An increase in mass causes a decrease in frequency, as indicted by the negative sign in Eq. (8.2). The Sauerbrey equation indicates that the change in frequency is proportional to the square of the fundamental frequency and inversely proportional to the surface area.

It should be noted that the Sauerbrey equation assumes a uniform distribution of mass over the active sensing area of the device, as the shear amplitude decreases from the center to the electrode edges [12]. Higher mass sensitivity may be achieved by operating the QCM at higher resonant frequencies corresponding to odd harmonics in the device structure [12].

Using the Rayleigh hypothesis, which assumes that the added mass layer does not affect the peak kinetic and potential energies [4], the mass sensitivity, S_m, of a QCM per unit area, A, of added mass can be defined as [24]:

$$S_m = \lim_{\Delta m \to 0} \frac{\Delta f / f_0}{\Delta m / A}, \tag{8.3}$$

where $f_0 \approx v_0 / 2\,h_0$, with v_0 and h_0 being the acoustic velocity and thickness of the unloaded resonator respectively, and $m_0 = \rho A h_0$ is the initial mass of the resonator.

This only holds true for total added mass that is less than 2% of the resonator's initial mass. The mass sensitivity of a QCM operating at 5 MHz has been found [3] to be approximately 0.057 Hz cm^2 ng^{-1}. A minimum detectable mass of 10 ng/cm^2 has been reported for a QCM structure [12].

For a mechanical system, resonance will take place at frequencies at which the peak kinetic energy exactly balances the peak potential energy. By definition, the series and parallel resonant frequencies, f_s and f_p respectively, occur when the reactance of the resonator is zero. They can be found from the Butterworth-Van Dyke (BVD) equivalent circuit of a TFR, shown in Fig. 8.6:

where R_m, C_m, and L_m are the resistive, capacitive, and inductive contributions, respectively, arising from the motional electromechanical coupling contributions of the piezoelectric. Similarly, C_0 and R_s are the clamped capacitance and series resistance respectively. The series resonance frequency occurs within the R_m–C_m–L_m part of the circuit, and the parallel resonance (or anti-resonance) is determined by the entire circuit, including C_0 and the parasitic capacitance from any test fixture. The series resonant frequency, f_s, is defined as the frequency

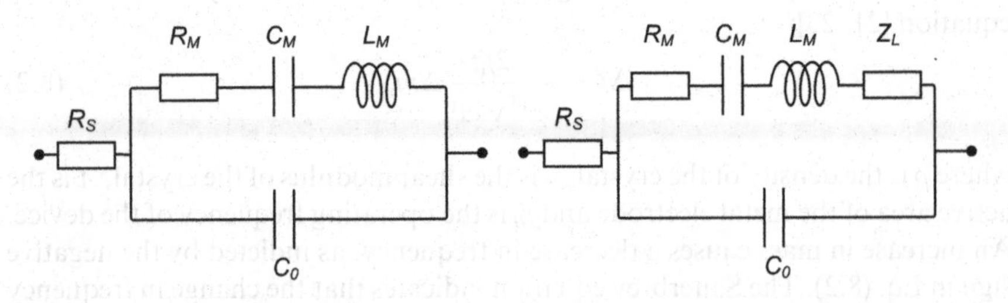

Fig. 8.6 Butterworth-Van Dyke (BVD) equivalent circuit of a TFR. *Left:* unloaded, *right:* loaded

at which the motional reactance is zero, whereas the parallel resonant frequency, f_p, is defined as the frequency at which the total reactance is zero, as follows [4]:

$$f_s = \frac{1}{2\pi\sqrt{L_m C_m}}, \tag{8.4}$$

$$f_p = \frac{1}{2\pi\sqrt{\frac{1}{L_m C_m} + \frac{1}{L_m C_0}}}. \tag{8.5}$$

Impedance analysis generally reveals sharp conductance peaks at the resonant frequency of the structure. The circuit has a capacitive response at frequencies below f_s and above f_p, whereas between f_s and f_p it has an inductive response, as shown by the frequency response in Fig. 8.7.

For bulk wave modes, such as the ones present in QCMs, sharp conductance peaks are indicative of a high quality factor, Q, which is a measure of energy stored to energy dissipated per cycle [12]:

$$Q = \omega \frac{energy\ stored}{power\ loss}, \tag{8.6}$$

where $\omega = 2\pi f$. The Q values for high quality quartz crystals can exceed 100,000. However, values ranging between 5,000 and 20,000 are typically employed for gas-phase analysis. The Q at the series and parallel resonant frequencies can be obtained by fitting the measured impedance to the Butterworth van Dyke model [25]:

$$Q_{s/p} = \frac{f_{s/p}}{2} \frac{d\phi_{s/p}}{df_{s/p}}, \tag{8.7}$$

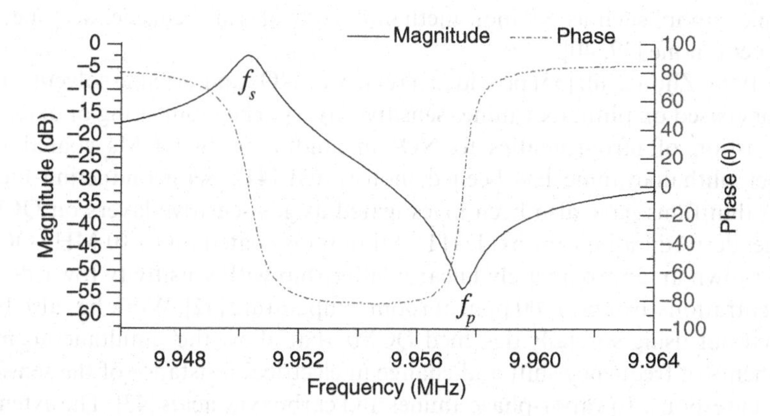

Fig. 8.7 Magnitude and phase of impedance vs. frequency of a 10 MHz AT-cut QCM

where $\phi_{s/p}$ is the phase of the impedance of the device at the series/parallel resonant frequency. QCMs can be fabricated to have higher operating frequencies; however in most cases Q decreases as the piezoelectric crystal is thinned.

8.2.1.2 QCM Gas Sensor Performance

QCM-based gas sensors are widely utilized as a result of their robust nature, availability and affordable interface electronics [3]. Originally employed for mass sensing in gas-phase deposition processes, they have evolved into a versatile platform for both liquid and gas-phase chemical sensing. Early QCM-based hydrocarbon sensors employed sensitive layers developed for gas chromatography [26]. Furthermore they enabled the first complete characterization of the triple-point wetting transition of light molecular gases on gold and rubidium surfaces [20, 27, 28] and have been extensively used in experiments in the field of chemistry and biology [29]. There is a significant amount of literature demonstrating their mass sensitivity for gas-sensing applications [30, 31, 32], and as a result a large number of QCM-based gas sensors have been reported [33, 34, 35, 36].

As with most acoustic-wave based sensors, the small fractional frequency (see Table 8.3) change when compared to the resonant frequency may pose signal-to-noise ratio issues with regards to electronic signal processing. Due to the relatively low operating frequency, a significant advantage of QCM-based sensors is that inexpensive drive electronics can be employed for routine sensing applications in the gas phase [37, 38].

The characteristics of QCM gas sensors invariably depends on the type of sensing films coated on their electrodes. However, factors such as the history/memory (or non-reversibility) of the sensitive layer, period of exposure, gas concentration and the coexistence of other molecules in the ambient sensing environment can significantly influence performance. In many cases, organic polymers are used as coating materials, yet they suffer from poor selectivity. PVC blended membrane coated QCMs have shown high sensitivity towards organic vapors such as ethanol, methanol, chloroform, benzene, acetone and cyclopentanone [39, 40].

In 1996, Zhou et al. [35] provided a review of NO_x and organic solvent vapor sensors based on phthalocyanines sensitive layers. The monitoring of oxidative combustion of niroaromatics to NO_2 in landfill using QCMs coated with copper phthalocyanine has been demonstrated [41]. Semiconducting metal-oxide thin films have also been investigated as gas sensitive layers on QCMs. Sol-gel derived indium-tin-oxide (ITO) thin films coated on to 10 MHz QCMs have shown an approximately linear relationship with sensitivity towards NO concentrations around 1000 ppm at room temperature [42]. Work has also been carried out using specially designed QCMs that allow the simultaneous measurements of frequency shift and change in electrical resistance of the sensitive layer in response to vapor-phase amines and carboxylic acids [43]. The extent of the responses was beyond that expected by mass uptake alone on the gas-sensing layer, which was composed of composites of conductive carbon black and insulating organic multifunctional polymers.

More recently, organic monolyers have gained prominence as sensitive layers. It has been demonstrated that QCMs with gas sensitive organic monolayers respond to ammonia in a relatively rapid fashion, with a 90% response occur within 90s [33]. Different polythiophenes prepared by the Langmuir-Blodgett (LB) technique for use as gas sensitive layers on 6 MHz gold-coated QCMs have been reported by Kim et al. [44], which produced frequency shift of over 130 Hz towards 550 ppm NO_2. Also, self assembled monolayers of alkanethiols with different functional groups have been studied by Okahata et al. [36] for the molecular recognition of acetic acid vapors. For a 9 MHz AT-cut crystal operating at the 7th overtone (63 MHz) a 300 Hz decrease in frequency was observed for a mass increase of 15 ng. These QCM platforms were found to be useful for in-situ studies of adsorption behavior of organic molecules on the monolayer surface [45].

Recent advances in molecular imprinted polymers (MIPs) have seen the emergence of new materials with molecular recognition capabilities [46]. Hiryama et al. [47] prepared a MIP sensing layer on a QCM for monitoring acetaldehyde in aqueous and gaseous samples. They reported very high selectivity for acetaldehyde when compared to the following potential interfering compounds: acetone, chloroform and trichloroethane. Similarly, Feng et al. [48] prepared a formaldehyde MIP sensitive layer on a 9 MHz QCM, reporting a ~90% increase in signal for formaldehyde detection when compared to a non-imprinted polymer.

Other approaches for increasing selectivity have made use of cavitands as receptors. These synthetic organic compounds have rigid cavities whose shape and dimension can be tailored to interact preferentially with certain types of gas molecules. Ferrari et al. [49] demonstrated that AT-cut QCMs employing cavitand coatings (Qx-Cav and Me-Cav) exposed to organic vapors such as toluene and acetonitrile were capable of monitoring concentrations down to the order of 100 ppm for environmental applications. Göpel et al. [50] described the assembly of a monomolecular cavitand film on QCMs that exhibitied strong sensitivity for perchloroethylene, and Hartman et al. [51, 52] investigated QCMs coated with different cavitand layers towards organic vapors such as cyclohexane, ethanol, acetontrile, benzene and chloroform.

Zeolite-coated QCMs have also been used for increasing selectivity towards specific analyte gases. Osada et al. [53] employed a faujasite-coated (zeolite) QCM to sense SO_2 gas at elevated operating temperatures of 150°C. They showed that 300 ppm of SO_2 could be sensed in the presence of 22,500 ppm of oxygen. Also, work regarding the sensing of H_2O, NO, and SO_2 using an array of QCMs with Zeolite A, silicalite-1 and sodalite sensitive layers has been carried out [54]. This work showed that response time, which is one of the most important factors, is significantly improved by using a differential signal instead monitoring the frequency shift alone. Additionally, other zeolite-coated QCMs have been employed for monitoring H_2O and organic molecules at room temperature [55, 56].

Recently, nanomaterials have gained much attention as sensitive layers towards a variety of gases and vapors. Their potential to increase response by one order of magnitude over those of conventional sensing layers is being widely investigated. For example, poly vinyl alcohol (PVA) and poly acrylic acid (PAA) nanofibrous membranes spun on QCMs have exhibited responses of 730 Hz towards NH_3 concentrations of 200 ppm [57]. Of all nanomaterials, carbon nanotubes have received the most attention for sensing applications. Multi-walled carbon nanotube (MWCNT) coated QCMs have been reported for relative humidity (RH) sensing applications, in which frequency shifts between 2.5 kHz and 10 kHz have been recorded for RH values between 5 and 97% [58]. Cusano et al. [59, 60] showed that single-walled carbon nanotube (SWCNT) gas-sensitive films deposited onto 10 MHz QCMs by the LB technique exhibited frequency shifts in excess of 200 Hz towards VOC concentrations less than 100 ppm. Penza et al. have researched SWCNTs deposited onto the active sensing area of QCMs for monitoring low concentrations of VOCs [59, 60, 61, 62, 63]: Fig. 8.8 shows the response towards ethanol vapor of a 10 MHz QCM whose active sensing area is coated with 20 monolayers of CdA, which is in turn covered with two monolayers of SWCNTs. These sensors possess a linear response characteristic towards ethanol concentrations below 200 ppm, with a frequency shift of 0.49 Hz/ppm. Figure 8.9 shows their response towards toluene vapors, exhibiting increased sensitivity with an increasing number of SWCNT monolayers.

QCM sensors are also well suited to mercury vapor-sensing applications. Such work dates back to 1975, where Scheide and Taylor [64] developed a meter for mercury vapor in air. As the 'sticking efficiency' of mercury vapor on a gold surface is infinite for the first monolayer [65], it is possible to use the QCM electrode itself as the sensitive layer. The mercury vapor adsorbs/amalgamates with the gold, increasing the mass of the electrode. Figure 8.10 shows the response of two 10 MHz QCM sensors towards 4.3 mg/m^3 of mercury vapor.

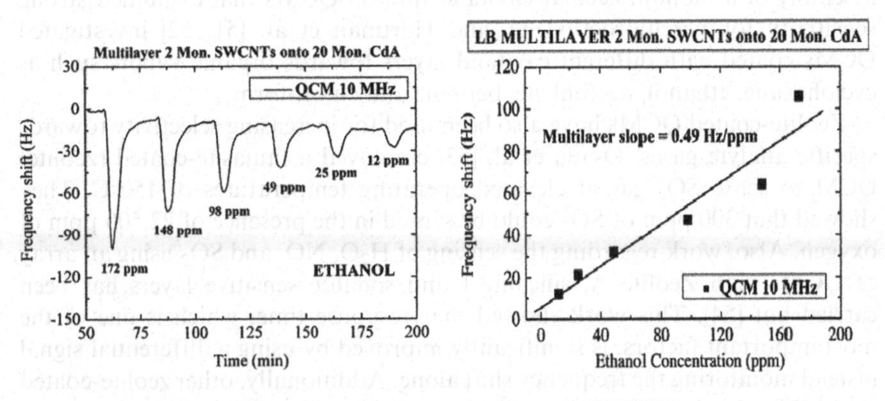

Fig. 8.8 A 10-MHz QCM sensor coated with 2 monolayers of SWCNTs on top of 20 monolayers of CdA, towards ethanol at room temperature. © 2006 IEEE [63]

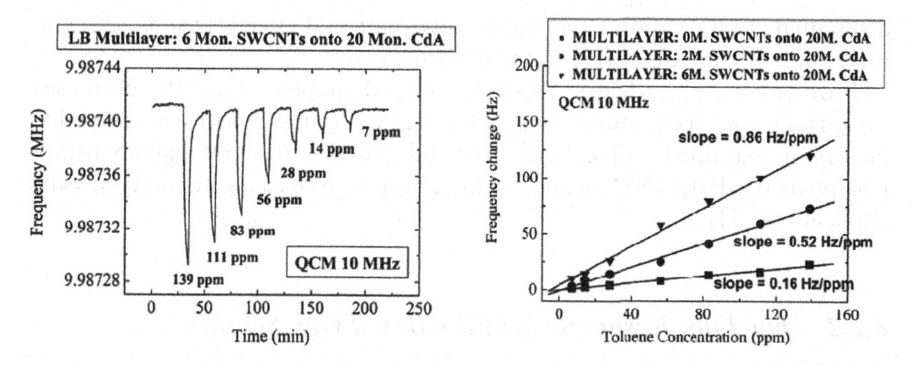

Fig. 8.9 A 10-MHz QCM sensor coated with 6 monolayers of SWCNTs on top of 20 monolayers of CdA, towards ethanol at room temperature. *Left*: dynamic response to different concentrations, *right:* calibration curve. © 2006 IEEE [63]

The top curve shows the response of the sensor using a non-modified evaporated gold electrode, while the second curve is the response of a QCM with augmented electrodes and a higher surface area to volume ratio [66]. The QCM with the augmented electrodes is shown to exhibit up to 144 Hz larger frequency shift than the non-modified QCM at an operating temperature of 37°C.

Since the selectivity of a single sensor is not specific to a single analyte, generally the analysis of the output pattern of the sensor array by neural networks or multivariate analysis may be employed to recognise specific gas species [67]. More recently, multichannel QCM structures have been developed [68, 69]. There are significant benefits to this approach, including cost and size reductions; however the structures must be carefully designed to minimize interference between the working devices. The structures can also be used in

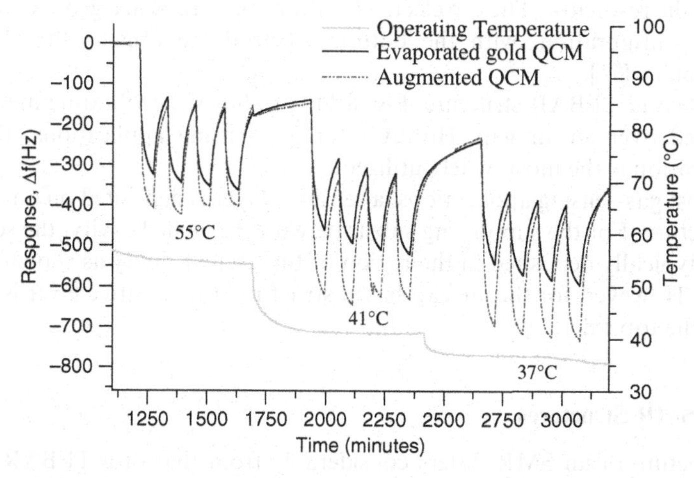

Fig. 8.10 Sensors responses to 4.3 mg/m^3 mercury vapor pulses at 55, 41 and 37°C [66]

deferential configuration to minimize environmental effects such as temperature, humidity and other interfering influences.

Furthermore, piezoelectric substrates other than quartz have also been used for gas-sensing applications, which have also been shown to have quality factors comparable to AT-cut quartz crystals [70]. For instance, gallium orthophosphate ($GaPO_4$) TSM resonators have been employed for humidity-sensing applications [71].

8.2.2 Thin-Film Resonator (TFR)-Based Gas Sensors

TFRs are capable of launching acoustic waves with frequencies up to 20 GHz [72, 73, 74], resulting in a much improved mass sensitivity over QCMs. The two main types of TFR structures that typically utilized for sensing applications are thin-film bulk acoustic resonators (TFBARs) and solidly mounted resonators (SMRs). Since they are BAW devices, depending on the crystal orientation of the piezoelectric materia they are capable of supporting several modes of operation, including longitudinal and shear horizontal modes. However for sensing applications the TSM is typically employed. The thickness of the piezoelectric layer is typically less than half the acoustic wavelength, and may vary between 0.5 to <10 μm, which is significantly less than in QCMs.

8.2.2.1 TFBAR Structures

Figure 8.11 shows two typical TFBAR structures. The first (a) shows the electrodes and thin piezoelectric film deposited on top of an optional insulating layer (typically silicon nitride) supported by a substrate. A portion of the substrate is removed, typically by a wet chemical etching process, thereby defining the resonator. The top electrode may be patterned in a ground–signal–-ground configuration, with the bottom electrode serving as the electrical ground plane [75].

The second TFBAR structure, Fig. 8.11 (b), shows the vibrating membrane suspended over an air gap. However for gas-sensing applications, the first configuration is the most widely utilized.

In most gas-sensing applications, a sensitive layer is deposited on one side of the structure. For the supporting substrate with the etched cavity, the sensitive layer is typically deposited in the region of this etched cavity as shown in Fig. 8.11 (a). However, for the air-gap-based structure, the sensitive layer is deposited on the top side.

8.2.2.2 SMR Structures

The structure of an SMR differs considerably from that of a TFBAR, and a typical cross-section is shown in Fig. 8.12. The piezoelectric layer is 'solidly'

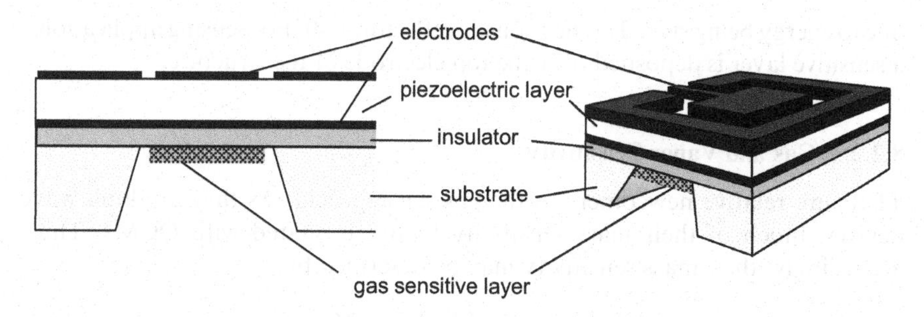

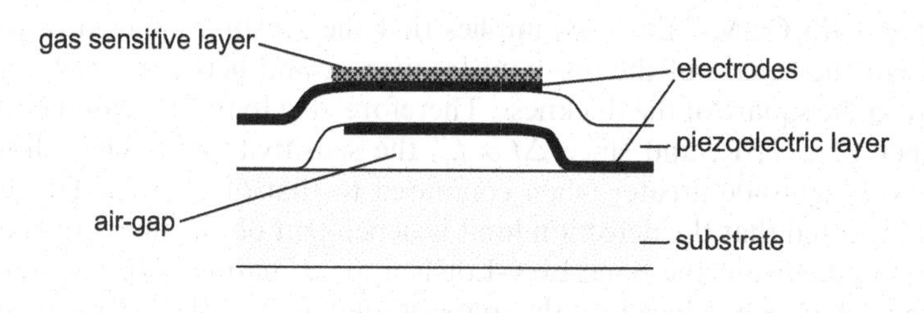

Fig. 8.11 (**a**) Cross-section and diagram of a TFBAR with etched supporting substrate and (**b**) diagram of TFBAR having the vibrating part suspended over an air-gap

mounted onto an acoustic reflector stack which alleviates the need for an etched supporting substrate or an air-gap. The reflector stack acts as a Bragg grating, which causes the wave amplitude to diminish within the depth of the reflector, hence confining the mechanical energy to the piezoelectric layer. First described by Newell [76], the SMRs comprise nominally quarter wavelength-thick layers of materials having alternating high/low acoustic impedance values that are stacked on top of each other. The number of layers required to obtain a satisfactory reflection coefficient strongly depends on the mechanical impedance contrast between layers, and to a lesser extent on the substrate [73]. Compared to TFBARs, SMRs suffer from a lower electromechanical coupling coefficient

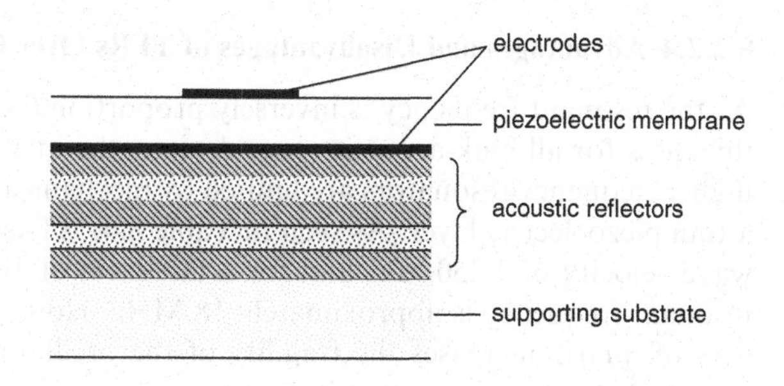

Fig. 8.12 Cross section of a typical SMR

due to energy being stored in the acoustic reflectors [74]. For sensing applications, a sensitive layer is deposited over the top electrode of the structure.

8.2.2.3 Gas and Vapor Sensitivity

TFRs are relative newcomers to the gas-sensing field. As they are bulk wave devices, much of their mass sensitivity theory is shared with QCMs. Thus, accordingly, their mass sensitivity may be described by:

$$S_m = -\frac{v_0}{2\rho}\left(\frac{1}{h_0}\right)^2 = -\frac{2f_0^2}{v_0\rho}. \tag{8.8}$$

Just as with QCMs, Eq. (8.8) implies that the sensitivity increases proportionally to the square of the resonant frequency, and hence inversely proportionally to the square of the thickness. Therefore, due to the increased operating frequency of a TFR, and since $\Delta f \propto f_0^2$, the sensitivity of a TFR should be orders of magnitude greater when compared to that of a QCM. However, it should be noted that the detection limit is dependent on the noise and stability of the system. From the Sauerbrey-Lotsis approximation [77], the fractional frequency change produced by the mass of analyte adsorbed per unit surface may be defined as:

$$\Delta f = -\frac{v_0}{2\rho}\left(\frac{1}{h_0}\right)^2\left(\frac{\Delta m}{A}\right). \tag{8.9}$$

This suggests that for a given change in mass per unit area, A, the change in frequency increases with acoustic wave velocity and with decreasing density, as well as with decreasing thickness, highlighting TFRs' improved performance over QCMs [78]. However in practice, the sensitivity has been found to depend on the Q factor of the TFR, which is generally lower than that of a QCM. It has been found that the sensitivity does not necessarily increase proportionally to the square of the resonant frequency. However, at the very least, sensitivity increases linearly with operating frequency [79, 80]. It has also been suggested that the detection limit is the product of the sensitivity and the Q factor of the structure [80].

8.2.2.4 Advantages and Disadvantages of TFRs Over QCMs

As the resonant frequency is inversely proportional to the piezoelectric layers thickness for all bulk acoustic wave devices, the inherent problems in making higher frequency resonators are limited by mechanical constraints in achieving a thin piezoelectric layer. As seen in Table 8.4, AT-cut quartz has an acoustic wave velocity of 3,750 m/s, and for a thickness of 100 μm the corresponding resonant frequency is approximately 18 MHz. However, decreasing the thickness of quartz increases the fragility of the oscillating membrane and hence

Table 8.4 Physical properties of some piezoelectric materials utilized in TFRs. Reused with permission from M. Benetti, D. Cannata, F. Di Pietrantonio, V. Foglietti, and E. Verona, Applied Physics Letters, 87, 173504 (2005). Copyright 2005, American Institute of Physics. [78]

Piezoelectric Material	v_p (m/s)	ρ_p (kg/m^3)	$v_p/2d^2\rho$ Hz(ng/cm^2)$^{-1}$
AT-Quartz (h_0 100 μm)	3750	2648	0.708
c-AlN (h_0 1 μm)	11345	3260	34,800
c-ZnO (h_0 1 μm)	6370	5665	11,244

increases the risk of damage. Conversely, piezoelectric materials such as ZnO and AlN, which have higher density and acoustic velocity, can have thickness of less than one micrometer without the risk of damage.

Therefore by utilizing a thinner piezoelectric layer, the resonant frequency and mass sensitivity of a TFR can be increased. Furthermore, sensitivity may also be increased by employing a piezoelectric material with a greater acoustic wave velocity.

A further consideration is the Q value of a TFR device, which strongly depends on the fabrication process. Devices operating at around 1 GHz for gas-phase analysis can exhibit Q values in the range of 200–10,000. This value may be less than those for QCMs that operate between 10 and 30 MHz, however QCMs cannot be fabricated for operation in the GHz range. Unfortunately, the lower Q of a TFR results in a decreased signal-to-noise ratio. Measures such as: removing the insulating layer, changing the electrode material or depositing a piezoelectric layer of better quality can be used to increase a TFR's Q value. Specifically, for an SMR, the reflector stack ensures that a resonance with a high Q is obtained. The number of layers required in the reflector stack is best determined by analyzing the resonance response as a function of the number of layers versus Q and electromechanical coupling coefficient [74]. In general, the performance of a TFR sensor will improve proportionally to the Q value, as it has been shown that the minimum detectable mass change for a TFR is comparable with that of a QCM that has a Q one hundred times higher [81]. Furthermore, one of the most significant advantages is that when compared with all acoustic-wave-based sensors, TFRs show the most promise for compatibility and integration with conventional IC circuits (CMOS technology) and fabrication procedures.

8.2.2.5 TFR Gas Sensor Performance

One of the earliest reports of a TFR-based gas sensor was that by O'Toole and coworkers [82]. The device had a resonant frequency of around 1 GHz, and employed a 5.5 μm thick piezoelectric A1N membrane situated between two gold electrodes in a TFBAR configuration. The TFBARs were coated with poly methy methacrylate (PMMA) layers, that were functionalized with thiol (SH-terminated) molecules of different lengths, such as $CH_3(CH_2)_nSH$ (where n = 3, 17), as well as $CF_3(CF_2)_7(CH_2)_2SH$ to demonstrate the TFBAR's mass sensitivity towards methanol vapor. Responses were measured as the ratio of the fractional changes in either the series or parallel resonant frequencies ($\Delta f_{s/p}/f_{s/p}$). Mass sensitivities of around –550 Hzcm^2/g were achieved and fractional change in frequency of –2.7 ppm for concentrations of 2.68% v/v. Furthermore, the relationship between the response and methanol concentration was found to be non-linear, suggesting the surface coverage of sorbed species approaches saturation with increasing methanol concentrations.

Many developments have occurred in the field of TFR's since their initial discovery. For example, 2 GHz SMR based on ZnO were developed by Gabl et al. [83, 84] for sensing humidity levels in gaseous N_2. The active vibrating region was spin-coated with a water vapor absorbing polyimide layer, whose thickness was varied from 50 to 650 nm. They observed mass sensitivity three orders of magnitude larger than for a typical QCM. It was determined that a strong non-linear relationship between humidity sensitivity and the thickness of the sensitive layer exists due to the complex acoustic response for thicker polymer layers. As seen from Fig. 8.13, for thinner layers the change in frequency in response to humidity was negative whereas a positive change in frequency was observed for thicker layers.

Additionally, it has been demonstrated that AlN-based SMR devices operating at 8 GHz can exhibit resonant frequency shifts as large as 15 MHz towards saturated vapor-phase acetone [80]. It was reported that a device with a 20-nm uncured PMMA sensitive layer was capable of detecting less than 1 pg on an electrode of area 900 μm^2. The porosity of the uncured polymer layer was believed to promote the absorption of acetone molecules.

TFBAR structures implemented on (001) Si wafers, using Si_3N_4/AlN membranes, comprising palladium (Pd) and Co-tetra-phenyl-porphyrin (Co-TTP) sensitive layers have been reported [78, 85]. They showed that a TFBAR based on 1-μm-thick c-axis oriented AlN with operating frequency of approximately 1.25 GHz has sensitivity of approximately 38.4 kHzcm^2/ng. When compared with QCMs, this equates to an increase of an order of magnitude. These TFBARs were subsequently employed for monitoring different concentrations

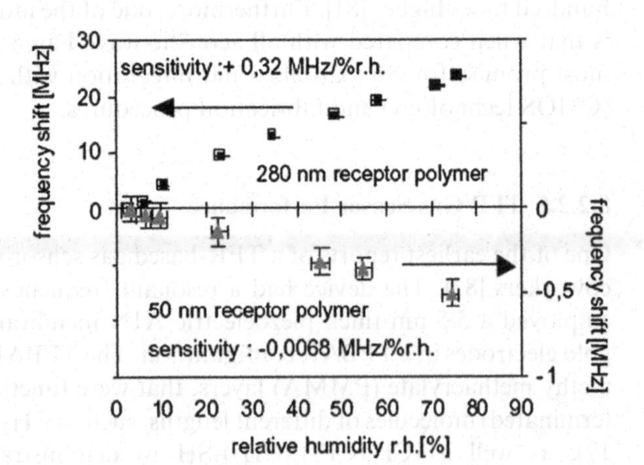

Fig. 8.13 Resonance frequency shift as function of relative humidity for two polymer layer thicknesses for a 2 GHz SMR based on ZnO. © 2004 IEEE [84]

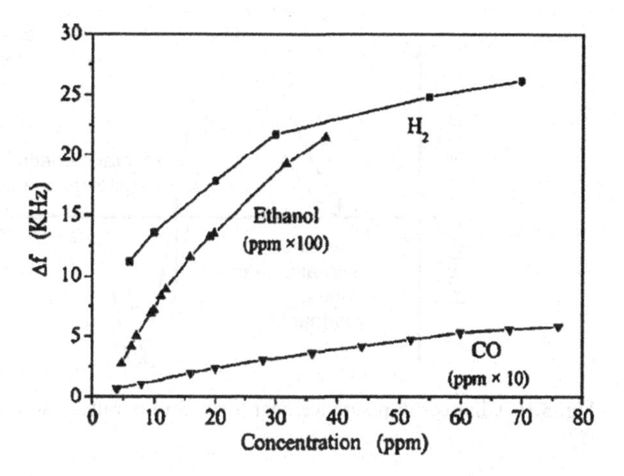

Fig. 8.14 Calibration curves for the TFBAR sensors upon exposure to H_2, Pd membrane, CO, and ethanol, Co-TPP membrane. Reused with permission from M. Benetti, D. Cannata, F. Di Pietrantonio, V. Foglietti, and E. Verona, Applied Physics Letters, 87, 173504 (2005). Copyright 2005, American Institute of Physics. [78]

of hydrogen and carbon monoxide gases balanced in nitrogen (see Fig. 8.14), where frequency shifts of approximately 24 kHz and 5 kHz were measured for 30 ppm H_2 and 600 ppm CO respectively.

More recently the use of nanomaterials as gas sensitive layers on TFBARs has been demonstrated by Penza et al. [86]. They employed SWCNTs) incorporated into an amphiphilic organic matrix of cadmium arachidate (CdA) for the monitoring of acetone vapors. These films were deposited by the Langmuir-Blodgett (LB) technique and their responses towards different acetone vapor concentrations can be seen in Fig. 8.15.

Zhang and Kim developed a Si_3N_4/ZnO TFBAR-based sensors for monitoring isopropanol vapor [81, 87], which was capable of detecting mass changes of 10^{-9} g/cm^2. The measurement technique used was able to track frequency shifts as small as 0.3 ppm with a network analyzer. Their technique involved determining the frequency near the series and parallel resonance, and then monitoring the change of the reactance at these frequencies upon the introduction of the

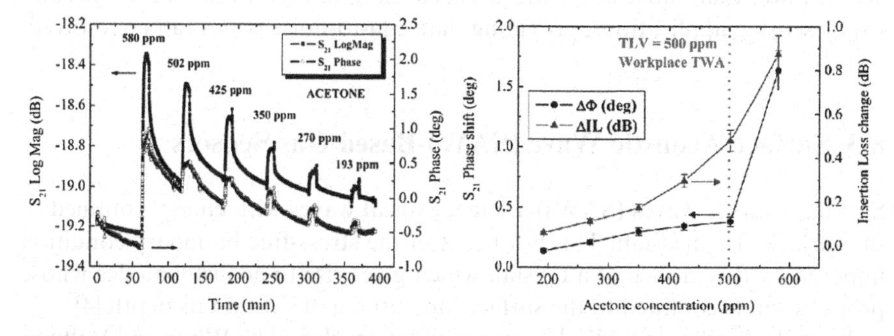

Fig. 8.15 Room-temperature response of TFBAR sensor coated with LB 75 wt.% nanocomposite film of SWCNTs-in-CdA towards acetone by measuring transmission phase and amplitude (S_{21}) by network analyzer (*left*) and calibration curve (*right*). © 2006 IEEE [86]

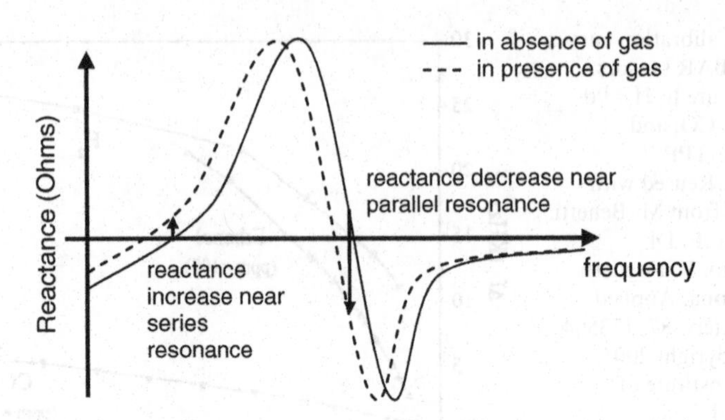

Fig. 8.16 Changes in reactance near series and parallel resonant frequencies. © 2003 IEEE [87]

analyte gas, as indicated in Fig. 8.16. This technique reduced the minimum detectable mass obtainable by a direct measurement of the frequency change by an order of magnitude. Their experimental results showed a mass sensitivity of 726 cm^2/g, which is about 50 times that of a typical QCM and that the minimum detectible mass change was improved for increasing Q value.

Typically, TFR-based gas sensors make use of a single resonator, however Lee and Yoon [86] demonstrated that differential-mode TFBARs are well suited for monitoring low-ppm concentrations of VOCs. The system was composed of two AlN-based TFBARs, one being exposed to the analyte gas and the other serving as a reference. Frequency shifts of 20 kHz towards 4 ppm of formaldehyde vapor were observed, and sensitivities for benzene, ethanol, and formaldehyde were found to be 2500 Hz/ppm, 84 Hz/ppm, and 5067 Hz/ppm, respectively.

Clearly a significant increase in mass sensitivity can be expected when the resonant frequency of an acoustic wave device increases well into the GHz range [80]. Therefore, TFRs with piezoelectric layer thickness several orders of magnitude lower than those of QCMs, are ideal candidates for mass-based gas and vapor-sensing applications, providing that sign-to-noise issues can be resolved.

8.3 Surface Acoustic Wave (SAW)-Based Gas Sensors

Surface acoustic waves (SAWs) are mechanical waves with energy confined to the surface of a medium. This is a result of the stress-free boundary condition imposed by the surface of a crystal, which gives rise to acoustic modes whose propagation is confined to the surface, and attenuates within its depth [4].

Piezoelectric-based SAW devices were first developed by White and Voltmer [88] in 1965. In their basic form, they consist of a piezoelectric substrate, on top of which two metallic interdigital transducers (IDTs) are patterned to form a

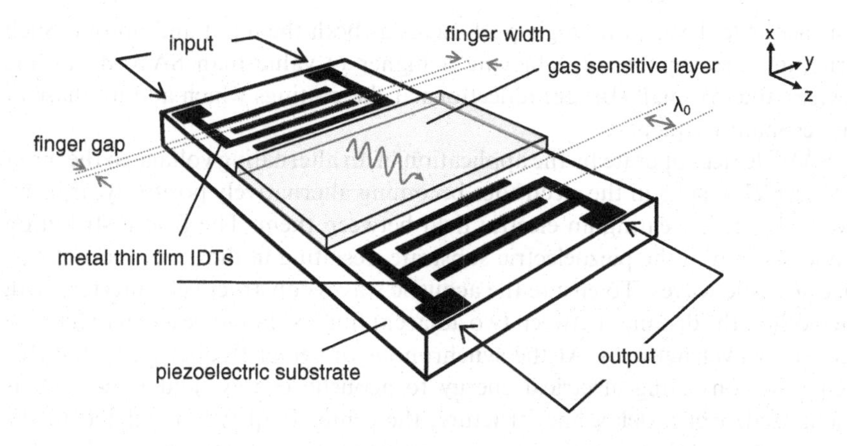

(a) 2-port delay line SAW device with gas sensitive layer

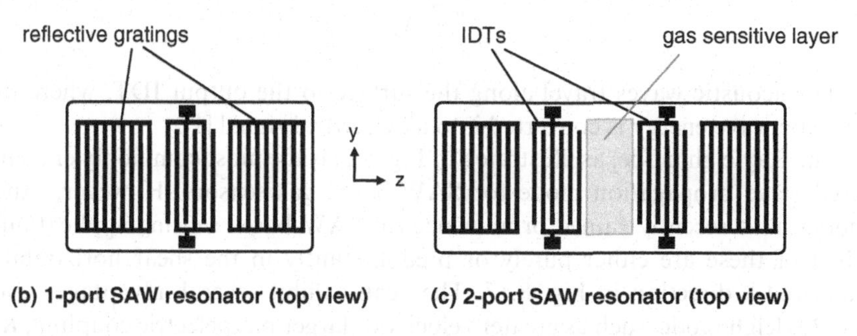

(b) 1-port SAW resonator (top view) (c) 2-port SAW resonator (top view)

Fig. 8.17 (**a**) Layout of a two-port delay line SAW device, (**b**) one-port resonator, and (**c**) two-port resonator

delay line structure, as shown in Fig. 8.17(a). This is the most commonly utilized structure for gas-sensing applications with the sensitive layer normally deposited in between the two IDT ports. However, if the layer is of low conductivity, it may be deposited over the entire device. The separation between the IDT ports determines the delay path between the transmission and reception of the surface wave. The primary factor in determining the acoustic wavelength (λ_0) of the structure is the IDT geometry patterned on the piezoelectric substrate. Generally, an equal finger width and spacing is used for sensing applications, however other arrangements are possible.

SAW resonator structures, as shown in Fig. 8.17 (b) and (c), can also be used for sensing applications. These may either be 1-port or 2-port devices having acoustic reflective electrode gratings that create a resonant cavity. With regards to two-port devices, one of the ports is used as the input port and the other as an output. The input signal excites an acoustic wave which propagates to the output port where it is converted to an electrical signal. Conversely for a one-port acoustic device, the single port serves as both the input and output. Such structures are capable of achieving a higher Q value than SAW delay line devices, due to small (Bragg) reflections at the gratings which add in phase at the resonant frequency.

SAW devices operate by the application of an alternating voltage to the input IDT, which results in the electrodes becoming alternatively positively or negatively charged, creating an electric field between them. The field distribution induces strain in the piezoelectric substrate, resulting in the generation of surface acoustic waves. To ensure the acoustic waves constructively interfere with each other, the distance between two adjacent fingers should be equal to half the acoustic wavelength, λ_0. At the synchronous or center frequency, f_0, the efficiency in converting electrical energy to acoustic energy, and vice-versa, is maximized. For a delay line structure, the center frequency is related to the wavelength and the acoustic wave velocity of the piezoelectric material, v_0, by:

$$f_0 = \frac{v_0}{\lambda_0}. \tag{8.10}$$

The acoustic waves travel along the surface to the output IDT, where the acoustic wave energy is converted into an electrical signal [1].

The Rayleigh mode, as illustrated in Fig. 8.2, is the most commonly encountered wave propagation mode for SAW-based gas sensors. However, other modes have recently gained prominence for SAW-based sensing applications. Most of these are either purely or predominantly in the shear horizontally polarized y-direction in Fig. 8.17. They can exhibit several advantages over the Rayleigh mode, such as greater velocity or larger piezoelectric coupling, K^2. They are of interest because they are able to propagate with minimal acoustic loss to the adjacent sensing medium or into the bulk of the substrate. Some of the different categories of surface acoustic waves are: Surface-skimming bulk wave (SSBW), leaky surface acoustic wave (LSAW) or pseudo surface acoustic wave (PSAW), shear horizontal surface acoustic wave (SH-SAW) or surface transverse wave (STW) and Love wave [1]. The material properties of the substrate are the primary parameters in determining which SAW modes are supported. Furthermore, the addition of a suitable intermediate or guiding layer deposited over the substrate can be utilized to favor a particular mode, such as in the case of Love wave devices [89, 90]. Predominately reported for liquid-phase analysis, Love modes can only exist if the shear wave propagation velocity in the guiding layer is less than that in the substrate [91].

SAW devices are well established in the electronics industry, and have been employed for gas-sensing applications for approximately 30 years [5]. They can be designed to offer high sensitivity with a reasonably large dynamic range, good linearity and low hysteresis. This makes them extremely suitable for sensing applications at parts-per-million (ppm) and parts-per-billion (ppb) concentrations.

There is an extensive body of knowledge regarding design and modeling of SAW devices [92]. Many combinations of materials that form new acoustic-based structures are being investigated for their sensing properties [93]. Advances in SAW sensor modeling have greatly facilitated the design and analysis of high performance multilayered structures [94, 95, 96, 97]. The choice of piezoelectric substrate and its crystal orientation, as well as the associated gas and vapor sensitive layers, has been investigated by different research groups to determine the supported SAW propagation modes, sensitivity towards mechanical or electrical perturbations and effect on overall device characteristics [1, 4, 98, 99]. The acoustic properties of commonly employed materials such as quartz, lithium tantalate ($LiTaO_3$) and lithium niobate ($LiNbO_3$) have been investigated in depth, and extensive and detailed material parameters have been compiled by Slobodnik [100, 101] and Auld [8]. Additionally, Hickernell [102] has also provided a thorough review of thin-film materials for SAW devices; however it primarily focuses on signal processing applications.

8.3.1 Conventional SAW Gas Sensors

The most common SAW-based gas sensor, referred to herein as a 'conventional SAW gas sensor', is one which has the sensitive layer deposited directly over the substrate between the IDT ports, as shown in Fig. 8.17 (a) and (c). The response to an analyte gas is determined by the shift in the propagating acoustic mode's phase velocity, due to the perturbations described by Eq. (8.1). The interaction between the sensitive layer and analyte gas occurs over the whole of the propagation path as opposed to TSM devices, where there is only interaction at the reflecting boundaries. Additionally, as most of the SAW energy is confined within one acoustic wavelength of the device's active surface [8], SAW-based sensors exhibit excellent sensitivity towards surface perturbations. Similar to TSM-based sensors, the two most common methods utilized for measuring changes in SAW velocity are via phase and frequency measurement techniques [103]. The change in resonant frequency of a SAW-based sensor can be related to the change in acoustic phase velocity (assuming negligible dispersion) by:

$$\frac{\Delta f}{f_0} \cong \kappa \frac{\Delta v}{v_0}, \tag{8.11}$$

where κ is the fractional coverage of the center-to-center distance between input and output IDTs affected by the perturbation [7], i.e. the area covered by the sensitive layer. Similarly, the device response can be obtained by several attenuation and phase measurement techniques [104, 105, 106]. In all cases, the measured quantities should correlate to the relative concentration of the gas or vapor analyte.

8.3.2 Multi-Layered SAW Gas Sensors

Recently multi-layered SAW-based gas sensors have emerged, as illustrated in Fig. 8.18. They consist of an intermediate thin-film layer deposited between the SAW substrate (over the IDT electrodes) and the gas sensitive layer. The intermediate layer is added for its acoustic, dielectric and/or protective properties, and as a result, layered SAW sensors offer several advantages over conventional SAW devices [6]. Typically employed for liquid-phase-sensing applications, they have recently gained attention for sensing: hydrogen [107, 108, 109, 110, 111], nitrogen dioxide (NO_2) [112, 113, 114], and ethanol vapor [115].

Depending on the materials selected, the advantages of multi-layered SAW devices over conventional SAW devices can include: increased electromechanical coupling coefficient; increased sensitivity; ability to generate several modes of propagation; improved temperature compensation; isolation of the substrate from the gas-sensing environment; and if the layer is piezoelectric, it may even allow the generation of acoustic waves on a non-piezoelectric substrate [6]. Furthermore, the active sensing surface area can be increased, as the sensitive layer is no longer limited to being deposited solely in the gap between the IDT ports, which may also lead to greater sensitivity.

8.3.3 Gas and Vapor Sensitivity

The response from a SAW sensor is based on measurable changes in the velocity and attenuation of the propagating wave. As a wave propagating in piezoelectric media generates both mechanical deformation and electrical potential, an analyte (gas or a vapor) can be sensed by the changes it induces in the mechanical and electrical coupling between the propagating wave and the sensitive layer. Hence, the interaction between the analyte and the sensitive layer causes either mechanical or electrical perturbations in the surface boundary conditions. Benes et al. have provided a comparison between BAW and SAW sensor principles [116].

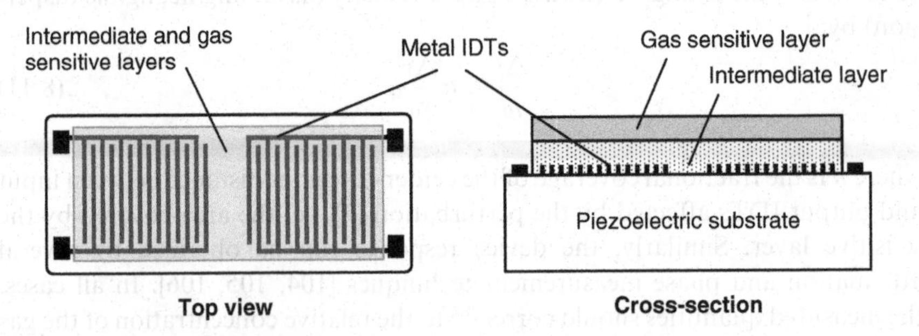

Fig. 8.18 Layout of a multilayered two-port layered delay line SAW device

8.3.3.1 Mechanical Perturbations

Mechanical perturbations at the sensor surface, which include changes in mass loading, elasticity, and viscoelasticity, are the most commonly utilized interactions for SAW-based gas sensors. Mass loading occurs when analyte molecules are adsorbed onto the surface, and typically cause a decrease in SAW phase velocity and/or an increase in attenuation. The most widely used definition of the mass sensitivity of a SAW mode, S_m^v, is [117]:

$$S_m^v = \lim_{\Delta m \to 0}\left(\frac{\Delta v / v_0}{\Delta m / A}\right). \tag{8.12}$$

The mass sensitivity of the SAW mode may be expressed in terms of frequency, S_m^f, as follows [24]:

$$S_m^f = \lim_{\Delta m \to 0}\left(\frac{\Delta f / f_0}{\Delta m / A}\right) = \left(\frac{v_g}{v_0}\right) S_m^v, \tag{8.13}$$

where v_g is the group velocity of the SAW mode, which can be calculated from the dispersion characteristics. Since SAW devices have operating frequencies of several hundred MHz or more, sensitivities in excess of 100 pg/cm^2 can be achieved [7]. In a non-layered device, the group and phase velocities are equal, thus S_m^f equals S_m^v. In multi-layered devices with significant velocity dispersion, this equality no longer holds, therefore the sensitivity can either be greater or worse depending on the degree of the dispersion.

Conversely, elasticity and viscoelastic effects arise from changes in density occurring on the sensing surface [4]. For example, SAW devices utilizing Pd sensitive layers undergo a change both in density and stiffness of the layer in response to H_2 adsorption and desorption [118]. Where the mass layer is very thin and itself has no effect on the elastic properties of the substrate, the mass loaded layer will not store any potential energy during the vibration cycle of the propagating wave [117]. However, as this is not always the case with sensitive layers, Wohltjen [119] has shown that for layers that are non-piezoelectric, lossless, insulating and isotropic, the Tiersten boundary condition can be applied:

$$\frac{\Delta f}{f} = (k_1 + k_2) f_0^2\, h \rho_f - k_2 f_0^2\, h \left\{ \frac{4\mu'}{v_0^2} \frac{\lambda' + \mu'}{\lambda' + 2\mu'} \right\}, \tag{8.14}$$

where k_1 and k_2 are the substrate material constants, ρ_f is the film density, h is the film thickness, and λ and μ are the first and second Lamé constants (c_{12} and c_{44} of the stiffness matrix), which are the bulk and film shear elastic moduli, respectively. Since k_1 and k_2 both are negative in sign, Eq. (8.14) implies increasing mass loading (first term) will cause a frequency decrease, while increasing shear modulus (second term) will cause a frequency increase [93].

In other words, increasing the mechanical stiffness has the opposite effect on the frequency shift than mass loading. For example, in the case of mercury amalgamation with a gold sensitive layer, which is commonly used for sensing applications, both effects occur [120]. This is also the case for many polymer sensitive layers employed for sensing VOCs. Therefore, it is quite possible that as the vapor concentration increases, non-linear calibration curves, positive frequency shifts and unusual attenuation may be observed. This has implications for the sorption and desorption profiles obtained during SAW sensor measurements, and is further discussed by Thompson and Stone [93]. Models have also been proposed by Grate et al. [121] and Martin et al. [122] to explain the combined effect of temperature and vapor sorption on sensor response.

8.3.3.2 Acoustoelectric Perturbations

SAW devices are also well suited for monitoring electrical perturbations at the sensor surface. When a SAW propagates along the surface of a piezoelectric material, the mechanical wave is accompanied by a traveling electric field that interacts with charge carriers on or adjacent to the surface [7]. Hence, the interaction between the gas molecules and the sensitive layer perturbs the electrical boundary conditions at the surface of the structure. As a result, the velocity and attenuation of the electromechanical waves are perturbed. The change in velocity for a given SAW mode due to change in surface conductivity can be calculated by perturbation theory [123]:

$$\frac{\Delta v}{v_0} = \frac{-K^2}{2} \frac{1}{1 + \left(\frac{v_0 \varepsilon_p}{\sigma_{sh}}\right)^2}, \tag{8.15}$$

where, σ_{sh} is the sheet conductivity of the layer, ε_p is the effective permittivity of the structure (including free space, ε_0), and K^2 is the electro-mechanical coupling coefficient of the SAW mode at the device surface. K^2 can be calculated by determining the perturbed wave velocity caused by a change in the electrical boundary conditions, and can be obtained theoretically or experimentally from [1]:

$$K^2 = 2\frac{v_f - v_m}{v_f}, \tag{8.16}$$

where v_f is the phase velocity with a free surface boundary condition, and v_m is the phase velocity of a metallized surface. Typically the metallized surface is modeled with an infinitely thin, perfectly conducting layer on the surface of the device. Table 8.5 lists the K^2 calculated from the free and metallized surface for common $LiNbO_3$ and $LiTaO_3$ substrate cuts.

It is important to note that in Eq. (8.15), the acoustoelectric response is a non-linear function of σ_{sh}, and it is largest when $\sigma_{sh} = v_f \varepsilon_p$. This indicates that maximum sensitivity is achieved when the sheet conductivity of the sensitive

Table 8.5 Substrate crystal cut properties calculated for a wavelength of 24 μm

Substrate material	Crystal cut	SAW axis	Velocity (m/s) free	Velocity (m/s) metallized	f_{free} (MHz)	K^2	Polarization
$LiNbO_3$	X	Y	3696.1	3639.5	154.00	3.06%	Generalized SAW
$LiNbO_3$	X	Z	3483.2	3404.2	145.13	4.54%	Rayleigh
$LiTaO_3$	36° Y	X	4226.3	4108.8	176.10	5.56%	Quasi shear-horizontal

layer is matched to the velocity–permittivity product of the SAW mode. Therefore, in the case of a multilayered SAW structure employing an intermediate layer, the thickness of the intermediate layer can be tailored to modify the velocity–permittivity product of the SAW mode such that a match is achieved. Thus, the sensitivity of the structure can be maximized for conductivity changes at the device surface, highlighting a significant advantage of using a multi-layered SAW-based structure for gas-sensing applications [115, 124]. Typical materials employed to form the intermediate layer of a multi-layered structure include: ZnO, SiO_2, Si_3N_4, phthalocyanines, and various polymers.

Figure 8.19 shows the frequency shift arising from the acoustoelectric response as a function of the layer's sheet conductivity. The acoustoelectric interaction is maximized for electrical perturbations when the conductivity of

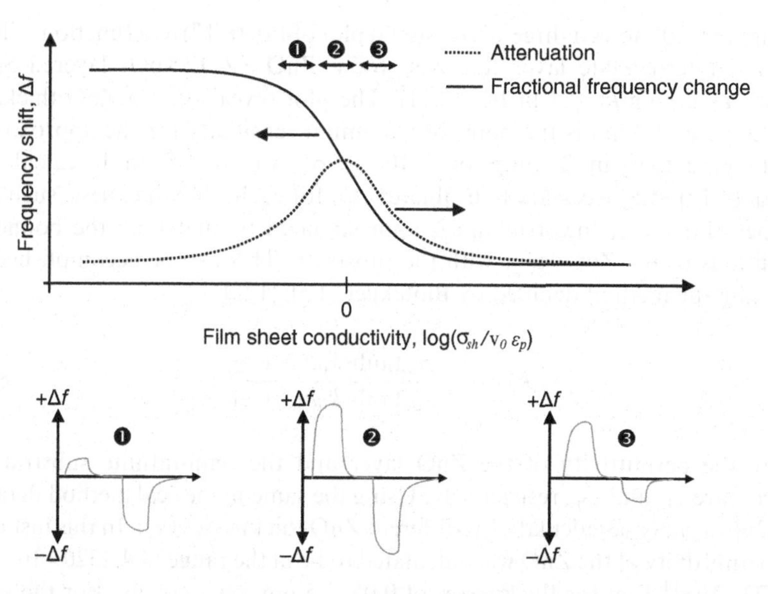

Fig. 8.19 Fractional change in frequency vs. film sheet conductivity with diagrammatic explanation of the sensor response for different operating regions (positive frequency shifts are towards reducing gases, and negative frequency shifts are towards oxidizing gases)

the sensitive layer falls within the range where the curve has a significantly large gradient (denoted by region 2).

The diagrammatic explanation in the bottom of Fig. 8.19 illustrates the assumed fractional frequency change towards reducing and oxidizing gases that arise from opposite but equal changes in sheet conductivity magnitude. At the point of maximum sensitivity (center of region 2), the frequency shifts are shown to be of opposite directions and of equal magnitude. However, operation in regions 1 and 3 is shown to produce non-proportional, opposing responses to the same reducing and oxidizing conditions. Unlike mass sensitivity, where the addition of mass on the sensor surface generally causes a decrease in frequency, sheet conductivity changes may cause either an increase or decrease in resonant frequency, depending on whether the analyte gas is oxidizing or reducing, respectively. Furthermore, Eq. (8.15) implies that for larger K^2 at the device surface, the influence of sheet conductive changes on the acoustic wave velocity is greater.

In the case of multi-layered structures, to appreciate the full effect of the intermediate layer thickness with regards to maximizing device acousto-electric response, the expression for sheet conductivity sensitivity can be determined by taking the derivative of Eq. (8.15):

$$\frac{d(\Delta v/v_0)}{d\sigma_{sh}} = \frac{-K^2 v_0^2 \varepsilon_p^2}{\left(1+(v_0^2\varepsilon_p^2/\sigma_{sh}^2)\right)^2 \sigma_{sh}^3}. \tag{8.17}$$

Figure 8.20 shows a three-dimensional plot of Eq. (8.17) as a function of both σ_{sh} and intermediate layer thickness for a ZnO/XZ $LiNbO_3$ layered SAW device, assuming $\kappa = 1$ in Eq. (8.11). The plot reveals that a ZnO thickness between 1 and 2 μm is the point of maximum sensitivity for the approximate sheet conductivity in the range of $1\times10^{-7} < \sigma_{sh} < 1\times10^{-6}$ S. In the calculation of Eq. (8.17), it is necessary to evaluate $\varepsilon_p(k)$ for each ZnO thickness, such that the permittivity of the structure is approximated by matching the boundary conditions of the ZnO layer with the substrate. This can be accomplished by adapting the method detailed by Blotekjær et al. [125]:

$$\varepsilon_p(k) = \varepsilon_{1p}\frac{\varepsilon_{1p}\tanh(k_0 h_1) + \varepsilon_{2p}}{\varepsilon_{2p}\tanh(k_0 h_1) + \varepsilon_{1p}}, \tag{8.18}$$

where the permittivity of the ZnO layer and the semi-infinite substrate of interest are ε_{1p} and ε_{2p}, respectively. Using the same numerical method detailed in [126], ε_{1p} may be calculated at different ZnO thicknesses (h_1). In this instance, the permittivity of the ZnO was calculated to be in the range of 4.2320×10^{-11} to 9.8973×10^{-11} F/m for thicknesses of 0.05–2.5 μm, respectively. For the semi-infinite XZ $LiNbO_3$ substrate, ε_{2p} is 4.1777×10^{-10} F/m. Similarly, a plot for a ZnO/36° YX $LiTaO_3$ structure can be found in [115].

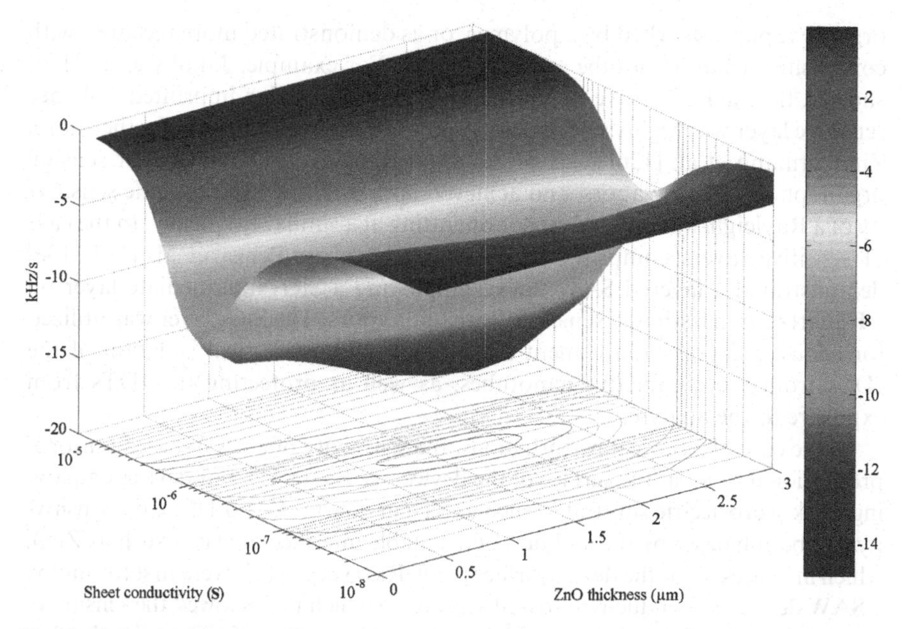

Fig. 8.20 Sensitivity of multilayered structure employing a ZnO intermediate layer on a XY $LiNbO_3$ substrate

8.3.4 SAW Device Gas Sensor Performance

The first report of a SAW device employed for gas sensing was in 1979 by Wohltjen and Dessy [127]. In their paper, they investigated the operation of a 2-port delay line SAW device utilizing a quartz and $LiNbO_3$ piezoelectric substrate. Following this work, reports emerged in literature on a variety of SAW-based sensors for monitoring gases such as: sulfur dioxide (SO_2) [16], hydrogen (H_2) [118, 128], humidity (H_2O), carbon dioxide (CO_2) [129], hydrogen sulfide (H_2S) [130], nitrogen oxide (NO) [131, 132], organophosphorous compounds [133], as well as many others. SAW-based vapor sensors have also been extensively investigated [134].

As the acoustic radiation from a SAW device into the surrounding gas media is very low, Rayleigh waves are commonly employed for gas-phase-sensing applications. Since Rayleigh waves are strongly guided by the surface, they are highly sensitive towards mass perturbations, and an additional layer does not generally improve mass sensitivity. However, the mass sensitivity of a SH-SAW or Love wave can be greater than that of a Rayleigh wave, typically due to the wave guiding properties provided by a well chosen intermediate layer. In many cases, greater confinement of the acoustic wave at the device surface is achieved for SH modes when compared to the non-layered structures supporting Rayleigh modes [6]. Thus, multi-layered SAW sensors supporting SH modes are advantageous for mass-sensing applications, such as the sensing of organic vapors absorbed by a polymer, or as demonstrated more recently, with composite carbon nanotube sensitive layers. For example, Jakoby et al. [135] showed that a Love wave device employing a molecular imprinted polymer sensitive layer was highly suitable for vapor-sensing applications. Furthermore, Zimmermann et al. [136] successfully applied a Love wave device for sensing organophosphorous vapors, and achieved approximately 10 times the sensitivity of a Rayleigh-wave-based sensor operating at a similar frequency. In the case of sensitive layers comprised of carbon nanotubes, Penza et al. [137, 138] demonstrated a layered SAW device employing a SiO_2 intermediate layer on ST quartz for mass-based sensing of organic vapors. The SiO_2 layer was utilized for acoustic energy confinement and for providing electrical isolation of the IDTs from the conductive nanotubes, as well as protecting the IDTs from exposure to the analyte.

In the case of electrical perturbations, the sensitivity of a SAW-based sensor is proportional to K^2 at the surface of the device, see Eq. (8.15). For devices employing weak piezoelectric substrates, such as ST-cut quartz ($K^2 = 0.11\%$), the sensitivity can be enhanced by the addition of a suitable intermediate layer such as ZnO, which increases K^2 at the device surface [139]. Ricco et al. [123] were first to employ a SAW device for conductivity-based sensing, in which they showed the sensitivity towards NO_2 was 1000 times greater, in terms of the number of NO_2 molecules that can be sensed by a similar SAW sensor functioning via mass loading. Later, Niewenhuizen et al. [140] developed a multi-layered device employing a Si_3N_4/ZnO/SiO_2/Si structure with a copper phthalocyanine sensitive layer for NO_2 sensing, where the Si_3N_4 acts as a passivation layer. Jakubik et al. [107, 108, 109] introduced a multi-layered structure targeting H_2, using metal free phthalocyanine and copper phthalocyanine intermediate layers with a palladium sensitive layer. The multi-layered structure was employed in an attempt to increase the sensitivity towards the analyte gas, by matching the operating region of the SAW mode to the conductivity of the palladium sensitive layer.

Other work carried out by Caron et al. employed a 261 MHz SAW delay line device based on a 24° rotated Y-cut quartz substrate with a rf sputtered ruthenium-doped tungsten trioxide sensitive layer for nitric oxide sensing [141]. The results were obtained at different relative humidity levels, and included a frequency shift of almost 50 kHz towards 3 ppm with no discernable interference from 100 ppm of NH_3 at an operating temperature of 250°C. Similarly, Penza investigated the sensitivity of a 128°XY $LiNbO_3$ SAW device with a WO_3 sensing layer towards NO and NO_2 in the presence of interfering gases such as CO and CH_4 [131].

Table 8.6 provides a summary of the performance of several other SAW-based gas and vapor sensors towards various analytes. Comprehensive surveys detailing the developments of SAW sensors have been compiled by D'Amico et al. [14] in 1989, Ballantine and Wohltjen [2] in 1989 and Grate et al. [16,17] in 1993. Since then, several other reviews of acoustic wave gas sensor technology have been completed by Martin et al. [142] in 1996, Vellekoop [143] in 1998, Cheeke and Wang [23] in 1999, and Drafts [144] in 2001.

Table 8.6 Responses of different SAW sensors towards different gases

Sensitive layer	Structure	Analyte gas	Operating frequency (MHz)	Frequency shift (Hz)	Ref
Conventional SAW-based gas sensors					
WO_3:Ru	24° rotated Y-cut quartz	3 ppm of NO	261	50,000	[141]
doped WO_3	YZ $LiNbO_3$	30 ppm of H_2S	60	10,000	[145]
WO_3:Au	YZ $LiNbO_3$	100 ppm H_2S	38.15	5,000	[130]
Metal-phthalocyanine	-	1.4 ppm of NO_2	600	8*	[132]
lead phthalocyanine	YZ $LiNbO_3$	10 ppm of NO_2	110	154*	[123]
catalysed lead phthalocyanine	YZ $LiNbO_3$	10 ppm of NO_2	110	330*	[123]
tetra-4-tert butyl silicon phthalocyanine dichloride	ST quartz	100 ppm of NO_2	98.6	420	[146]
lead phthalocyanine	YZ $LiNbO_3$	3.1 ppm of NO_2	~43	9,500*	[147]
Au	28° rotated Y-cut quartz	500 ppb of Hg	261	4,000	[148]
WO_3	27° quartz	10 ppm of H_2S	260	4,000*	[149]
WO_3	YZ $LiNbO_3$	27 ppm of H_2S	60	9,900*	[145]
Pt/WO_3	langasite	1000 ppm of H_2S	167.14	6,000*	[150]
Multilayered SAW-based gas sensors					
TiO_2	SiO_2/ ST quartz	10,000 ppm and 100 ppm of O_2	63.04	45,000 3,000	[151]
ZnO	ZnO/90° rotated ST quartz	50 ppm of O_2	90	11,000	[152]
ZnO	ZnO/90° rotated ST quartz	100 ppm of C_2H_8	90	11,000	[152]
palladium	copper phthalocyanine/ YZ $LiNbO_3$	2% H_2	-	1,000*	[109]
palladium	phthalocyanine/ YZ $LiNbO_3$	2% H_2	-	5,000*	[108]
copper phthalocyanine	SiO_2/90° rotated ST quartz	1 ppm of NO_2	100	1,400	[114]
WO_3	ZnO/36° YX $LiTaO_3$	500 ppm of ethanol	167.75	119,000	[115]
WO_3	ZnO/36° YX $LiTaO_3$	1% H_2	164.07	24,700	[111]
Au-WO_3	ZnO/36° YX $LiTaO_3$	1% H_2	165.7	755,000	[110]
SWCNTs	SiO_2/ST-cut X-propagating quartz	50 ppm of Isopropanol	433	8,000	[153]

* Interpreted or calculated from information provided

Experimental studies on Love mode devices based on SiO_2/ST-cut quartz have also investigated the mass sensitivity, velocity, insertion loss, oscillation frequency stability and temperature coefficient as a function of SiO_2 layer thickness [90]. Du and Harding [154] also studied multi-layered structures employing PMMA/SiO_2/quartz. The structures were observed to exhibit higher mass sensitivity when compared to devices having only the SiO_2 or PMMA layers. Harding [155] further demonstrated that the acoustic properties of the SiO_2 layer can be modified by introducing CF_4 gas during the deposition process. This was shown to reduce the propagation velocity in the SiO_2 guiding layer, resulting in increased mass sensitivity. Although the application of this work was for liquid-phase analysis, it clearly demonstrates that there are significant sensitivity advantages of multi-layered SAW structures when compared to their non-layered counterparts. Recently, Chang et al. [156] completed a study of Love modes in ZnO/quartz and ZnO/$LiTaO_3$ structures with the preferred orientation of doped ZnO films on the 36° YX $LiTaO_3$ substrates [157] for application in both gas and liquid media.

Sauberlich et al. [158] proposed a SAW-based mercury sensor, taking advantage of the high sensitivity of SAWs towards mechanical perturbations. Following this, mercury vapor sensors based on dual delay line SAW structures coated with a gold sensitive layer were investigated by Caron et al. [148, 159]. In this work, the low concentration sensing of gaseous mercury in the range of 100 to 500 ppb was achieved. Responses towards mercury at room temperature were accumulative, which required periodic regeneration of the gold sensitive layer in-between successive measurements. However, at an elevated operating temperature of 200°C the response was found to become a direct measure of the instantaneous concentration. Similarly, Haskell et al. [120] investigated the effect of Au sensitive layer thickness on sensitivity for mercury concentrations between 20 and 100 ppb. The largest response magnitude was obtained for a gold film thickness of 75 Å. It was modeled and demonstrated experimentally that the net contribution is a result of both mass loading and stiffening effects of mercury vapor adsorbed/amalgamated on to gold sensitive layer.

Ricco et al. [160] demonstrated the use of SAW chemical sensor arrays with sensitive layers comprising of self-assembled mono layers of organic materials and plasma processed films, combined with novel cluster analysis for highly selective monitoring of VOCs. Lu and Zellers [161] utilized a 520 MHz ST-cut quartz-based SAW device with a sensitive layer of poly(dimethylsiloxane) having a hexafluorobisphenol A-molecules incorporated along the backbone. Using chromographic separation by employing a pre-concentrator, the sensing of low-ppb analyte concentrations in sample mixtures containing up to 20 different VOCs was conducted.

In the case of layered structures employing strong piezoelectric substrates such as XZ $LiNbO_3$ ($K^2 = 4.54\%$), the addition of a piezoelectric intermediate layer may not necessarily increase the K^2 at the device surface, which is an important quantity for measuring electrical perturbations. For example, the use of a ZnO intermediate layer will reduce the electromechanical coupling at the device surface. For a device with an electrode period of 24 μm and a 1.2-μm-thick ZnO layer, K^2 at the device surface will be reduced from 4.54 to 2.81% [162]. Despite this decrease, the sensitivity towards electrical perturbations is increased as a result of matching the velocity–permittivity product of the SAW mode to the sheet conductivity, as shown in Fig. 8.20. For instance, a ZnO/XZ $LiNbO_3$ SAW device employing a 40 nm thick InO_x sensitive layer exhibits a negative frequency shift towards H_2, as shown in the Fig. 8.21. The frequency shifts toward concentrations of 0.06–1.00% are in the range of 78.5–319.4 kHz, at an operating temperature of 246°C. Similarly, the same sensor shows a positive frequency shift towards NO_2. A frequency shift in excess of 91 kHz is shown for 2.125 ppm of NO_2, with response magnitude saturation occurring at this concentration [112].

The use of catalysts and dopants can also significantly improve the sensitivity and selectivity of SAW-based gas sensors. For example, conductive polyaniline nanofibers as gas-sensitive layers has been synthesized with a template-free rapidly mixed polymerization of aniline using HCl and CSA as the dopant acids [163]. In this work, room temperature responses towards 1% of H_2 were 3 and 14.6 kHz for HCl and CSA doped polyaniline nanofiber sensors, respectively. Additionally, response magnitudes can be as much as 50 times larger for catalytically activated sensitive layers. For example, SAW delay line structures based on ZnO/36° YX $LiTaO_3$ consisting of a WO_3-sensitive layer activated with an Au catalyst resulted in a frequency shift of approximately 755 kHz towards 1% H_2 in air [110]. By contrast, a similar device consisting of only the WO_3 sensitive layer exhibits a frequency shift 25.8 kHz towards the same H_2 concentration [111].

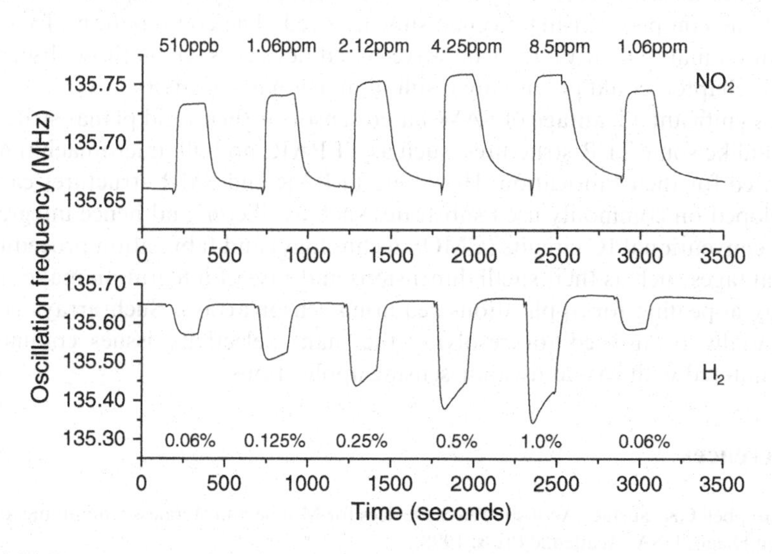

Fig. 8.21 Response towards H_2 and NO_2 for a InO_x/ZnO/XZ $LiNbO_3$ SAW sensor at an operating temperature of 246°C [112]

8.4 Concluding Remarks

To date, acoustic-wave-based sensors have been developed for numerous gas and vapor analytes including: CO_x, H_2, H_2O, NO_x, VOCs, hydrocarbons, and many others. Current research efforts focus on optimizing the transducers structure as well as developing gas sensitive layers.

Acoustic-wave-based sensors have been, and are still, strongly influenced by advances in devices developed for signal processing and communications applications. Both QCM- and SAW-based sensors are quite mature, thus only incremental changes are likely to occur in these areas. When considering new developments that may have a significant impact in the field of acoustic-wave-based sensors, it is most likely that TFRs will provide the greatest impetus for significant new advances. TFRs are still relatively new compared to QCM and SAW devices, and offer much higher operational frequency that extends well into the GHz range. However, despite the lower sensitivity offered by QCMs when compared to TFR or SAW-based sensors, their robust nature, availability and affordable electronics makes them highly viable sensing platforms [3].

Another major area of development in acoustic-wave-based gas and vapor sensing is the fabrication and synthesis of novel sensitive layers. The rapidly advancing research in nanotechnology has a far-reaching impact on gas and vapor-sensing technology. It is likely that new functionalized materials with high surface area to volume ratio will be discovered and exploited in future acoustic-wave-based sensors. For example, recently developed carbon nanowires, mesoporous single crystal nanowires and nanobelts appear very promising. However, further research investigating the interactions between nanomaterial sensitive layers and the gas or vapor analyte is required to understand the complex sensing mechanisms involved. The combination of nanostructured materials with acoustic-wave-based devices, such as those discussed in this chapter, could potentially result in higher gas sensitivity.

A significant advantage of SAW-based sensors is their rigid planar structure and unlike some TFR structures, such as TFBAR, no bulk micromachining is required for their fabrication. However, TFBAR and SMR structures can be developed on commonly used substrates such as silicon, and hence integrated with conventional IC circuits (CMOS technology) and fabrication procedures. Advantages such as their small dimensions and ease of integration make them highly appealing for applications requiring sensor arrays. Such arrays could potentially be utilized for resolving the many selectivity issues commonly encountered with gas- and vapor-sensing applications.

References

1. Campbell CK. Surface Acoustic Wave Devices for Mobile and Wireless Communications. San Diego, USA: Academic Press; 1998.
2. Ballantine DS, Wohltjen H. Surface acoustic wave devices for chemical analysis. Anal Chem. 1989;61:704–715.

3. Janshoff A, Galla H-J and Steinem C. Piezoelectric mass-sensing devices as biosensors – an alternative to optical biosensors? Angew Chem Int Edit. 2000;39:4004–4032.
4. Ballantine DS, White RM, Martin SJ, Ricco AJ, Zellers ET, Frye G C et al. Acoustic Wave Sensors: Theory, Design, & Physico-Chemical Applications (Applications of Modern Acoustics). San Diego, USA: Academic Press; 1996.
5. Wohltjen H, Dessy R. Surface acoustic-wave probe for chemical-analysis .1. Introduction and Instrument Description. Anal Chem. 1979;51:1458–1464.
6. Powell DA, Kalantar-zadeh K, Wlodarski W, Ippolito SJ. Layered surface acoustic wave chemical and Bio-Sensors. In: CA Grimes, EC Dickey, MV Pishko, editors. Encyclopedia of Sensors. California, USA: American Scientific Publishers, Stevenson Ranch; 2006.
7. Ricco AJ, Martin SJ. Thin metal-film characterization and chemical sensors: Monitoring electronic conductivity, mass loading and mechanical properties with surface acoustic wave devices. Thin Solid Films. 1991;206:94–101.
8. Auld BA. Acoustic Fields and Waves in Solids. Malabar, Florida, USA: Krieger Publishing Company; 1990.
9. Morgan DP. Surface-Wave Devices for Signal Processing. Amsterdam, The Netherlands: Elsevier; 1991.
10. Cheeke JDN. Fundamentals and Applications of Ultrasonics Waves. Boca Raton, USA: CRC Press; 2001.
11. Ballantine DS, White RM, Martin SJ, Ricco AJ, Frye GC, Zellers E T et al. Acoustic Wave Sensors: theory, design and Physico-Chemical Applications. New York, USA: Academic Press; 1997.
12. Ward MD, Buttry DA. Insitu interfacial mass detection with piezoelectric transducers. Science. 1990;249:1000–1007.
13. Di Natale C, Davide F A M and D'Amico A. A self-organizing system for pattern classification: Time varying statistics and sensor drift effects. Sens Actuators B. 1995;27:237–241.
14. D'Amico A and Verona E. SAW Sensors. Sens Actuators. 1989;17:55–66.
15. Grate JW, Rosepehrsson SL, Venezky DL, Klusty M, Wohltjen H. Smart sensor system for trace organophosphorus and organosulfur vapor detection employing a temperature-controlled array of surface-acoustic-wave sensors, automated sample preconcentration, and pattern-recognition. Anal Chem. 1993;65:1868–1881.
16. Grate JW, Martin SJ, White RM. Acoustic-wave microsensors. Anal Chem. 1993a;65: A940–A948.
17. Grate JW, Martin SJ, White RM. Acoustic-wave microsensors. Anal Chem. 1993b; 65:A987–A996
18. Fernandez MJ, Fontecha JL, Sayago I, Aleixandre M, Lozano J, Gutierrez J et al. Discrimination of volatile ccompounds through an electronic nose based on ZnO SAW sensors. Sens Actuators B. 2007;127:277–283.
19. Joo BS, Huh JS, Lee DD. Fabrication of polymer SAW sensor array to classify chemical warfare agents. Sens Actuators B. 2007;121:47–53.
20. Mistura G, Lee HC, Chan MHW. Hydrogen adsorption on alkali-metal surfaces – wetting, prewetting and triple-point wetting. J Low Temp Phys. 1994;96:221–244.
21. Sauerbrey GZ. The use of quartz oscillators for weighing thin layers and for microweighing. Z Physik. 1959;155:206–222.
22. Buck RP, Lindner E, Kutner W, Inzelt G. Piezoelectric chemical sensors. Pure Appl Chem. 2004;76:1139–1160.
23. Cheeke JDN, Wang Z. Acoustic wave gas sensors. Sens Actuators B. 1999;59:146–153.
24. Wenzel SW, White RM. Analytic comparison of the sensitivities of bulk-wave, surface-wave, and flexural plate-wave ultrasonic gravimetric sensors. Appl Phys Lett. 1989;54:1976–1978.
25. Lakin KM, Wang JS. Acoustic bulk wave composite resonators. Appl Phys Lett. 1981;38: 125–127.
26. King WH. Piezoelectric sorption detector. Anal Chem. 1964;36:1735.

27. Krim J, Dash JG. Suzanne J. Triple-point wetting of light molecular gases on Au(111) surfaces. Phys Rev Lett. 1984;52:640–643.
28. Mistura G, Lee HC, Chan MHW. Quartz microbalance study of hydrogen and helium adsorbed on a rubidium surface. Physica B. 1994;194:661–662.
29. Lu C, Cazanderna AW. Applications of piezoelectric quartz crystal microbalances. Amsterdam, The Netherlands: North-Holland; 1984.
30. Henkel K, Oprea A, Paloumpa I, Appel G, Schmeisser D, Kamieth P. Selective polypyrrole electrodes for quartz microbalances: NO_2 and gas flux sensitivities. Sens Actuators B. 2001;76:124–129.
31. Nanto H, Dougami N, Mukai T, Habara M, Kusano E, Kinbara A et al. A smart gas sensor using Polymer-Film-Coated quartz resonator microbalance. Sens Actuators B. 2000;66:16–18.
32. Nanto H, Tsubakino S, Habara M, Kondo K, Morita T, Douguchi Y et al. A novel chemical sensor using $CH_3Si(OCH_3)_3$ sol-gel thin film coated quartz-resonator microbalance. Sens Actuators B. 1996;34:312–316.
33. Brousseau LC, Mallouk TE. Molecular design of intercalation based sensors. 1. Ammonia sensing with quartz crystal microbalances modified by copper biphenylbis(Phosphonate) thin films. Anal Chem. 1997;69:679–687.
34. Dickert FL, Baumler UPA, Stathopulos H. Mass-sensitive solvent vapor detection with Calix 4 resorcinarenes: tuning sensitivity and predicting sensor effects. Anal Chem. 1997;69:1000–1005.
35. Zhou R, Josse F, Gopel W, Ozturk ZZ, Bekaroglu O. Phthalocyanines as sensitive materials for chemical sensors. Appl Organomet Chem. 1996;10:557–577.
36. Okahata Y, Matsuura K, Ito K, Ebara Y. Gas-phase selective adsorption on functional monolayers immobilized on a highly sensitive quartz-crystal microbalance. Langmuir. 1996;12:1023–1026.
37. Schroder J, Borngraber R, Eichelbaum F, Hauptmann P. Advanced interface electronics and methods for QCM. Sens Actuators A. 2002;97–8:543–547.
38. Bruschi L, Delfitto G, Mistura G. Inexpensive but accurate drving circuits for quartz crystal microbalances. Rev Sci Instr. 1999;70:153–157.
39. Nakamura K, Nakamoto T and Moriizumi T. Classification and evaluation of sensing films for QCM odor sensors by steady-state sensor response measurement. Sens Actuators B. 2000;69:295–301.
40. Islam A, Ismail Z, Ahmad MN, Saad B, Othman AR, Shakaff AYM et al. Transient parameters of a coated quartz-crystal microbalance sensor for the detection of Volatile Organic Compounds (VOCs). Sens Actuators B. 2005;109:238–243.
41. Rocha-Santos TAP, Gomes M, Duarte AC, Oliveira J. A quartz crystal microbalance sensor for the determination of nitroaromatics in landfill gas. Talanta. 2000;51: 1149–1153.
42. Zhang J, Hu J, Zhu ZQ, Gong H, O'Shea SJ. Quartz crystal microbalance coated with sol-gel-derived indium-tin oxide thin films as gas sensor for NO detection. Colloid Surf A. 2004;236:23–30.
43. Sartore L, Penco M, Della Sciucca S, Borsarini G, Ferrari V. New carbon black composite vapor detectors based on multifunctional polymers. Sens Actuators B. 2005;111:160–165.
44. Kim SR, Choi SA, Kim JD, Kim KJ, Lee C, Rhee SB. Preparation of polythiophene LB films and their gas sensitivities by the suartz-crystal microbalance. Synth Met. 1995;71:2027–2028.
45. Matsuura K, Ebara Y, Okahata Y. Guest selective adsorption from the gas phase onto a functional self-assembled monolayer immobilized on a super-sensitive quartz-crystal microbalance. Thin Solid Films. 1996;273:61–65.
46. Alexander C, Andersson HS, Andersson LI, Ansell RJ, Kirsch N, Nicholls IA et al. Molecular imprinting science and technology: A survey of the literature for the years up to and including 2003. J Mol Recognit. 2006;19:106–180.
47. Hirayama K, Sakai Y, Kameoka K, Noda K, Naganawa R. Preparation of a sensor device with specific recognition sites for acetaldehyde by molecular imprinting technique. Sens Actuators B. 2002;86:20–25.

48. Feng L, Liu YJ, Zhou XD, Hu JM. The fabrication and characterization of a formaldehyde odor sensor using molecularly imprinted polymers. J Colloid Interface Sci. 2005;284:378–382.
49. Ferrari M, Ferrari V, Marioli D, Taroni A, Suman M, Dalcanale E. Cavitand-coated PZT resonant piezo-layer sensors: Properties, structure, and comparison with QCM sensors at different temperatures under exposure to organic vapors. Sens Actuators B. 2004;103:240–246.
50. Schierbaum KD, Weiss T, Vanvelzen EUT, Engbersen JFJ, Reinhoudt DN, Gopel W. Molecular recognition by self-assembled monolayers of cavitand receptors. Science. 1994;265:1413–1415.
51. Hartmann J, Hauptmann P, Levi S, Dalcanale E. Chemical sensing with cavitands: Influence of cavity shape and dimensions on the detection of solvent vapors. Sens Actuators B. 1996;35:154–157.
52. Dalcanale E, Hartmann J. Selective detection of organic-compounds by means of cavitand-coated QCM transducers. Sens Actuators B. 1995;24:39–42.
53. Osada M, Sasaki I, Nishioka M, Sadakata M, Okubo T. Synthesis of a faujasite thin layer and its application for SO_2 sensing at elevated temperatures. Microporous Mesoporous Mat. 1998;23:287–294.
54. Sasaki I, Tsuchiya H, Nishioka M, Sadakata M, Okubo T. Gas sensing with zeolite-coated quartz crystal microbalances-principal component analysis approach. Sens Actuators B. 2002;86:26–33.
55. Yan YG, Bein T. Molecular recognition on acoustic-wave devices – sorption in chemically anchored zeolite monolayers. J Phys Chem. 1992;96:9387–9393.
56. Bein T, Brown K, Frye GC, Brinker CJ. Molecular-sieve sensors for selective detection at the nanogram level. J Am Chem Soc. 1989;111:7640–7641.
57. Ding B, Kim JH, Miyazaki Y, Shiratori SM. Electrospun nanofibrous membranes coated quartz crystal microbalance as gas sensor for NH_3 detection. Sens Actuators B. 2004;101:373–380.
58. Zhang YS, Yu K, Xu RL, Jiang DS, Luo LQ, Zhu ZQ. Quartz crystal microbalance coated with carbon nanotube films used as humidity sensor. Sens Actuators A. 2005;120:142–146.
59. Consales M, Campopiano S, Cutolo A, Penza M, Aversa P, Cassano G et al. Carbon nanotubes thin films fiber optic and acoustic VOCs sensors: Performances analysis. Sens Actuators B. 2006;118:232–242.
60. Consales M, Campopiano S, Cutolo A, Penza M, Aversa P, Cassano G et al. Sensing properties of buffered and not buffered carbon nanotubes by fibre optic and acoustic sensors. Meas Sci Technol. 2006;17:1220–1228.
61. Penza M, Cassano G, Aversa P, Antolini F, Cusano A, Consales M et al. Carbon nanotubes-coated multi-transducing sensors for VOCs detection. Sens Actuators B. 2005;111:171–180.
62. Penza M, Cassano G, Aversa P, Antolini F, Cusano A, Cutolo A et al. Alcohol detection using carbon nanotubes acoustic and optical sensors. Appl Phys Lett. 2004;85:2379–2381.
63. Penza M, Cassano G, Aversa P, Cusano A, Consales M, Giordano M et al. Acoustic and optical VOCs sensors incorporating carbon nanotubes. IEEE Sens J. 2006;6:867–875.
64. Scheide EP, Taylor JK. Piezoelectric crystal dosimeter for monitoring mercury-vapor in industrial atmospheres. Am Ind Hyg Assoc J. 1975;36:897–901.
65. Schroeder WH, Munthe J. Atmospheric mercury – an overview. Atmos Environ. 1998;32:809–822.
66. Dhawan D, Bhargava S, Tardio J, Wlodarski W. Kalantar-zadeh K. Gold coated nanostructured molybdenum oxide mercury vapour quartz crystal microbalance sensor. Sens Lett. 2008;6(1):231–236.
67. Nakamoto T, Moriizumi T. Artificial olfactory system using neural network. In: H Yamazaki, editor. Handbook of Sensors and Actuators. Amsterdam, The Netherlands: Elsevier;1996.

68. Shen F, Lee KH, O'Shea SJ, Lu P, Ng TY. Frequency interference between two quartz crystal microbalances. IEEE Sens J. 2003;3:274–281.
69. Abe T, Esashi M. One-chip multichannel Quartz Crystal Microbalance (QCM) fabricated by deep RIE. Sens Actuators A. 2000;82:139–143.
70. Worsch PM, Koppelhuber-Bitschnau B, Mautner FA, Krempl PW, Wallnofer W, Doppler P et al. High temperature X-Ray powder diffraction study on the phase transition between the alpha-quartz and the beta-cristobalite-like phase of $GaPO_4$. Mater Sci Forum. 2000;321–323:914–917.
71. Pedarnig JD, Peruzzi M, Salhofer H, Schwodiauer R, Reichl W, Runck J. F_2-laser patterning of $GaPO_4$ resonators for humidity sensing. Appl Phys A. 2005;80:1401–1404.
72. Lakin KM. Thin film resonators and filters. IEEE Ultrasonics Symposium. 1999;895–906.
73. Lakin KM. A review of thin-film resonator technology. IEEE Microw Mag. 2003; 4:61–67.
74. Lakin KM. Thin film resonator technology. IEEE Trans Ultrason Ferr. 2005;52:707–716.
75. Penza M, Cassano G, Aversa P, Suriano D, Verona E, Benetti M et al. Thin film bulk acoustic resonator vapor sensors with single-walled carbon nanotubes-based nanocomposite layer. Proceedings of the 2007 IEEE Sensors Conference. 2007.
76. Newell WE. Face-mounted piezoelectric resonators. Proc IEEE. 1965;53:575–581.
77. Lostis P. The study, production and control of thin films giving a chosen path difference between perpendicularly polarized components. Rev Opt Theor Instr. 1959;38:1.
78. Benetti M, Cannata D, Di Pietrantonio F, Foglietti V, Verona E. Microbalance chemical sensor based on thin-film bulk acoustic wave resonators. Appl Phys Lett. 2005;87:173504.
79. Gizeli E. Acoustic transducers. In: E. Gizeli, C.R. Lowe editors. Biomolecular Sensors. London, UK: Taylor & Francis; 2002.
80. Rey-Mermet S, Lanz R, Muralt P. Bulk acoustic wave resonator operating at 8 GHz for gravimetric sensing of organic films. Sens Actuators B. 2006;114:681–686.
81. Zhang H, Kim ES, Micromachined acoustic resonant mass sensor. J Microelectromech Syst. 2005;14:699–706.
82. O'Toole RP, Burns SG, Bastiaans GJ, Porter MD. Thin aluminum nitride film resonators – miniaturized high-sensitivity mass sensors. Anal Chem. 1992;64:1289–1294.
83. Gabl R, Green E, Schreiter M, Feucht HD, Zeininger H, Primig R et al. Novel integrated FBAR sensors: A universal technology platform for Bio- and Gas-Detection. Proceedings of the 2003 IEEE Sensors Conference. 2003;1184–1188.
84. Reichl W, Runck J, Schreiter M, Greert E, Gabl R. Novel gas sensors based on thin film bulk acoustic resonators. Proceedings of the 2004 IEEE Sensors Conference. 2004;1504–1505.
85. Benetti M, Cannata D, D'Amico A, Di Pietrantonio F, Foglietti V, Verona E. Thin Film Bulk Acoustic Wave Resonator (TFBAR) gas sensor. Proceeding of the IEEE Ultrasonics Symposium. 2004;1581–1584;
86. Lee H-M, Kim H-T, Choi H-K, Lee H-C, Hong H-K, Lee D-H et al. A highly-sensitive differential-mode microchemical sensor using TFBARs with on-chip microheater for Volatile Organic Compound (VOC) detection. Proceeding on the IEEE International Conference on Micro Electro Mechanical Systems, Istanbul, Turkey; 2006. pp. 490–493.
87. Zhang H, Kim ES. Vapor and liquid mass sensing by micromachined acoustic resonator. Proceeding of the 16th Annual IEEE International Conference on Micro Electro Mechanical Systems – MEMS-03, Kyoto, Japan; 2003. pp. 470–473.
88. White RM, Voltmer FW. Direct piezoelectric coupling to surface elastic waves. Appl Phys Lett. 1965;12:314.
89. Harding GL, Du J. Design and properties of quartz-based love wave acoustic sensors incorporating silicon dioxide and PMMA guiding layers. Smart Mater Struct. 1997;6:716–720.

90. Du J, Harding GL, Ogilvy JA, Dencher PR, Lake M. A study of love-wave acoustic sensors. Sens Actuators A. 1996;56:211–219.
91. Kovacs G, Vellekoop MJ, Haueis R, Lubking GW, Venema A. A love wave sensor for (Bio)Chemical Sensing in Liquids. Sens Actuator A-Phys. 1994;43:38–43.
92. Morgan DP. A history of surface acoustic wave devices. In: CCW Ruppel, TA Fjeldly, editors. Advances in surface acoustic wave technology, systems and applications. Singapore: World Scientific Publishing; 2000.
93. Thompson M, Stone DC. Surface-launched acoustic wave sensors. chemical sensing and thin-film characterization. In: JDWinefordner, editor. Chemical Analysis: A Series of Monographs on Analytical Chemistry and its Applications, New York, USA: John Wiley & Sons; 1997.
94. Adler EL. SAW and pseudo-SAW properties using matrix-methods. IEEE Trans Ultrason Ferr. 1994;41:876–882.
95. Powell DA, Kalantar-zadeh K, Wlodarski W. Numerical calculation of SAW sensitivity: Application to $ZnO/LiTaO_3$ transducers. Sens Actuators A. 2004;115:456–461.
96. Powell DA. PhD Thesis: Modelling of Layered Surface Acoustic Wave Resonators for Liquid Media Sensing Applications, Melbourne, Australia: RMIT University;2006.
97. Ippolito SJ, Kalantar-Zadeh K, Powell DA, Wlodarski W. A 3-Dimensional finite element approach for simulating acoustic wave propagation in layered SAW devices. Proceedings of the IEEE Symposium on Ultrasonics, 2003;303–306.
98. Caron JJ, Andle JC, Vetelino JF. Surface acoustic wave substrates for high temperature applications. Proceedings of the IEEE Frequency Control Symposium. 1996;222–227.
99. Caron JJ, Andle JC, Vetelino JF. Surface acoustic wave substrates for gas sensing applications. Proceedings of the IEEE Ultrasonics Symposium, 1995;461–466.
100. Slobodnik Jr AJ. Materials and Their Influence on Performance. In: AA Oliner, editor. Acoustic Surface Waves, Berlin, Germany: Springer-Verlag; 1978.
101. Slobodnik Jr AJ, Conway ED, Delmonico RT. Volume 1A. Surface wave velocities. In: Microwave Acoustic Handbook, National Technical Information Service, US Dept of Commerce. 1973.
102. Hickernell FS. Thin films for SAW devices. In: CCW Ruppel, TA Fjeldly, editors. Advances in Surface Acoustic Wave Technology, Systems and Applications, Singapore: World Scientific Publishing; 2000.
103. Galipeau DW, Story PR, Vetelino KA, Mileham RD. Surface acoustic wave microsensors and applications. Smart Mater Struct. 1997;6:658–667.
104. Rugemer A, Reiss S, Geyer A, von Schickfus M, Hunklinger S. Surface acoustic wave NO_2 sensing using attenuation as the measured quantity. Sens Actuators B. 1999;56:45–49.
105. Becker H, vonSchickfus M. Hunklinger S. A new sensor principle based on the reflection of surface acoustic waves. Sens Actuators A. 1996;54:618–621.
106. Martin SJ, Ricco AJ. Effective utilization of acoustic wave sensor responses: Simultaneous measurement of velocity and attenuation. Proceedings of the IEEE Ultrasonics Symposium, 1989;621–625.
107. Jakubik WP, Urbanczyk MW, Bodzenta J, Pietrzyk MA. Investigations on the resistance of the metal-free phthalocyanine and palladium bilayer sensor structure influenced by hydrogen. Sens Actuators B. 2005;105:340–345.
108. Jakubik WP, Urbanczyk MW, Kochowski S, Bodzenta J. Palladium and phthalocyanine bilayer films for hydrogen detection in a surface acoustic wave sensor system. Sens Actuators B. 2003;96:321–328.
109. Jakubik WP, Urbanczyk MW, Kochowski S, Bodzenta J. Bilayer structure for hydrogen detection in a surface acoustic wave sensor system. Sens Actuators B. 2002;82:265–271.
110. Ippolito SJ, Kandasamy S, Kalantar-Zadeh K, Wlodarski W. Layered SAW hydrogen sensor with modified tungsten trioxide selective layer. Sens Actuators B. 2005;108: 553–557.

111. Ippolito SJ, Kandasamy S, Kalantar-Zadeh K, Trinchi A, Wlodarski W. A layered surface acoustic wave ZnO/$LiTaO_3$ structure with a WO_3 selective layer for hydrogen sensing. Sens Lett. 2003;1:33–36.
112. Ippolito SJ, Kandasamy S, Kalantar-Zadeh K, Wlodarski W, Galatsis K, Kiriakidis G et al. Highly sensitive layered ZnO/$LiNbO_3$ SAW device with InO_x selective layer for NO_2 and H_2 gas sensing. Sens Actuators B. 2005;111:207–212.
113. Ippolito SJ, Kalantar-Zadeh K, Trinchi A, Wlodarski W, Tobar M. Layered SAW nitrogen dioxide sensor based on a ZnO/36°YX $LiTaO_3$ structure with WO_3 selective layer. Proceedings of the 2003 IEEE International Frequency Control Sympposium. 2003;931–934.
114. Ippolito SJ, Kalantar-Zadeh K, Wlodarski W, Galatsis K, Fischer WJ, Berger O et al. A layered SAW based NO_2 sensor with a copper phthalocyanine selective layer. Proceedings of the Conference on Optoelectronic and Microelectronic Materials and Devices. 2002;165–168.
115. Ippolito SJ, Ponzoni A, Kalantar-zadeh K, Wlodarski W, Comini E, Faglia G et al. Ethanol sensor based on layered WO_3/ZnO/36°YX $LiTaO_3$ SAW devices. Proceedings of TRANSDUCERS '05. 2005;1915–1918.
116. Benes E, Groschl M, Seifert F and Pohl A. Comparison between BAW and SAW sensor principles. IEEE Trans Ultrason. 1998;45:1314–1330.
117. Wang Z, Cheeke JDN, Jen CK. Perturbation method for analyzing mass sensitivity of planar multilayer acoustic sensors. Ultrasonics, Ferroelectrics and Frequency Control, IEEE Transactions on 1996;43:844–851.
118. D'Amico A, Palma A, Verona E. Surface acoustic-wave hydrogen sensor. Sens Actuators. 1982;3:31–39.
119. Wohltjen H, Ballantine DS, Jarvis NL. Vapor detection with surface acoustic-wave microsensors. ACS Symposium Series. 1989;403:157–175.
120. Haskell RLB, Caron JJ, Duptisea MA, Ouellette JJ, Vetelino JF. Effects of film thickness on sensitivity of SAW mercury sensors. Proceedings of the IEEE Ultrasonics Symposium. 1999;429–434.
121. Grate JW, Klusty M, McGill RA, Abraham MH, Whiting G, Andonianhaftvan J. The predominant role of swelling-induced modulus changes of the sorbent phase in determining the responses of polymer-coated surface acoustic-wave vapor sensors. Anal Chem. 1992;64:610–624.
122. Martin SJ, Frye GC, Senturia SD. Dynamics and response of polymer-coated surface-acoustic-wave devices – effect of viscoelastic properties and film resonance. Anal Chem. 1994;66:2201–2219.
123. Ricco AJ, Martin SJ, Zipperian TE. Surface acoustic-wave gas sensor based on film conductivity changes. Sens Actuators. 1985;8:319–333.
124. Kalantar-zadeh K, Powell DA, Ippolito S, Wlodarski W. Study of layered SAW devices operating at different modes for gas sensing applications. Proceeding of the IEEE Ultrasonics Symposium, 2004;191–194.
125. Blotekjær K, Ingebrig Ka, Skeie H. Method for analyzing waves in structures consisting of metal strips on dispersive media. IEEE Trans Electron Devices ED. 1973;20:1133–1138.
126. Powell DA, Kalantar-zadeh K, Ippolito S, Wlodarski W. Comparison of conductometric gas sensitivity of surface acoustic wave modes in layered structures. Sens Lett. 2005;3:66–70.
127. Wohltjen H, Dessy R. Surface acoustic-wave probes for chemical-analysis. 2. Gas-Chromatography Detector. Anal Chem. 1979;51:1465–1470.
128. D'Amico A, Palma A, Verona E. Hydrogen sensor using a palladium coated surface acoustic wave delay-line. Proceedings of the IEEE Ultrasonics Symposium. 1982;308–311.
129. Nieuwenhuizen MS, Nederlof AJ. A saw gas sensor for carbon-dioxide and water – preliminary experiments. Sens Actuators B. 1990;2:97–101.

130. Galipeau JD, Falconer RS, Vetelino JF, Caron JJ, Wittman EL, Schweyer MG et al. Theory, design and operation of a surface-acoustic-wave hydrogen-sulfide microsensor. Sens Actuators B. 1995;24:49–53.
131. Penza M, Vasanelli L. SAW NO_x gas sensor using WO_3 thin-film sensitive coating. Sens Actuators B. 1997;41:31–36.
132. Rapp M, Stanzel R, Vonschickfus M, Hunklinger S, Fuchs H, Schrepp W et al. Gas-detection in the ppb-range with a high-frequency, high-sensitivity surface acoustic-wave device. Thin Solid Films. 1992;210:474–476.
133. Rebiere D, Dejous C, Pistre J, Lipskier JF, Planade R. Synthesis and evaluation of fluoropolyol isomers as SAW microsensor coatings: Role of humidity and temperature. Sens Actuators B. 1998;49:139–145.
134. Wohltjen H. Mechanism of operation and design considerations for surface acoustic-wave device vapor sensors. Sens Actuators. 1984;5:307–325.
135. Jakoby B, Ismail GM, Byfield MP, Vellekoop MJ. A novel molecularly imprinted thin film applied to a love wave gas sensor. Sens Actuator A-Phys. 1999;76:93–97.
136. Zimmermann C, Rebiere D, Dejous C, Pistre J, Chastaing E, Planade R. A love-wave gas sensor coated with functionalized polysiloxane for sensing organophosphorus compounds. Sens Actuators B. 2001;76:86–94.
137. Penza M, Antolini F, Antisari MV. Carbon nanotubes as SAW chemical sensors materials. Sens Actuators B. 2004;100:47–59.
138. Penza M, Aversa P, Cassano G, Wlodarski W, Kalantar-Zadeh K. Layered SAW gas sensor with single-walled carbon nanotube-based nanocomposite coating. Sens Actuators B. 2007;127:168–178.
139. Kalantar-Zadeh K, Trinchi A, Wlodarski W, Holland A. A novel love-mode device based on a ZnO/ST-Cut quartz crystal structure for sensing applications. Sens Actuators A. 2002;100:135–143.
140. Nieuwenhuizen MS, Nederlof AJ. A silicon-based SAW chemical sensor for NO_2 by applying a silicon-nitride passivation layer. Sens Actuators B. 1992;9:171–176.
141. Caron JJ, Kenny TD, LeGore LJ, Libby DG, Freeman CJ, Vetelino JF. A surface acoustic wave nitric oxide sensor. Proceedings of the IEEE Frequency Control Symposium. 1997;156–162.
142. Martin SJ, Frye GC, Spates JJ, Butler MA. Gas sensing with acoustic devices. Proceedings of the IEEE Ultrasonics Symposium. 1996;423–434.
143. Vellekoop MJ. Acoustic wave sensors and their technology. Ultrasonics. 1998;36:7–14.
144. Drafts B. Acoustic wave technology sensors. IEEE Trans Microw Theory. 2001;49:795–802.
145. Lec R, Vetelino JF, Falconer RS, Xu A. Macroscopic theory of surface acoustic wave gas microsensors. Proceedings of the IEEE Ultrasonics Symposium. 1988;581:585–589.
146. Holcroft B, Roberts GG. Surface acoustic-wave sensors incorporating langmuir-blodgett films. Thin Solid Films. 1988;160:445–452.
147. Jakubik W, Urbanczyk M, Opilski A. Sensor properties of lead phthalocyanine in a surface acoustic wave system. Ultrasonics. 2001;39:227–232.
148. Caron JJ, Haskell RB, Benoit P, Vetelino JF. A surface acoustic wave mercury vapor sensor. IEEE Trans Ultrason Ferr. 1998;45:1393–1398.
149. Galipeau JD, LeGore LJ, Snow K, Caron JJ, Vetelino JF, Andle JC. The integration of a chemiresistive film overlay with a surface acoustic wave microsensor. Sens Actuators B. 1996;35:158–163.
150. Thiele JA, Pereira da Cunha M. High temperature LGS SAW devices with Pt/WO_3 and Pd sensing films. Proceedings of the IEEE Ultrasonics Symposium, 2003;1752: 1750–1753.
151. Kalantar-zadeh K, Li YX, Wlodarski W. Brennan F. A layered structure surface acoustic wave for oxygen sensing. Proceedings of the Conference on Optoelectronic and Microelectronic Materials and Devices, 2000;202–205

152. Kalantar-zadeh K, Trinchi A, Wlodarski W, Holland A, Atashbar MZ. A novel love mode device with nanocrystalline ZnO film for gas sensing applications. Proceedings of the IEEE Nanotechnology Conference. 2001;556–561.
153. Penza M, Tagliente MA, Aversa P, Re M, Cassano G. The effect of purification of single-walled carbon nanotube bundles on the alcohol sensitivity of nanocomposite langmuir-blodgett films for SAW sensing applications. Nanotechnology. 2007;18: 185502 (12pp).
154. Du J, Harding GL. A multilayer structure for love-mode acoustic sensors. Sens Actuators A. 1998;65:152–159.
155. Harding GL. Mass sensitivity of love-mode acoustic sensors incorporating silicon dioxide and silicon-oxy-fluoride guiding layers. Sens Actuators A. 2001;88:20–28.
156. Chang RC, Chu SY, Hong CS, Chuang YT. A study of love wave devices in ZnO/Quartz and ZnO/$LiTaO_3$ structures. Thin Solid Films. 2006;498:146–151.
157. Chang RC, Chu SY, Hong CS, Chuang YT. An investigation of preferred orientation of doped ZnO films on the 36° YX-$LiTaO_3$ substrates and fabrications of love-mode devices. Surf Coat Technol. 2006;200:3235–3240.
158. Sauberlich R, Petter P, Bu W, Wall B, Nindel N. Method and device for the concentration of mercury in gases. Deutsche Demokratische Republik Patent. 1989;No: 00268530A1.
159. Caron JJ, Haskell RB, Libby DG, Freeman CJ, Vetelino JF. A surface acoustic wave mercury vapor sensor. Proceedings of the IEEE Frequency Control Symposium. 1997; 133–139.
160. Ricco AJ, Crooks RM, Osbourn GC. Surface acoustic wave chemical sensor arrays: New chemically sensitive interfaces combined with novel cluster analysis to detect volatile organic compounds and mixtures. Accounts Chem Res. 1998;31:289–296.
161. Lu CJ, Zellers ET. A dual-adsorbent preconcentrator for a portable indoor-VOC microsensor system. Anal Chem. 2001;73:3449–3457.
162. Ippolito SJ. PhD Thesis: Investigation Of Multilayered Surface Acoustic Wave Devices For Gas Sensing Applications: Employing Piezoelectric Intermediate and Nanocrystalline Metal Oxide Sensitive Layers, Melbourne, Australia: RMIT University; 2006.
163. Sadek AZ, Wlodarski W, Li Y, Yu W, Li X, Yu X et al. A ZnO nanorod based layered ZnO/64° YX $LiNbO_3$ SAW hydrogen gas sensor. Thin Solid Films. 2007;515:8705–8708.

Chapter 9
Cantilever-Based Gas Sensing

Hans Peter Lang

9.1 Introduction to Microcantilever-Based Sensing

9.1.1 Early Approaches to Mechanical Sensing

Male individuals of certain animal species like the large domestic silk moth *Bombyx mori*, who is the adult of the silk-thread-producing silkworm, are able to detect pheromones emitted by the female over several miles by means of their antennas. Such high sensitivity is achieved by evolution-driven optimization of chemical detection aimed for the survival of a species. A single pheromone molecule already triggers perception. However, a change of behavior only occurs at higher concentration. Highly specific receptors for certain chemical compounds are often based on the geometrical conformation of the target molecule, supported by chemical affinities and binding between specific functional groups. The adsorption process is frequently related to local conformational changes, which are of mechanical nature. Thin membranes and beams also possess mechanical properties that render them suitable for detection of small forces. This fact can be easily demonstrated when water adsorbs on thin membranes, where the large effect of surface tension of water leads to deformation of such thin membranes.

Detecting adsorption by measurement of bending or change in resonance frequency using beams of silicon as sensors was already described by Wilfinger et al. [72], who detected resonances in large silicon cantilever structures of 50 mm × 30 mm × 8 mm. Actuation was performed by localized thermal expansion in diffused resistors (piezoresistors) located near the cantilever support to create a temperature gradient that drives the cantilever at its resonance frequency. Similarly, the piezoresistors could also be used to monitor mechanical deflection of the cantilever. Heng [25] fabricated gold microcantilevers capacitively coupled to microstrip lines for mechanical trimming of high-

H.P. Lang
National Center of Competence for Research in Nanoscale Science, Institute of Physics of the University of Basel, Klingelbergstrasse 82, 4056 Basel, Switzerland
e-mail: Hans-Peter.Lang@unibas.ch

E. Comini et al. (eds.), *Solid State Gas Sensing*,
DOI: 10.1007/978-0-387-09665-0_9, © Springer Science+Business Media, LLC 2009

frequency oscillator circuits. Petersen [48] constructed cantilever-type micromechanical membrane switches made from silicon designed to bridge the gap between silicon transistors and mechanical electromagnetic relays. Kolesar [31] suggested the use of cantilever structures as electronic detectors for nerve agents. The breakthrough for microcantilevers came with the advent of atomic force microscopy (AFM) (Binnig et al. [7]), as micro-fabricated cantilevers have become easily commercially available. The technical development also triggered research reports of microcantilever use as sensors. Itoh et al. [29] presented a microcantilever coated with a thin film of zinc oxide with piezoresistive deflection readout. Cleveland et al. [13] reported the tracking of cantilever resonance frequency to detect nanogram changes in mass loading when small particles are deposited onto AFM probe tips. Gimzewski et al. [21] showed first chemical-sensing applications, in which static cantilever bending revealed chemical reactions, such as the platinum-assisted catalytic conversion of hydrogen and oxygen into water at very high sensitivity. Thundat et al. [68] demonstrated that the resonance frequency as well as static bending of microcantilevers is influenced by changing ambient conditions, such as moisture adsorption. Furthermore they found that deflection of metal-coated cantilevers is also thermally influenced (bimetallic effect). Later Thundat et al. [69] observed changes in the resonance frequency of microcantilevers due to adsorption of analyte vapor on exposed surfaces. The frequency changes are caused through mass loading or adsorption-induced changes in cantilever spring constant. By coating cantilever surfaces with hygroscopic materials, such as phosphoric acid or gelatin, the cantilever can sense water vapor at picogram mass resolution.

9.1.2 Cantilever Sensors

For the use of microcantilevers as sensors, neither a tip at the cantilever apex nor a sample surface is required. The microcantilever surfaces represent the platform to sense adsorption of molecules. Such processes involve generation of surface stress, resulting in bending of the microcantilever, provided adsorption preferentially occurs on one cantilever surface. Selective adsorption on one surface only is controlled by coating typically the upper surface with a thin layer showing affinity to the molecules in the environment to be detected. This surface will be called sensor surface or functionalized surface of the microcantilever (see Fig. 9.1a). The other surface, typically the lower surface, may be left uncoated or be coated with a passivation layer being inert or not exhibiting substantial affinity to the molecules that are to be detected. To establish functionalized surfaces, often a metal layer is evaporated onto the surface designed as sensor surface. Metal surfaces, such as gold, are frequently used to covalently bind a monolayer representing the actual detection layer, e.g., a thiol monolayer with defined surface chemistry. The molecules to be detected bind then to the thiol layer. The underlying gold coating also serves as reflection layer for optical readout of the cantilever.

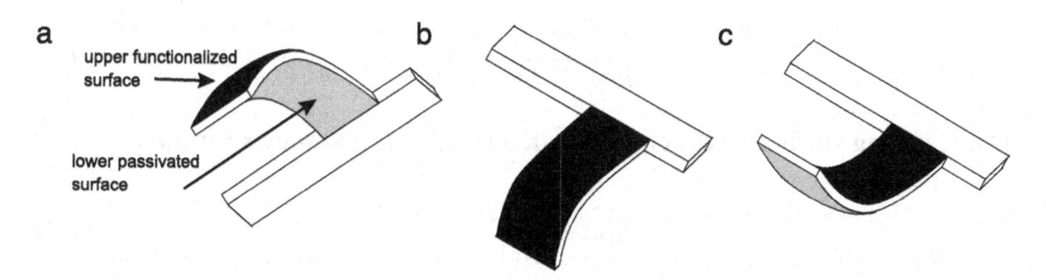

Fig. 9.1 (**a**) Schematic drawing of a microcantilever with its *lower surface* passivated and it *upper surface* functionalized for recognition of target molecules, (**b**) *downward bending* of a microcantilever due to compressive surface stress, (**c**) *upward bending* due to tensile surface stress

Adsorption of molecules on the upper (functionalized) surface will produce a downward bending of the microcantilever due to formation of surface stress. This process is called formation of compressive surface stress (see Fig. 9.1b), because the adsorbed layer of molecules (e.g., a monolayer of alkylthiols) causes downward bending of the microcantilever away from its functionalized side. If the opposite situation occurs, i.e., when the microcantilever bends upwards, tensile surface stress is produced (see Fig. 9.1c). A mixing of influences will take place, if both the upper and the lower surface of the microcantilever are prone to surface stress change effects. In this case, e.g., predominant compressive stress formation on the lower microcantilever surface might appear like tensile stress on the upper surface. Therefore, it is extremely important to properly passivate the lower surface so that no processes should take place on the lower surface of the microcantilever, facilitating evaluation and discussion of cantilever experiments.

9.1.3 Deflection Measurement

Adsorption of molecules onto the functional layer causes stress formation at the interface between functional layer and the forming molecular layer, resulting in bending of the microcantilever, as the forces within the functional layer try to keep the distance between molecules constant. The cantilever beam bends due to its extreme flexibility. This property is reflected by the spring constant k of the cantilever. For a rectangular microcantilever of length l, thickness t, and width w the spring constant is k is calculated as follows:

$$k = \frac{Ewt^3}{4l^3}, \tag{9.1}$$

where E is Young's modulus ($E_{Si} = 1.3 \times 10^{11}$ N/m^2 for Si(100)).

In first approximation the shape of the bent microcantilever is described as part of a circle with radius R. This radius of curvature is given by (Ibach [28]; Stoney [65])

$$\frac{1}{R} = \frac{6(1-\nu)}{Et^2} \Delta\sigma, \tag{9.2}$$

The resulting surface stress change is described using Stoney's formula :

$$\Delta\sigma = \frac{Et^2}{6R(1-\nu)}, \tag{9.3}$$

where E is Young's modulus, t the thickness of the cantilever, ν the Poisson's ratio ($\nu_{Si} = 0.24$), and R the bending radius of the cantilever.

The deflection of microcantilever sensors can be measured in various ways. They differ in sensitivity, effort for alignment and setup, robustness, and ease of readout as well as in potential for miniaturization.

9.1.3.1 Piezoresistive Readout

Piezoresistive microcantilevers (Itoh et al. [29]; Berger et al. [4]) are usually U-shaped and have diffused piezoresistors in both of the legs near to the fixed end. The resistance in the piezoresistors is measured using a Wheatstone bridge circuit composed of three reference resistors, of which one is adjustable. The current flowing between the two branches of the Wheatstone bridge is first nulled by changing the resistance of the adjustable resistor. If the microcantilever bends, the piezoresistor changes its value and a current will flow between the two branches of the Wheatstone bridge. The current is converted via a differential amplifier into a voltage, which is proportional to the deflection value. For dynamic-mode measurement, the piezoresistive microcantilever is externally actuated via a piezocrystal driven by a frequency generator. The ac actuation voltage is fed as reference voltage into a lock-in amplifier and compared with the response of the Wheatstone bridge circuit allowing to sweep resonance curves and to determine shifts in resonance frequency.

9.1.3.2 Piezoelectric Readout

Piezoelectric microcantilevers (Lee et al. [34]) are driven via the inverse piezoelectric effect (self-excitation) by applying an electric ac voltage to the piezoelectric material (lead zirconium titanate PhO-ZrO_2-T_iO_2 (PZT) or ZnO). Sensing of bending is performed by recording the piezoelectric current change taking advantage of the fact that the PZT layer produces a sensitive field response to weak stress through the direct piezoelectric effect. Piezoelectric microcantilevers are multilayer structures consisting of a SiO_2 cantilever and the PZT piezoelectric layer. Two electrode layers, insulated from each other, provide electrical contact. The entire structure is protected using passivation layers. An identical structure is usually integrated into the rigid chip body to provide a reference for the piezoelectric signals from the cantilever.

9.1.3.3 Capacitive Readout

Microcantilevers structures for capacitive readout are composed of a rigid beam with an electrode and a flexible cantilever with another electrode (Göddenhenrich et al. [22]; Brugger et al. [10]). Both electrodes are insulated from each other. When the flexible microcantilever is being bent, the capacitance between the two electrodes changes. From the capacitance change, the cantilever deflection can be determined. Both measurement of static bending as well as determination of resonance frequency are possible.

9.1.3.4 Beam-Deflection Optical Readout

The most frequently used approach to read out microcantilever deflections is optical beam deflection (Meyer and Amer [40]), because it is a comparatively simple method, which has an excellent resolution.

The actual cantilever deflection Δx scales with the cantilever dimensions; therefore deflection responses should be expressed in terms of surface stress $\Delta\sigma$ in N/m to be able to compare cantilever responses acquired with different setups. Surface stress takes into account the cantilever material properties, such as Poisson ratio ν, Young's modulus E, and the cantilever thickness t. The radius of curvature R of the cantilever characterizes bending, see Eq. (9.2). As shown in the drawing in Fig. 9.2, the actual cantilever displacement Δx is transformed into a displacement Δd on the position-sensitive detector (PSD), a lateral photodiode, used to determine the location of an incident light beam. The position of a light spot on a PSD is determined by measuring the photocurrents from the two facing electrodes. The movement of the light spot on the linear PSD is calculated from the two currents I_1 and I_2 and the size L of the PSD by

$$\Delta d = \frac{I_1 - I_2}{I_1 + I_2} \times \frac{L}{2}. \tag{9.4}$$

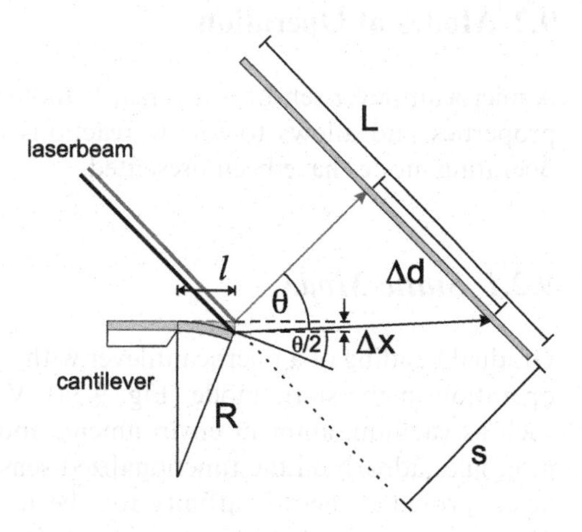

Fig. 9.2 Beam-deflection concept to determine microcantilever bending with an accuracy of one nanometer

As all angles are very small, it can be assumed that the bending angle of the cantilever is equal to half of the angle θ of the deflected laser beam, i.e., $\theta/2$. Therefore, the bending angle of the cantilever can be calculated to be

$$\frac{\theta}{2} = \frac{\Delta d}{2s}, \quad (9.5)$$

where s is the distance between the PSD and the cantilever. The actual cantilever deflection Δx is calculated from the cantilever length l and the bending angle $\theta/2$ by

$$\Delta x = \frac{\theta/2}{2} \times l. \quad (9.6)$$

Combination of Eqs. (9.5) and (9.6) relates the actual cantilever deflection Δx to the PSD signal:

$$\Delta x = \frac{l \times \Delta d}{4s}. \quad (9.7)$$

The relation between the radius of curvature and the deflection angle is

$$\frac{\theta}{2} = \frac{l}{R}, \quad (9.8)$$

and after substitution becomes

$$R = \frac{2ls}{\Delta d} \text{ or } R = \frac{2\Delta x}{l^2}. \quad (9.9)$$

9.2 Modes of Operation

A microcantilever sensor is a versatile tool for investigation of various sample properties and allows to follow reactions occurring on its surface. Various operating modes have been presented.

9.2.1 Static Mode

Gradual bending of a microcantilever with molecular coverage is referred to as operation in the static mode (Fig. 9.3a). Various environments are possible, such as vacuum, ambient environment, and liquids. In gaseous environment molecules adsorb on the functionalized sensing surface and form a molecular layer, provided there is affinity for the molecules to adhere to the surface.

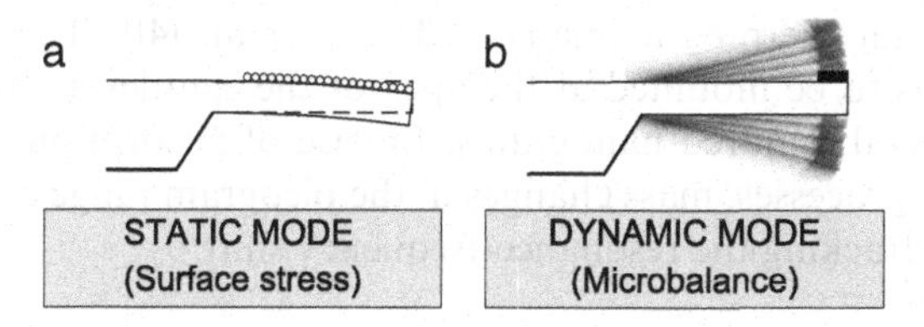

Fig. 9.3 Major operating modes of microcantilever sensors: (**a**) static-mode exploiting surface stress changes, (**b**) dynamic mode to extract mass changes

Polymer sensing layers show a partial sensitivity, because molecules from the environment diffuse into the polymer layer at different rates, mainly depending on the size and solubility of the molecules in the polymer layer. By selecting polymers among a wide range of hydrophilic/hydrophobic ligands, the chemical affinity of the surface can be influenced, because different polymers vary in diffusion suitability for polar/unpolar molecules. Thus, for detection in gas phase, the polymers can be chosen according to the detection problem, i.e., what the applications demand. Typical chemicals to be detected are volatile organic compounds (VOCs).

9.2.2 Dynamic Mode

By oscillating a microcantilever at its eigenfrequency, information on the amount of molecules adsorbed can be obtained. However, the surface coverage is basically not known. Furthermore, molecules on the surface might be exchanged with molecules from the environment in a dynamic equilibrium.

In contrast, mass changes can be determined accurately by tracking the eigenfrequency of the microcantilever during mass adsorption or desorption. The eigenfrequency equals the resonance frequency of an oscillating microcantilever if its elastic properties remain unchanged during the molecule adsorption/desorption process and damping effects are negligible. This operation mode is called the dynamic mode (Fig. 9.3b) . The microcantilever is used as a microbalance , as with mass addition on the cantilever surface, the cantilever's eigenfrequency will shift to a lower value. The mass change on a rectangular cantilever is calculated (Thundat et al. [68]) according to

$$\Delta m = (k/4\pi^2) \times (1/f_1^2 - 1/f_0^2), \tag{9.10}$$

where f_0 is the eigenfrequency before the mass change occurs, and f_1 the eigenfrequency after the mass change. For calculation of the spring constant k of the cantilever see Eq. (9.1).

Mass-change determination can be combined with varying environment temperature conditions to obtain a method introduced in the literature as "micromechanical thermogravimetry" (Berger et al. [4]). The sample to be investigated has to be mounted at the apex of the cantilever. Its mass should not exceed several hundred nanograms. In case of adsorption, desorption or decomposition processes, mass changes in the picogram range can be observed in real time by tracking the resonance-frequency shift.

9.3 Functionalization

It is essential that the surfaces of the cantilever are coated in a proper way to provide suitable receptor surfaces for the molecules to be detected. Such coatings should be specific, homogeneous, stable, reproducible, and either reusable or designed for single use only. For static mode measurements using the beam deflection technique, the cantilever's upper side, the sensor side, is coated with a 20-nm-thick layer of gold as reflection layer. The coating method of choice should be fast, reproducible, reliable, and allow one or both cantilever surfaces to be coated separately.

For convenient coating with polymer layers, inkjet spotting is used, as it is possible to coat only the upper or lower surface. The method is also appropriate for coating many cantilever sensor arrays in a rapid and reliable way (Bietsch et al. [5]; Bietsch et al. [6]), see Fig. 9.4. An x-y-z positioning system allows a fine nozzle (capillary diameter: 70 μm) to be positioned with an accuracy of approximately 10 μm over a cantilever (Fig. 9.4a). Individual droplets (diameter: 60–80 μm, volume 0.1–0.3 nl) can be dispensed individually by means of a piezo-driven ejection system in the inkjet nozzle. When the droplets are spotted with a pitch smaller than 0.1 mm, they merge and form continuous films. By adjusting the number of droplets deposited on cantilevers, the resulting film thickness can be controlled precisely (Fig. 9.4b). The

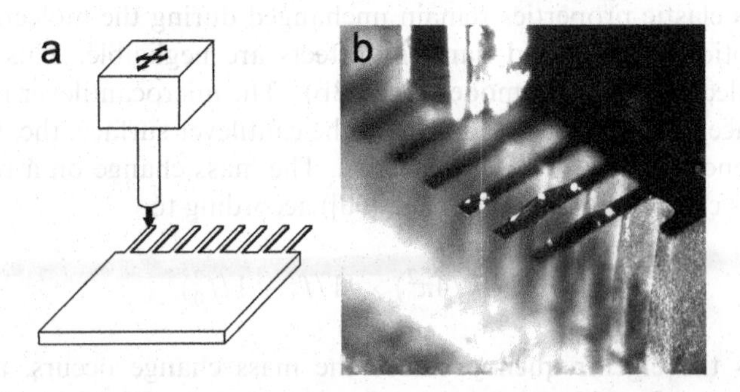

Fig. 9.4 (**a**) Schematic drawing of single-sided microcantilever coating using an inkjet spotter. (**b**) Optical micrograph of the inkjet spotter nozzle. The amount of liquid containing the probe molecules can be dosed accurately by choosing the number of drops being ejected from the nozzle

inkjet-spotting technique allows a cantilever to be coated within seconds and yields very homogeneous, reproducibly deposited layers of well-controlled thickness. Successful coating of self-assembled alkanethiol monolayers, polymer solutions, self-assembled DNA single-stranded oligonucleotides (Bietsch et al. [6]), and protein layers has been demonstrated. This technique has been applied to functionalize a polymer-coated microcantilever array for the chemical vapor detection experiments as described in the following section. In conclusion, inkjet spotting has turned out to be a very efficient and versatile method for functionalization that can even be used to coat arbitrarily-shaped sensors reproducibly and reliably (Lange et al. [33]; Savran et al. [60]).

9.4 Example of an Optical Beam-Deflection Setup

9.4.1 General Description

A measurement setup for microcantilever arrays consists of four main parts: (1) the measurement chamber hosting the microcantilever array, (2) an optical or piezoresistive system to detect the cantilever deflection (e.g., laser sources, collimation lenses, and a PSD), (3) electronics to amplify, process, and acquire the signals from the PSD, and (4) a gas handling system to reproducibly inject samples into the measurement chamber and purge the chamber.

Figure 9.5 shows a realization of a setup for experiments performed in gaseous environment. The microcantilever sensor array is hosted in an analysis chamber of 20 μl in volume, with inlet and outlet ports for gases. The bending of the microcantilevers is measured using an array of eight vertical-cavity surface-emitting lasers (VCSELs) arranged at a linear pitch of 250 μm that emit at a wavelength of 760 nm into a narrow cone of 5–10°.

The light of each VCSEL is collimated and focused onto the apex of the corresponding microcantilever by a pair of achromatic doublet lenses, 12.5 mm in diameter. This size was selected in order to make sure that all eight laser beams pass through the lenses close to its center in order to minimize scattering, and chromatic and spherical aberration artifacts. The light is then reflected off the gold-coated surface of the cantilever and hits the surface of a PSD. As only a single PSD is used, the eight lasers cannot be switched on simultaneously. Therefore, a time-multiplexing procedure is used to switch the lasers on and off sequentially at typical intervals of 10–100 ms. The resulting deflection signal is digitized and stored together with time information on a personal computer (PC), which also controls the multiplexing of the VCSELs as well as the switching of the valves and mass flow controllers used for setting the composition ratio of the analyte mixture.

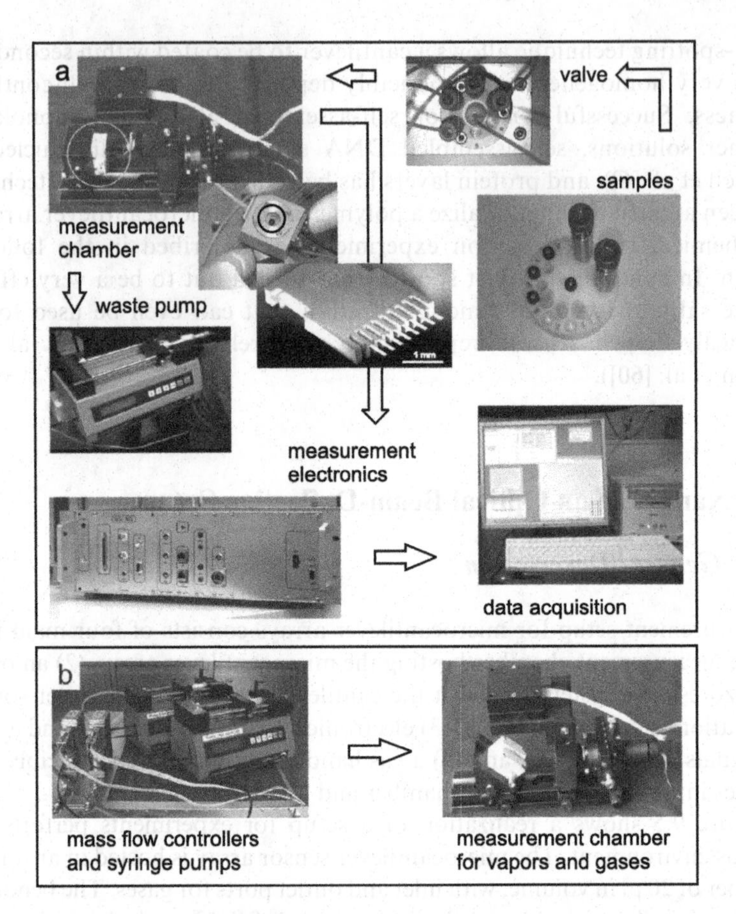

Fig. 9.5 (**a**) Realization of a static mode beam-deflection microcantilever-array sensor setup for measurements in gas. A six-way valve is used to select vapor from the headspace of a vial filled with liquid sample. The vapor is transported by nitrogen carrier gas to the measurement chamber hosting the microcantilever array. The microcantilever bending is measured using optical beam deflection involving VCSELs and a PSD. The location of the laser spot indicating microcantilever bending is processed and digitized using measurement electronics and a data acquisition card in a personal computer. (**b**) Gas handling system consisting of mass flow controllers and syringe pumps for well-defined flow rates

9.4.2 Cantilever-Based Electronic Nose Application

Besides specific receptor analyte binding, also the response pattern of a range of several differently coated sensors, e.g., coated with polymer layers, can be used to identify chemical vapors (artificial or electronic nose application). Using a cantilever array sensor setup consisting of 8 polymer-coated cantilevers and automated gas handling system, several solvent vapors can be distinguished. 0.1 ml of various solvents was placed in vials, and the vapor from the headspace above the liquid was sampled using microcantilever sensors, operated in static

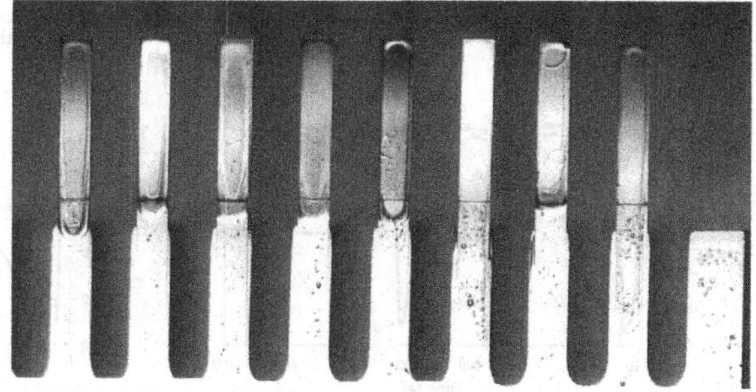

Fig. 9.6 Optical microscopy image of a polymer-coated microcantilever array for application as an artificial nose for solvents. The pitch between microcantilevers is 250 μm

deflection mode as a kind of artificial nose. Detection of vapors takes place via diffusion of the vapor molecules into the polymer, resulting in a swelling of the polymer and bending of the cantilever. Each cantilever is coated with a different polymer (see Fig. 9.6 and Table 9.1). The bending is specific for the interaction between solvent vapor and polymer time- and magnitude-wise.

Cantilever deflection traces upon subsequent injection of solvent vapor for 30 s and purging with dry nitrogen for 150 s are shown in Fig. 9.7 for (a) water, (b) methanol, and (c) ethanol.

The cantilever deflections at 10, 20, 30, and 40 s after completion of solvent vapor injection are extracted. They describe the time-development of the curves in a reduced data set, i.e. $8 \times 4 = 32$ cantilever deflection amplitudes ("fingerprint") that account for a measurement data set. This data set is then evaluated using principal component analysis (PCA) techniques, extracting the most dominant deviations in the responses for the various sample vapors. The axes refer to projections of the multidimensional datasets into two dimensions (principal components). The symbols in the PCA plot (Fig. 9.7d) indicate the individual measurements. The PCA plot shows well separated clusters of measurements indicating clear identification of vapor samples.

Table 9.1 Polymer coatings

Cantilever	Polymer	Full name of compound
1	PSS	Poly(sodium 4-styrenesulfonate)
2	PEI	Polyethylenimine
3	PAAM	Poly(allylamine hydrochloride)
4	CMC	Carboxymethylcellulose sodium salt
5	Dextran	Dextran from Leuconostoc spp.
6	HPC	Hydroxypropyl cellulose
7	PVP	Polyvinylpyrrolidone
8	PEO	Poly(2-ethyl-2-oxazoline)

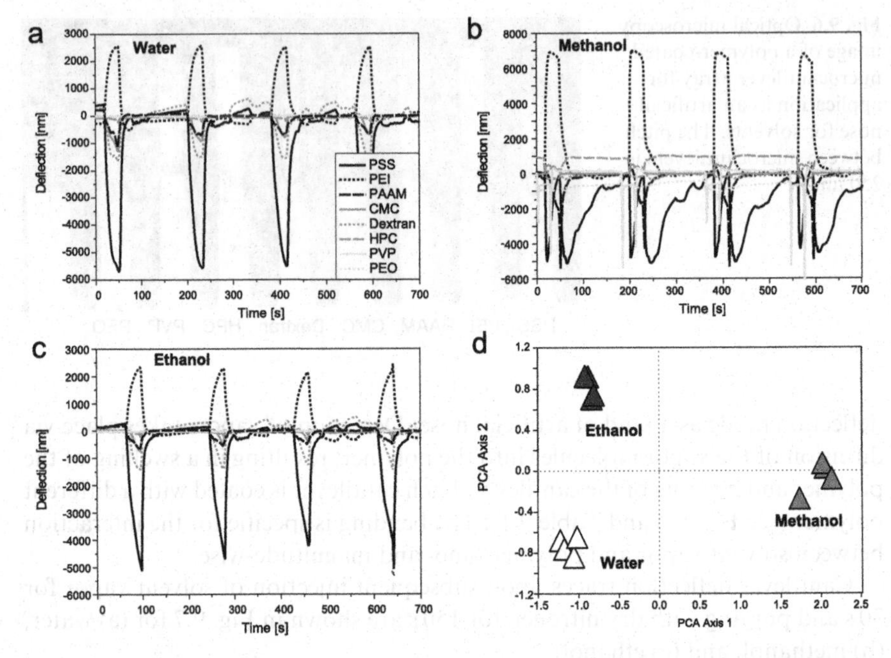

Fig. 9.7 Application of microcantilever array sensors as artificial electronic nose. Measurement traces of microcantilevers coated with polymers during the detection of **(a)** water, **(b)** methanol, **(c)** ethanol. For every solvent, four consecutive injections of vapor saturated with solvent are shown. Upon injection of solvent vapor, the microcantilevers deflect in a specific way due to the swelling of the polymers layer on exposure to the solvent vapor. Subsequent purging of the measurement chamber with dry nitrogen gas (flow rate: 100 ml/min) promotes diffusion of the solvent molecules out of the polymer layer, resulting in bending back of the microcantilevers to the baseline. **(d)** Principal component analysis (PCA) of the response patterns of all eight microcantilevers upon exposure to the three different solvent vapors. Clear separation of the clusters proves the excellent distinction capability of the artificial nose setup. Each symbol in the PCA plot corresponds to one of the injections in **(a)**–**(c)**

9.5 Applications of Cantilever-Based Gas Sensors

The application fields of microcantilever gas sensors are diverse: gas sensing, vapor sensing (artificial nose application), explosives detection, gas pressure, and flow sensing. The following sections will review some recent publications in these fields, complemented by theory, simulation, and fabrication aspects.

9.5.1 Gas Sensing

Several publications investigate the detection of hydrogen using Pd-coated cantilever sensors. Hu et al. [27] have studied bulk adsorption of hydrogen on palladium using differential stress formation on a bimaterial cantilever. The adsorption of hydrogen into palladium results in film expansion, whereby the bending magnitude is proportional to hydrogen partial pressure. Bending also depends on the thickness of the Pd film and is reversible. Unlike hydrogen adsorption, mercury adsorption on a gold film is irreversible, as it is a surface adsorption process. Baselt et al. [3] describe the design of a micro-electro-mechanical (MEMS) hydrogen sensor consisting of an array of 10 micromachined cantilever beams. The deflection is measured using the capacitive method with a dedicated readout circuit including wireless transmitters capable of transmitting data over a distance of 30 m. The sensitive coating is composed of 90% Pd and 10% Ni and can easily detect concentrations of 0.4% hydrogen. It operates reversibly. Influences of relative humidity and temperature have also been studied. Fabre et al. [18] present a hydrogen sensor based on Pd-coated microcantilevers. It was found that the surface stress response depends strongly on the hydrogen dissociation into the bulk of the Pd layer, whereby the influences of cantilever shape and surface roughness play also an important role. Ono et al. [44] report resonating cantilever sensors to investigate the hydrogen storage capacity of carbon nanotubes (CNT). The storage capacity was found to be up to 6% and the mass resolution was on the order of attogram. Zhou et al. [76] use self-excited piezoelectric microcantilevers with a layer of MFI zeolites to detect Freon down to a concentration of 10 ppm, whereby ethanol vapor shows no effect up to the largest concentration of 500 ppm that was measured (Zhou et al. [77]).

Mertens et al. [38] presented results on detection of HF, a decomposition product of nerve gas agents, using SiO_2-coated Si_3N_4 microcantilevers. Detection based on etching of SiO_2 was achieved in the range from 0.26 to 13 ppm. Embedded piezoresistive cantilevers have been used by Kooser et al. [32] for detection of carbon monoxide using a nickel-containing polyethylene oxide (PEO) layer for detection. Ni has been added to the PEO solution as nickel acetate and serves for resistivity measurement. After drying for 24 h and exposure for 1 h to dry nitrogen, the sensor was exposed to carbon monoxide, resulting in a resistance drop. For concentration ratios up to 1:10 (CO:N_2) the sensor response was fully reversible. Adams et al. [1] report on piezoelectric cantilevers for mercury detection down to a concentration of 50 ppb. Again, after three subsequent exposure steps, the cantilever was irreversibly saturated. Porter et al. [57] demonstrate HCN detection using embedded piezoresistive

Table 9.2 Detection limits of gaseous analytes

Analyte	Detection limit	Reference
H_2	20 ppm	Hu et al. [27]
Hg	0.4 ppb	Hu et al. [27]
Hg	50 ppb	Adams et al. [1]
Hg	4000 ppm	Baselt et al. [3]
Freon	10 ppm	Zhou et al. [76]
HF	0.26 ppm	Mertens et al. [38]
HCN	150 ppm	Porter et al. [57]

microcantilever sensors down to a concentration of 150 ppm. HCN was generated by reacting KCN with sulfuric acid. No response was observed for H_2SO_4 vapors only. Sensitivities are summarized in Table 9.2

9.5.2 Chemical Vapor Detection

Jensenius et al. [30] describe a microcantilever-based alcohol vapor sensor using the piezoresistive technique and polymer coating. They also present a simple evaporation model that allows determining the concentration. The detection limit found amounts to 10 ppm for methanol, ethanol, and 2-propanol. Hierlemann et al. [26] present an integrated complementary metal oxide semiconductor (CMOS) chemical microsensor with piezoresistive detection (Wheatstone bridge configuration) using poly(etherurethane) (PEUT) as the sensor layer. They are able to reversibly detect volatile organic compounds (VOCs) such as toluene, *n*-octane, ethyl acetate, and ethanol with a sensitivity level down to 200 ppm. An improved version of that device is described by Lange et al. [33]. The sensitivity could be increased to 5 ppm for *n*-octane. Later the technique has been refined by using electromagnetic rather than electrothermal actuation and transistor-based readout reducing power dissipation on the cantilever [70]. Fadel et al. [19] describe piezoelectric readout in dynamic mode and electromagnetic actuation of cantilevers spray-coated with PEUT, achieving a sensitivity of 14 ppm for ethanol. Wright et al. [75] published a study how to prepare polyethylene glycol (PEG) coated microcantilever sensors using a microcapillary pipette assisted method. PEG coating is suitable for ethanol sensing as ethanol quickly forms hydrogen bonds with the OH groups of the PEG. Sensor operation is reported to be reversible and reproducible. Senesac et al. [61] use artificial neural networks for analyte species and concentration identification with polymer coated optically read-out microcantilevers. The analytes detected are carbon dioxide, dichloromethane, diisopropylmethylphosphonate (DIMP), dioxane, ethanol, water, 2-propanol, methanol, trichloroethylene, and trichloromethylene. Lochon et al. [36] investigate the chemical-sensing performance of a silicon resonant microcantilever sensor in dependence on the thickness of the sensitive coating. For a coating thickness of 1, 4, and 21 μm of PEUT a limit of detection of 30 ppm was found for ethanol. Satyanarayana et al. [59] present a new concept of parylene micromembrane array for chemical sensing using the capacitive method. The parylene membrane is suspended over a metal pad patterned on the substrate. The pad and part of the membrane that is metal-coated serves as electrodes for capacitive measurement. The top electrode located on the membrane is chemically modified by applying a gold layer and self-assembled thiol monolayers (–COOH, $-CH_3$, and –OH) for detection of analyte molecules. Successful detection of 2-propanol and toluene is reported. Then et al. [67] describe a sensitive self-oscillating cantilever array for quantitative and qualitative analysis of organic vapor mixtures. The cantilevers are electromagnetically actuated and the resonance

Table 9.3 Detection limits of vapors

Analyte	Detection limit	Reference
Methanol, ethanol, 2-propanol	10 ppm	Jensenius et al. [30]
Toluene, n-octane, ethyl acetate	10 ppm	Jensenius et al. [30]
Ethanol	200 ppm	Hierlemann et al. [26]
n-octane	5 ppm	Lange et al. [33]
Ethanol	14 ppm	Fadel et al. [19]
Ethanol	30 ppm	Lochon et al. [36]
Butanol	0.19 ppm	Then et al. [67]
n-octane	0.6 ppm	Then et al. [67]
Toluene	0.38 ppm	Then et al. [67]

frequency is measured using a frequency counter. Sensor response is reproducible and reversible. Using PEUT coating the smallest measured concentration is 400 ppm, but the limit of detection is well below 1 ppm. Chapman et al. [11] report on a combination of gas chromatography with a microcantilever sensor array for enhanced selectivity. Test VOC mixtures composed of acetone, ethanol, and trichloroethylene in pentane, as well as methanol with acetonitrile in pentane were first separated in a gas chromatography column and then detected using micocantilevers coated with responsive phases such as 3-aminopropyltriethoxy silane, copper phtalocyanine, and methyl-β-cyclodextrin. Analytes detected include pentane, methanol, acetonitrile, acetone, ethanol, and trichloroethylene. Archibald et al. [2] present results on independent component analysis (ICA) of ethanol, propanol, and DIMP using cantilever coated with molecular recognition phases (MRP), whereby ICA has proven its feature extraction ability for components in mixtures. Sensitivities are summarized in Table 9.3

9.5.3 Explosives Detection

Preventive counter measures against terrorism threats require inexpensive, highly selective, and very sensitive small sensors that can be mass-produced and micro-fabricated. Such low-cost sensors could be arranged as a sensor grid for large area coverage of sensitive infrastructure, like airports, public buildings, or traffic infrastructure. Threats can be of chemical, biological, radioactive or explosive nature. Micro-fabricated detectors for explosives might be very useful for passenger baggage-screening, as compact versions of established technologies like the ion mobility spectrometer [17] or nuclear quadrupole resonance [20] have been developed, but are not likely to be miniaturized further. Another application area would be detection of landmines.

Microcantilever sensors offer sensitivities more than two orders of magnitude better than quartz crystal microbalances [43], flexural plate wave oscillators [15], and surface acoustic wave devices [23]. Several approaches to detect dangerous chemicals are already described in literature: photomechanical chemical microsensors based on adsorption-induced and photo-induced stress changes due to the presence of diisopropyl methyl phosphonate (DIMP), which is a model compound for phosphorous-containing chemical warfare agents, and trinitrotoluene (TNT), an explosive [16]. Further explosives frequently used include pentaerythritol tetranitrate (PETN) and hexahydro-1,3,5-triazine (RDX), often also with plastic fillers [49]. These compounds are very stable, if no detonator is present. Their explosive power, however, is very large, and moreover, the vapor pressures of PETN and RDX are very low, in the range of ppb and ppt. By functionalizing microcantilevers with self-assembled monolayers of 4-mercaptobeonzoic acid (4-MBA) PETN was detected at a level of 1400 ppt and RDX at a level of 290 ppt [50]. TNT was found to readily stick to Si surfaces, suggesting the use of microcantilevers for TNT detection, taking advantage of the respective adsorption/desorption kinetics [42, 53, 54]. Detection of TNT via deflagration on a microcantilever is described by Pinnaduwage et al. [53]. They used piezoresistive microcantilevers where the cantilever deflection was measured optically via beam deflection. TNT vapor from a generator placed 5 mm away from the microcantilever was observed to adsorb on its surface resulting in a decrease of resonance frequency. Application of an electrical pulse (10 V, 10 ms) to the piezoresistive cantilever resulted in deflagration of the TNT vapor and a bump in the cantilever bending signal. This bump was found to be related to the heat produced during deflagration. The amount of heat released is proportional to the area of the bump in the time vs. bending signal diagram of the process. The deflagration was found to be complete, as the same resonance frequency as before the experiment was observed. The amount of TNT mass involved was determined as 50 pg. The technique was later extended to the detection of PETN and RDX, where much slower reaction kinetics was observed [50, 51]. Traces of 2,4-dinitrotoluene (DNT) in TNT can also be used for detection of TNT, because it is the major impurity in production grade TNT. Furthermore DNT is a decomposition product of TNT. The saturation concentration of DNT in air at 20°C is 25 times higher than that of TNT. DNT was reported detectable at the 300 ppt level using polysiloxane polymer layers [52]. Micro-fabrication of electrostatically actuated resonant microcantilever beams in CMOS technology for detection of the nerve agent stimulant dimethylmethylphosphonate (DMMP) using polycarbosilane-coated beams [71] is an important step towards an integrated platform based on silicon microcantilevers, which besides compactness might also include telemetry [55]. Sensitivities are summarized in Table 9.4.

Table 9.4 Detection limits of explosives

Analyte	Detection limit	Reference
PETN	1400 ppt	Pinnaduwage et al. [49]
RDX	290 ppt	Pinnaduwage et al. [49]
DNT	300 ppt	Pinnaduwage et al. [54]

9.5.4 Gas Pressure and Flow Sensing

Gas sensing does not only involve chemical detection, but also pressure and flow sensing . Brown et al. [8] have studied the behavior of magnetically actuated oscillating microcantilevers at large deflections and have found hysteresis behavior at resonance. The amplitude at the actuation frequency changes depending on pressure due to damping. The authors have used cantilever in cantilever (CIC) structures, and have observed changes in deflection as gas pressure is varied. At atmospheric pressure, damping is large and the oscillation amplitude is relative small and hysteresis effects are absent. At lower pressure, abrupt changes in the oscillation amplitude occur with changes in the driving frequency. Since the change of amplitude and driving frequency, at which they occur are pressure dependent, these quantities can be used for accurate determination of gas pressure, demonstrated in the range between 10^{-3} and 10^{2} Torr. Brown et al. [9] emphasize that microelectromechanical system pressure sensors will have a wide range of applications, especially in the automotive industry. Piezoresistive cantilever-based deflection measurement has major advantages over diaphragms. The pressure range has been extended to 15–1450 Torr by means of design geometry adaptation. Su et al. [66] present highly sensitive ultra-thin piezoresistive silicon microcantilevers for gas velocity sensing, whereby the deflection increases with airflow distribution in a steel pipe. The detection principle is based on normal pressure drag producing bending of the cantilever. The minimum flow speed measured was 7.0 cm/s, which is comparable to classical hot-wire anemometers. Mertens et al. [39] have investigated the effects of temperature and pressure on microcantilever resonance response in helium and nitrogen. Resonance response as a function of pressure showed three different regimes, which correspond to molecular flow, transition regimes, and viscous flow, whereby the frequency variation of the cantilever is mainly due to changes in the mean free path of gas molecules. Effects observed allow measurement of pressures between 10^{-2} and 10^{6} Pa. Mortet et al. [41] present a pressure sensor based on a piezoelectric bimorph microcantilever with a measurement range between 0.1 and 8.5 bar. The resonance frequency shift is constant for pressures below 0.5 bar. For higher pressures the sensitivity is typically a few ppm/mbar,

Table 9.5 Detection limits of gas pressure and flow

Technique	Measured	Reference
Cantilever in cantilever, pressure	1.33–133 mbar	Brown et al. [8]
Cantilever in cantilever, pressure	20–1933 mbar	Brown et al. [9]
Resonant response, pressure	10^{-5}–1000 mbar	Mertens et al. [39]
Piezoelectric bimorph, pressure	0.1–8.5 bar	Mortet et al. [41]
Acousto-optical, pressure	0.1 mbar	Sievilä et al. [63]
Bending beam, flow	0.07 m/s	Su et al. [66]

but depends on the mode number. Sievilä et al. [63] present a cantilever paddle within a frame operating like a moving mirror to detect the displacements in the oscillating cantilever using a He/Ne laser in a Michelson interferometer configuration, whereby the cantilever acts as moving mirror element in one path of the interferometer. A fixed mirror serves a reference in the other arm of the interferometer. Sensitivities are summarized in Table 9.5.

9.6 Other Techniques

9.6.1 Metal Oxide Gas Sensors

Tin dioxide is the most investigated material for metal-oxide-based gas sensors, capable for detection of a wide range of gases [12]. The technique takes advantage of the fact that the resistance of the semiconductor changes with the ambient gas concentration and therefore the rate of redox reactions is measured. Sensor elements are typically operated at temperatures between 300 and 550°C. The sintered polycrystalline surface is responsible for a large resistance change, oxygen absorption increases the potential barrier between grain boundaries. By doping of SnO_2 with silicon the selectivity can be enhanced [14]. Doping of SnO_2 with 0.1 wt % of Ca and Pt allows detecting various volatile organic compounds [35]. Sputtered WO_3 thin films containing Pd, Au, Bi or Sb catalyst have been recognized as promising candidates for sensing elements in gas analysis [45]. Sensitivities are summarized in Table 9.6

Table 9.6 Metal oxide sensors

Gas	Detection limit	Reference
CO	15 ppm	Comini et al. [14]
NO_2	200 ppb	Comini et al. [14]
NO_2	5 ppm	Penza et al. [45]
H_2S	1 ppm	Penza et al. [45]
NO	10 ppm	Penza et al. [45]
CH_4	1 ppm	Penza et al. [45]
SO_2	1 ppm	Penza et al. [45]
Ethanol	1000 ppm	Lee et al. [35]
Methanol	200 ppm	Lee et al. [35]
Toluene	100 ppm	Lee et al. [35]
Acetone	200 ppm	Lee et al. [35]
Benzene	10 ppm	Lee et al. [35]

9.6.2 *Quartz Crystal Microbalance*

Quartz crystal microbalance (QMB) sensors (see also Chapter 8) are based on a piezoelectric substrate (quartz). When actuated by an alternating field, elastic waves are generated in the quartz crystal (typically at 10 MHz). Temporarily absorbed molecules perturb the propagation of acoustic waves because of the added mass (microbalance) and by changing the viscoelastic properties. From the shift of the fundamental frequency, the adsorbed mass can be measured at ppm sensitivity [24]. Specific realizations of resonance-based sensors are surface acoustic sensors (SAW), thickness-shear mode sensors (TSM), and flexural plate wave (FPW) devices.

9.6.3 *Conducting Polymer Sensors*

Conducting polymer sensors are based on chemoresistors (see also Chapter 5) that are manufactured by dissolving a chemically sensitive polymer in an appropriate solvent and mixing the dissolved polymer with conductive carbon particles [62]. This mixture (ink) is then deposited and dried onto a solid substrate on which a metal electrode has been applied previously. The upper surface of the polymer layer is also coated partially with a metal electrode. When chemical vapors come into contact with the polymers, the vapor will diffuse into the polymer layer, producing swelling of the polymer. The swelling changes the resistance between the electrodes, and a response signal can be measured and recorded. The amount of swelling is proportional to the concentration of the vapor present. A variety of volatile organic compounds can be detected in this way. By purging with e.g. dry nitrogen gas the swelling process can be reversed. For trichloroethylene concentrations between 1000 and 10,000 ppm have been detected with polyisobutylene or polyepichlorohydrine modified with carbon black [58]. Using poly-*N*-vinyl-pyrrolidone and carbon black, methanol, and ethanol have been detected at the ppt level [37]. A different possibility to apply conductive polymers for detection of volatile organic compounds is the use of piezoresistive microcantilevers embedded in conducting polymers [56]. This design has the advantage of higher stability and ease of use, since only an ohmmeter connected to the piezoresistors is required for measurements.

9.6.4 *Surface Acoustic Waves*

The principle of using the surface acoustic wave (SAW) effect for chemical vapor sensing is pioneered by Wohltjen and Dessy [73] and described in detail by Wohltjen [74] (see also Chapter 8). SAW devices are frequently used for chemical vapor detection because of the ruggedness, low cost, electronic output,

Table 9.7 Surface acoustic wave sensors

Vapor	Detection range	Reference
Methanol	15–130 ppm	Penza et al. [46]
Acetone	50–250 ppm	Penza et al. [46]
Propanol	5–70 ppm	Penza and Cassano [47]

sensitivity, and adaptibility to a wider variety of vapors. The substrate for a SAW device must be piezoelectric (e.g. quartz) to permit the easy generation of radio frequency Rayleigh waves. As chemical sensor often a polymer layer is applied. The resonance frequency of the device is measured during exposure to the analyte vapor, allowing to deduce the mass change caused by the adsorbing vapor molecules. Using SAW multisensors and pattern recognition techniques vapors can be detected routinely in the ppm range [46, 47], see Table 9.7

9.6.5 Field Effect Transistor Sensors Devices

The detection principle of field effect transistor (FET) sensors with catalytic metal gates is based on the change of the electric field in the insulator, into which gases diffuse (see also Chapter 4). Therewith the number of mobile carriers in the semiconductor is changed. Spetz et al. [64] use SiC FET sensors to monitor combustion processes at temperatures up to 1000°C. By use of different gate electrode materials the selectivity towards specific gases can be optimized, as well as by choice of the operation temperature. Measured concentrations range from a few 10 ppm to 12%, as seen from Table 9.8

Acknowledgments I thank R. McKendry (University College London, London, U.K.), M. Hegner, W. Grange, Th. Braun (CRANN Dublin), Ch. Gerber, J. Zhang, A. Bietsch, V. Barwich, M. Ghatkesar, F. Huber, N. Backmann, G. Yoshikawa, J.-P. Ramseyer, A. Tonin, H.R. Hidber, E. Meyer, and H.-J. Güntherodt (University of Basel, Basel, Switzerland) for valuable contributions and discussions, as well as U. Drechsler, M. Despont, H. Schmid, E. Delamarche, H. Wolf, R. Stutz, R. Allenspach, and P.F. Seidler (IBM Research, Zurich Research Laboratory, Rüschlikon, Switzerland). I also thank the European Union FP 6 Network of Excellence FRONTIERS for support. This project is funded partially by the National Center of Competence in Research in Nanoscience (Basel, Switzerland), the Swiss National Science Foundation and the Commission for Technology and Innovation (Bern, Switzerland).

Table 9.8 Detection of gases with FET sensors (after Spetz [64])

Gas	Low concentration	High concentration
CO_2	3%	9%
CO	50 ppm	250 ppm
O_2	6%	12%
NO	200 ppm	1000 ppm
C_3H_6	130 ppm	260 ppm
H_2O vapor	2%	2%

References

1. Adams JD, Rogers B, Manning L, Hu Z, Thundat T, Cavazos H, Minne SC. Piezoelectric self-sensing of adsorption-induced microcantilever bending. Sens Act A. 2005;121:457–461.
2. Archibald R, Datskos P, Devault G, Lamberti V, Lavrik N, Noid D, Sepianiak M, Dutta P. Independent component analysis of Nanomechanical responses of cantilever arrays. Anal Chim Acta. 2007;584:101–105.
3. Balselt DR, Fruhberger B, Klassen E, Cemalovic S, Britton Jr CL, Patel SV, Mlsna TE, McCorkle D, Warmack B. Design and performance of a microcantilever-based hydrogen sensor. Sens Act B. 2003;88:120–131.
4. Berger R, Lang HP, Gerber C, Gimzewski JK, Fabian JH, Scandella L, Meyer E, Güntherodt HJ. Micromechanical thermogravimetry. Chem Phys Lett. 1998;294:363–369.
5. Bietsch A, Hegner M, Lang HP, Gerber C. Inkjet deposition of alkanethiolate monolayers and DNA oligonucleotides on gold: Evaluation of spot uniformity by wet etching. Langmuir. 2004;20:5119–5122.
6. Bietsch A, Zhang J, Hegner M, Lang HP, Gerber C. Rapid functionalization of cantilever array sensors by inkjet printing. Nanotechnology. 2004;15:873–880.
7. Binnig G, Quate CF, Gerber C. Atomic force microscope. Phys Rev Lett. 1986;56:930–933.
8. Brown KB, Ma Y, Allegretto W, Lawson RPW, Vermeulen FE, Robinson AM. Microstructural pressure sensor based on an enhanced resonant mode hysteresis effect. J Vac Sci Technol B. 2001;19:1628–1632.
9. Brown KB, Allegretto W, Vermeulen FE, Robinson AM. Simple resonating microstructures for gas pressure measurement. J Micromech Microeng. 2002;12:204–210.
10. Brugger J, Buser RA, de Rooij NF. Micromachined atomic force microprobe with integrated capacitive read-out. J Micromech Microeng. 1992;2:218–220.
11. Chapman PJ, Vogt F, Dutta P, Datskos PG, Devault GL, Sepianiak MJ. Facile hyphenation of gas chromatography and a microcantilever array sensor for enhanced selectivity. Anal Chem. 2007;79:364–370.
12. Chiorino A, Ghiotti G, Prinetto F, Carotta MC, Gnani D, Martinelli G. Preparation, characterization of SnO_2, MoO_x-SnO_2 nano-sized powders for thick film gas sensors. Sens Act B. 1999;58:338–349.
13. Cleveland JP, Manne S, Bocek D, Hansma PK. A nondestructive method for determining the spring constant of cantilevers for scanning force microscopy. Rev Sci Instrum. 1993;64:403–405.
14. Comini E, Faglia G, Sberveglieri G. CO and NO_2 response of tin oxide silicon doped thin films. Sens Act B. 2001;76:270–274.
15. Cunningham B, Weinberg M, Pepper J, Clapp C, Bousquet R, Hugh B, Kant R, Daly C, Hauser E. Design, fabrication and vapor characterization of a microfabricated flexural plate resonator sensor and application to integrated sensor arrays. Sens Actuators B. 2001;73:112–123.
16. Datskos PG, Sepaniak MJ, Tipple CA, Lavrik N. Photomechanical chemical microsensors. Sens Act B. 2001;76:393–402.
17. Ewing RG, Miller CJ. Detection of volatile vapors emitted from explosives with a handheld ion mobility spectrometer. Field Anal Chem Technol. 2001;5:215–221.
18. Fabre A, Finot E, Demoment J, Contreras S. Monitoring the chemical changes in Pd induced by hydrogen absorption using microcantilevers. Ultramicroscopy. 2002;97:425–432.
19. Fadel L, Lochon F, Dufour I, Français O. Chemical sensing: Millimeter size resonant microcantilever performance. J Mcromech Microeng. 2004;14:S23–S30.

20. Garroway AN, Buess ML, Miller JB, Suits BH, Hibbs AD, Barrall GA, Matthews R, Burnett LJ. Remote sensing by nuclear quadrupole resonance. IEEE Trans. Geosci. Remote Sens. 2001;39:1108–1118.
21. Gimzewski JK, Gerber C, Meyer E, Schlittler RR. Observation of a chemical-reaction using a micromechanical sensor. Chem Phys Lett. 1994;217:589–594.
22. Göddenhenrich T, Lemke H, Hartmann U, Heiden C. Force microscope with capacitive displacement detection. J Vac Sci Technol A. 1990;8:383–387.
23. Grate JW. Acoustic wave microsensor arrays for vapor sensing. Chem Rev (Washington, DC) 2000;100:2627–2647.
24. Guan S. Frequency encoding of resonant mass sensors for chemical vapor detection. Anal Chem. 2003;75:4551–4557.
25. Heng TMS. Trimming of microstrip circuits utilizing microcantilever air gaps. IEEE Trans Microw Theor Techn. 1971;19:652–654.
26. Hierlemann A, Lange D, Hagleitner C, Kerness N, Koll A, Brand O, Baltes H. Application-specific sensor systems based on CMOS chemical microsensors. Sens Act B. 2000;70:2–11.
27. Hu Z, Thundat T, Warmack RJ. Investigation of adsorption and adsorption-induced stresses using microcantilever sensors. J Appl Phys. 2001;90:427–431.
28. Ibach H. Adsorbate-induced surface stress. J Vac Sci Technol A. 1994;12:2240–2243.
29. Itoh T, Suga T. Force sensing microcantilevers using sputtered zinc-oxide thin-film. Appl Phys Lett. 1994;64:37–39.
30. Jensenius H, Thaysen J, Rasmussen AA, Veje LH, Hansen O, Boisen A. A microcantilever-based alcohol vapor sensor-application and response model. Appl Phys Lett. 2000;76:2815–2817.
31. Kolesar ES. United States Patent No. 4,549,427, filed Sept 19, 1983.
32. Kooser A, Gunter RL, Delinger WD, Porter TL, Eastman MP. Gas sensing using embedded piezoresistive microcantilever sensors. Sens Act B. 2004;99:474–479.
33. Lange D, Hagleitner C, Hierlemann A, Brand O, Baltes H. Complementary metal oxide semiconductor cantilever arrays on a single chip: Mass-sensitive detection of volatile organic compounds. Anal Chem. 2002;74:3084–3095.
34. Lee C, Itoh T, Ohashi T, Maeda R, Suga T. Development of a piezoelectric self-excitation and self-detection mechanism in PZT microcantilevers for dynamic scanning force microscopy in liquid. J Vac Sci Technol B. 1997;15:1559–1563.
35. Lee DS, Jung JK, Lim JW, Huh JS, Lee DD. Recognition of volatile organic compounds using SnO_2 sensor array and pattern recognition analysis. Sens Act B. 2001; 77:228–236.
36. Lochon F, Fadel L, Dufour I, Rebière D, Pistré J. Silicon made resonant microcantilever: Dependence of the chemical sensing performances on the sensitive coating thickness. Mat Sci Eng C. 2006;26:348–353.
37. Longeran MC, Severin EJ, Doleman BJ, Beaber SA, Grubbs RH, Lewis NS. Array-based vapor sensing using chemically sensitive, carbon black-polymer resistors. Chem Mater. 1996;8:2298–2312.
38. Mertens J, Finot E, Nadal MH, Eyraud V, Heintz O, Bourillot E. Detection of gas trace of hydrofluoric acid using microcantilever. Sens Act B. 2003;99:58–65.
39. Mertens J, Finot E, Thundat T, Fabre A, Nadal MH, Eyraud V, Bourillot E. Effects of temperature and pressure on microcantilever resonance response. Ultramicroscopy. 2003;97:119–126.
40. Meyer G, Amer NM. Novel optical approach to atomic force microscopy. Appl Phys Lett. 1988;53:2400–2402.
41. Mortet V, Petersen R, Haenen K, D'Olieslaeger M. Wide range pressure sensor based on a piezoelectric bimorph microcantilever. Appl Phys Lett. 2006;88:133511.
42. Muralidharan G, Wig A, Pinnaduwage LA, Hedden D, Thundat T, Lareau RT. Absorption-desorption characteristics of explosive vapors investigated with microcantilevers. Ultramicroscopy. 2003;97:433–439.

43. O'Sullivan CK, Guilbault GG. Commercial quartz crystal microbalances – theory and applications. Biosens Bioelectron. 1999;14:663–670.
44. Ono T, Li X, Miyashita H, Esashi M. Mass sensing of adsorbed molecules in sub-picogram sample with ultrathin silicon resonator. Rev Sci Instrum. 2003; 74:1240–1243.
45. Penza M, Cassano G, Tortorella F. Gas recognition by activated WO3 thin-film sensors array. Sens Act B. 2001;81:115–121.
46. Penza M, Cassano G, Tortorella F. Identification and quantification of individual volatile organic compounds in a binary mixture by SAW multisensor array and pattern recognition analysis. Meas Sci Technol. 2002;13:846–858.
47. Penza M, Cassano G. Application of principal component analysis and artificial neural networks to recognize the individual VOCs of methanol/2-propanol in a binary mixture by SAW multi-sensor array. Sens Act B. 2003;89:269–284.
48. Petersen KE. Micromechanical membrane switches on silicon. IBM J Res Develop. 1979;23:376–385.
49. Pinnaduwage LA, Boiadjiev V, Hawk JE, Thundat T. Sensitive detection of plastic explosives with self-assembled monolayer-coated microcantilevers. Appl Phys Lett. 2003;83:1471–1473.
50. Pinnaduwage LA, Gehl A, Hedden DL, Muralidharan G, Thundat T, Lareau RT, Sulchek T, Manning L, Rogers B, Jones M, Adams JD. A microsensor for trinitrotoluene vapour. Nature. 2003;425:474.
51. Pinnaduwage LA, Thundat T, Gehl A, Wilson SD, Hedden DL, Lareau RT. Desorption characteristics, of uncoated silicon microcantilever surfaces for explosive and common nonexplosive vapors. Ultramicroscopy. 2004;100:211–216.
52. Pinnaduwage LA, Thundat T, Hawk JE, Hedden DL, Britt R, Houser EJ, Stepnowski S, McGill RA, Bubb D. Detection of 2,4-dinitrotoluene using microcantilever sensors. Sens Act B. 2004;99:223–229.
53. Pinnaduwage LA, Wig A, Hedden DL, Gehl A, Yi D, Thundat T, Lareau RT. Detection of trinitrotoluene via deflagration on a microcantilever. J Appl Phys. 2004;95:5871–5875.
54. Pinnaduwage LA, Yi D, Tian F, Thundat T, Lareau RT. Adsorption of trinitrotoluene on uncoated silicon microcantilever surfaces. Langmuir. 2004;20:2690–2694.
55. Pinnaduwage LA, Ji HF, Thundat T. Moore's law in homeland defense: An integrated sensor platform based on silicon microcantilevers. IEEE Sens J. 2005;5:774–785.
56. Porter TL, Eastman MP, Macomber C, Delinger WG, Zhine R. An embedded polymer piezoresistive microcantilever sensor. Ultramicroscopy. 2003;97:365–369.
57. Porter TL, Vail TL, Eastman MP, Stewart R, Reed J, Venedam R, Delinger W. A solid-state sensor platform for the detection of hydrogen cyanide gas. Sens Act B. 2007;123:313–317.
58. Rivera D, Alam MK, Davis CE, Ho CK. Characterization of the ability of polymeric chemiresistor arrays to quantitate trichloroethylene using partial least squares (PLS): Effects of experimental design, humidity, and temperature. Sens Act B. 2003;92:110–120.
59. Satyanarayana S, McCormick DT, Majumdar A. Parylene micro membrane capacitive sensor array for chemical and biological sensing. Sens Act B. 2006;115:494–502.
60. Savran CA, Burg TP, Fritz J, Manalis SR. Microfabricated mechanical biosensor with inherently differential readout. Appl Phys Lett. 2003;83:1659–1661.
61. Senesac LR, Dutta P, Datskos PG, Sepianiak MJ. Analyte species and concentration identification using differentially functionalized microcantilever arrays and artificial neural networks. Anal Chim Acta. 2006;558:94–101.
62. Severin EJ, Lewis NS. Relationships among resonant frequency changes on a coated quartz crystal microbalance, thickness changes, and resistance responses of polymer-carbon black composite chemiresistors. Anal Chem. 2000;72:2008–2015.
63. Sievilä P, Rytkönen VP, Hahtela O, Chekurov N, Kauppinen J, Tittonen I. Fabrication and characterization of an ultrasensitive acousto-optical cantilever. J Micromech Microeng. 2007;17:852–859.

64. Spetz AL, Tobias P, Uneus L, Svenningstorp H, Ekedahl LG, Lundstrom I. High temperature catalytic metal field effect transistors for industrial applications. Sens Act B. 2000;70:67–76.
65. Stoney GG. The tension of thin metallic films deposited by electrolysis. Proc R Soc London Ser A. 1909;82:172–175.
66. Su Y, Evans AGR, Brunnschweiler A, Ensell G. Characterization of a highly sensitive ultra-thin piezoresistive silicon cantilever probe and its application in gas flow velocity sensing. J Micromech Microeng. 2002;12:780–785.
67. Then D, Vidic A, Ziegler C. A highly sensitive self-oscillating cantilever array for the quantitative and qualitative analysis of organic vapor mixtures. Sens Act B. 2006; 117:1–9.
68. Thundat T, Warmack RJ, Chen GY, Allison DP. Thermal and ambient-induced deflections of scanning force microscope cantilevers. Appl Phys Lett. 1994;64:2894–2896.
69. Thundat T, Chen GY, Warmack RJ, Allison DP, Wachter EA. Vapor detection using resonating microcantilevers. Anal Chem. 1995;67:519–521.
70. Vancura C, Rüegg M, Li Y, Hagleitner C, Hierlemann A. Magnetically actuated complementar metal oxide semiconductor resonant cantilever gas sensor systems. Anal Chem. 2005;77:2690–2699.
71. Voiculescu I, Zaghloul ME, McGill RA, Houser EJ, Fedder GK. Electrostatically actuated resonant microcantilever beam in CMOS technology for the detection of chemical weapons. IEEE Sens J. 2005;5:641–647.
72. Wilfinger RJ, Bardell PH, Chhabra DS. Resonistor – a frequency selective device utilizing mechanical resonance of a silicon substrate, IBM J Res Develop. 1968;12:113–118.
73. Wohltjen H, Dessy RE. Surface acoustic probe for chemical analysis I. Introduction and instrument description. Anal Chem. 1979;51:1458–1475.
74. Wohltjen H. Mechanism of operation and design considerations for surface acoustic wave device vapour sensors. Sens Act. 1984;5:307–325.
75. Wright YJ, Kar AK, Kim YW, Scholz C, George MA. Study of microcapillary pipette-assisted method to prepare polyethylene glycol-coated microcantilever sensors. Sens Act B. 2005;107:242–251.
76. Zhou J, Li P, Zhang S, Huang Y, Yang P, Bao M, Ruan G. Self excited piezoelectric microcantilever for gas detection. Microelectr Eng. 2003;69:37–46.
77. Zhou J, Li P, Zhang S, Long Y, Zhou F, Huang Y, Yang P, Bao M. Zeolite-modifies microcantilever gas sensor for indoor air quality control. Sens Act B. 2003;94:337–342.

Index

Y

Z